# Basic Environmental Data Analysis for Scientists and Engineers

T0138794

# Basic Environmental Data Analysis for Scientists and Engineers

**Ralph R.B. von Frese**
Academy Professor, School of Earth Sciences
The Ohio State University, Columbus, OH, USA

CRC Press is an imprint of the
Taylor & Francis Group, an **informa** business

The cover shows the 160-foot (160') high Leaning Tower of Pisa, Italy with some travel-time observations on the free fall of a mass (green ball) at one-second (1s) intervals that helped spawn modern data analysis.

CRC Press
Taylor & Francis Group
6000 Broken Sound Parkway NW, Suite 300
Boca Raton, FL 33487-2742

First issued in paperback 2022

© 2020 by Taylor & Francis Group, LLC
CRC Press is an imprint of Taylor & Francis Group, an Informa business

No claim to original U.S. Government works

ISBN 13: 978-1-03-247506-6 (pbk)
ISBN 13: 978-1-138-62778-9 (hbk)

DOI: 10.1201/9780429291210

This book contains information obtained from authentic and highly regarded sources. Reasonable efforts have been made to publish reliable data and information, but the author and publisher cannot assume responsibility for the validity of all materials or the consequences of their use. The authors and publishers have attempted to trace the copyright holders of all material reproduced in this publication and apologize to copyright holders if permission to publish in this form has not been obtained. If any copyright material has not been acknowledged please write and let us know so we may rectify in any future reprint.

Except as permitted under U.S. Copyright Law, no part of this book may be reprinted, reproduced, transmitted, or utilized in any form by any electronic, mechanical, or other means, now known or hereafter invented, including photocopying, microfilming, and recording, or in any information storage or retrieval system, without written permission from the publishers.

For permission to photocopy or use material electronically from this work, please access www.copyright.com (http://www.copyright.com/) or contact the Copyright Clearance Center, Inc. (CCC), 222 Rosewood Drive, Danvers, MA 01923, 978-750-8400. CCC is a not-for-profit organization that provides licenses and registration for a variety of users. For organizations that have been granted a photocopy license by the CCC, a separate system of payment has been arranged.

**Trademark Notice:** Product or corporate names may be trademarks or registered trademarks, and are used only for identification and explanation without intent to infringe.

Publisher's Note
The publisher has gone to great lengths to ensure the quality of this reprint but points out that some imperfections in the original copies may be apparent.

**Library of Congress Cataloging-in-Publication Data**

Names: Von Frese, R., author.
Title: Basic environmental data analysis for scientists and engineers / Ralph R.B. von Frese.
Description: Boca Raton, FL : CRC Press, Taylor & Francis Group, 2019. | Includes bibliographical references and index.
Identifiers: LCCN 2019008283| ISBN 9781138627789 (hardback : alk. paper) | ISBN 9780429291210 (ebook)
Subjects: LCSH: Environmental sciences--Data processing. | Environmental sciences--Statistical methods.
Classification: LCC GE45.D37 V66 2019 | DDC 363.7001/5195--dc23
LC record available at https://lccn.loc.gov/2019008283

**Visit the Taylor & Francis Web site at**
**http://www.taylorandfrancis.com**

**and the CRC Press Web site at**
**http://www.crcpress.com**

# Dedication

---

*To my wife, Janet Kay Shaw,*
*and*
*data analysis students*

# Contents

# PART II Digital Data Inversion, Spectral Analysis, Interrogation, and Graphics

# Preface

## 0.1 BOOK OBJECTIVES AND ORGANIZATION

This book is designed to make university undergraduate and beginning graduate students in the environmental sciences and engineering proficient in modern, electronically driven digital data analysis. It seeks to demonstrate the fundamental elegance of data analysis that ultimately depends on only the simple inverse arithmetic of sums and differences. The material is organized into two main parts with each part constituting a single-semester university course offering.

**Part I** for beginning undergraduate students introduces the basic approaches for quantifying data variations in terms of environmental parameters. They include using simple numerical three-data point derivatives and two-data point integrals to establish the respective differential and integral calculus properties of the data, which are the ultimate analytical attributes of any signal. A second approach uses the data's measurement units or dimensions to quantify data behavior, and develop effective scale modeling studies of difficult-to-quantify environmental phenomena [e.g., tectonic basin evolution, mountain building, storms, and ground erosion, shaking, excavation, blasting, etc.]. A third basic approach invokes non-statistical and statistical data errors and their propagation to characterize data variations.

These approaches are presented in the context of the data array or matrix, which is the fundamental data and mathematical processing format of modern electronic computing. Data sampling issues are also considered for designing the array to represent critical data variations for analysis, address data accessibility, and survey logistical and other sampling limitations that affect the success of environmental studies. **Part I** concludes with the introduction of the inverse problem for estimating the intercept and slope coefficients of the least squares straight line fit of data.

**Part II** for advanced undergraduate and beginning graduate students extends the inverse problem to least squares solutions involving more than two unknowns. It generalizes my gravity and magnetic anomaly-specific processing topics in *Appendix A* of [**43**] to the broad array of digital data analysis topics in environmental sciences and engineering. The sensitivity analysis is further developed for assessing solution performance to satisfy observed data with unavoidable measurement errors using models subject to parameter and computational uncertainties. The fast Fourier transform is presented as the dominant inverse problem in electronic data analysis because of its computational efficiency and accuracy in modeling data arrays with minimal assumptions. Procedures for filtering or interrogating the spectral model for the data's derivative and integral properties are also introduced. Finally, effective presentation graphics to display the data and data processing results are presented as essential elements of modern data analysis.

As a study aid to the reader, each chapter is introduced with an *Overview* and concluded with a summary section listing the *Key Concepts* of the chapter. The overview is not an abstract, but rather provides the reader with a broad, generalized summary of the chapter that is a useful guide in reading and studying the chapter. The concluding key concepts draw the reader's attention to the more important concepts that are presented. They are not a listing of what should be known upon reading the chapter, but rather guide the reader in reviewing what has been emphasized in the chapter.

The book uses mathematics up to and including basic differential equations, but develops the methods in simple digital array operations with minimal use of the arcane integral and differential calculus notation. Thus, this book is more computationally oriented than most previous works because the reader can readily implement and explore analytical results with electronic equation solvers like MATLAB, Mathematica, Mathcad, etc.

To facilitate this hands-on approach, the book's numerous numerical examples are alpha-numerically labelled and catalogued in the index for easy access by the reader. The directional filter example labelled ***Example 11.4.2.2f***, for instance, is the ***sixth [= f] example*** in ***sub-sub-section*** 2 of ***sub-section*** 2 in ***section*** 4 of ***Chapter*** 11. For instructors with in-class access to an electronic equation solver, the labelling convention also facilitates bringing up quickly one or more of the examples for students to work through during class time. In addition, the website **www.crcpress.com/9781138627789** offers exercises and other supplemental materials to foster further understanding of the topics in the book.

In general, the text and related on-line exerecises and other materials are well suited to serve the learning objectives of both the reader and instructor. The book is not only a textbook, but also a handy reference for environmental practitioners [e.g., agronomists, archaeologists, soil scientists, ecologists, biologists, geologists, geophysicists, geodesists, engineers, etc.]. Thus, the book serves a wide audience in the environmental sciences and engineering ranging from undergraduate and graduate students to professionals in academia, government, and industry.

## 0.2 RELATED BOOKS

This comprehensive book is unique in the approach and breadth of the content. However, several books pertaining to environmental data analysis are available that the reader may find as useful supplements to this book. A selected list of these books includes *Data Reduction and Error Analysis for the Physical Sciences* [**6**], *Statistics and Data Analysis in Geology* [**20**], *Methods of Environmental Data Analysis* [**41**], *Spatial Data Analysis in the Social and Environmental Sciences* [**39**], *Environmental Statistics and Data Analysis* [**72**], *Analysis and Modelling of Spatial Environmental Data* [**53**], *Statistical Data Analysis Explained: Applied Environmental Statistics with R* [**79**], and *Environmental Data Analysis with MATLAB* [**68**].

## 0.3 ACKNOWLEDGMENTS

I thank the students, faculty, and administration of the School of Earth Sciences at The Ohio State University for their encouragement and assistance in preparing this book. I also thank John W. Olesik, Mohammad F. Asgharzadeh, and Lawrence W. Braile for reviewing various technical elements of this book. They provided invaluable and constructive advice, but the responsibiliy for errors of omission or commission in the final manuscript are solely mine.

I'm very grateful to Yuriy Yeremenko for his significant expertise and assistance with the electronic production of the book's graphics. I also thank Irma Britton and Rebecca Pringle of CRC Press for their able editorial assistance and patience with me and my sundry commitments that complicated the completion of the book.

The book's organization of technical topics was inspired by Prof. Braile's applied data analysis course for advanced undergraduate and beginning graduate students at the Dept. of Geosciences, Purdue University, which I taught for a term as a visiting assistant professor. At Ohio State's School of Earth Sciences, I expanded the topics into two single-semester courses, which include a beginning undergraduate 4-credit hour course required of the School's majors that covers **Part I** of the book, and a popular 3-credit hour elective course for advanced undergraduate and beginning graduate students that covers the topics from **Part II**. I thank the many students who participated in these courses for their good humor and probing questions that constantly pushed me to improve my understanding and presentation of the materials in this book.

## 0.4 AUTHOR BIOGRAPHY

Ralph R.B. von Frese is an Academy Professor in the School of Earth Sciences, The Ohio State University, where, from 1982 to 2017, he taught undergraduate courses in introductory Earth sciences, geophysics, and environmental geoscience and data analysis, as well as graduate courses in exploration geophysics, Earth and planetary physics, and geomathematics. His research has focused mainly on archaeological and planetary applications of geophysics. He has authored or co-authored more than 125 journal articles and 1 book, co-edited 4 special issues of scientific journals, and served on several government and scientific panels. Prof. von Frese co-founded in 1995 and chaired through 2017 the Antarctic Digital Magnetic Anomaly Project (ADMAP), an international collaboration of the Scientific Committee for Antarctic Research (SCAR) and the International Association of Geomagnetism and Aeronomy (IAGA). He currently serves on ADMAP's steering committee, and is a member of the Society of Exploration Geophysicists, the American Geophysical Union, and the Geological Society of America.

# *Part I*

## *Digital Data Calculus, Units, Errors, Statistics, and Linear Regression*

# 1 Computing Trends

## 1.1 OVERVIEW

*Modern data analysis incorporates the internationally recognized Arabric numerals* 1, 2, ..., 9 *supplemented by the symbol* 0 *for the Hindu invention of zero. These* 10 *numerals are used to represent any number no matter how large or small.*

*The plural of 'datum' is 'data,' which are discrete numbers with or without attached units of measurement. Analog data are measured against continuous quantities [e.g., lengths of a ruler, heights of mercury in a tube, slide rule calculations, etc.], whereas digital data are referenced against discrete quantities [e.g., clock ticks, fingers, abacus beads, open and closed electronic circuits, etc.]. Digital data, as well as digital mathematics and graphics dominate modern data analysis.*

*Numerical derivatives and integrals of successive* 3- *and* 2-*data point sequences, respectively, form a closed system that completely defines the data's analytical attributes. Thus, numerically evaluating and graphing the inverse derivative and integral properties of data fully describe their analytical attributes. The rise-to-run ratios of successive* 3-*data point sequences define* $[n-2]$ *of the derivatives of the n-point signal. The defined derivative estimates involve minimal assumptions in application because they are constrained by the observed datum at the center of each sequence. The two undefined derivatives at each end of the signal can be determined using the First Fundamental Theorem of Calculus, which states that the signal is the integral to within a constant of its derivative. Here, the integral values are just the cumulative areas under the derivative curve established from the equivalent areas of simple rectangles with vertically extending triangles [i.e., trapezoids]. The integral values are defined everywhere except at the first data point where the undefined value can be established by the Second Fundamental Theorem of Calculus, which holds that the signal is also the derivative of its integral.*

*In general, the signal's numerical derivatives and integrals to higher p orders are just the numerical derivatives and integrals of the lower* $[p-1]$ *order derivatives and intregals. These numerical derivatives involve* $[n-2p]$ *undefined values at the two boundaries of the signal, whereas the numerical integration involves p undefined values at the signal's beginning. However, the undefined boundary values may all be established using the two fundamental calculus theorems to reveal the signal's complete derivative and integral properties for all p orders consistent with the working precision of the data and computations. These results, in turn, can provide considerable insight for quantifying data in terms of applicable environmental parameters.*

*The mechanics of computation provide practical simplifying insights on computational theory. Computing devices developed to date have been mostly digital, including the abacus that originated over* 3,000 *years ago and reigns today as perhaps the most popular computing aid of all time. Modern electronic computing is based on Leibniz's* 17*th century fomulation of the binary number system and Babbage's early* 19*th century development of a programmable mechanical computer. The latter device established the elements of the modern computer system that include a central processing unit [CPU] to perform input and output, arithmetic, and logical or Boolean operations.*

*Computing technology is advancing so rapidly at present that whatever can be purchased commercially is only several years behind the technology under development.*

## 1.2 NUMERATION

To better consume and exploit the environment, humans invented numbers and discovered effective recipes and fomulae, as well as devices, for analyzing numbers. The first numbers were whole numbers or integers like the Roman numerals $I, II, III, IV$, etc. or the equivalent Arabic numerals $1, 2, 3, 4$, etc. However, Arabic numerals also include the symbol 0 for zero, which was a Hindu invention. This may have been the most revolutionary step in mathematical history because its inclusion yielded 10 Arabic numerals for representing any number no matter how large or small. Thus, no new symbols like the Roman numerals $X[= 10], L[= 50], C[= 100], D[= 500], M[= 1000]$, etc. were needed to accommodate changing multiples of 10. In addition, numerals have symbols that are the same in all languages, and thus are internationally recognized.

Measurements provide discrete, finite samples of phenomena that in theory may exhibit an infinite continuum of infinitesimally small variations. These discrete numbers with or without attached units of measurement are called *data*, which is the Latin plural for *datum*. Data are called *analog* when measured from lengths of a ruler, heights of mercury in a tube, slide rule calculations, or other continuous quantities, and *digital* when determined by counting clock ticks, heart beats, fingers, toes, bones, stones, abacus beads, or other discrete quantities.

## 1.3 NUMERICAL CALCULUS

With the advent of widespread electronic computing in the latter half of the 20th century, the counting of open and closed electronic circuits in pocket calculators, and personal, mainframe, and super computers became the norm for representing and analyzing data. Thus, modern analysis has regressed to the pre-calculus era that existed roughly up through the early 17th century when the more intuitive linear mathematics of digital data series prevailed [**Figure 1.1**]. However, the remarkable advances of electronic computing in the current *information age* allow us to explore and visualize the analytical

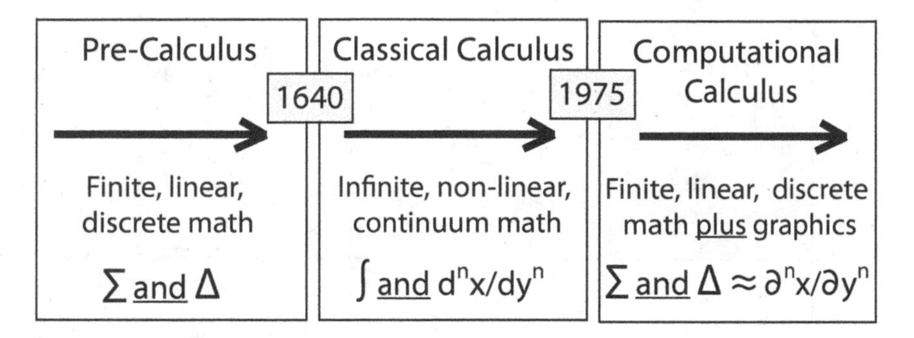


**Figure 1.1** Historical trends in computing range from the pre-calculus era of ancient times when analysis was essentially based on linear discrete mathematics, through the classical calculus era that began roughly with Newton and Leibniz who incorporated continuum mathematics and non-linear systems, to the present computational calculus era that was inaugurated for the masses by the arrival of the electronic pocket calculator. All data analyses are fundamentally based on taking data sums [$\Sigma$] and differences [$\Delta$].

properties of data far more readily and comprehensively than any previous generation of scientists and engineers.

In general, all analysis is based on the elementary inverse *addition* and *subtraction* operations of arithmetic that define basic sums and differences in data. These fundamental operations are combined into the inverse operations of *multiplication* and *division* to describe the more complex arithmetic of data products and quotients, respectively. With the algebra, the arithmetic operations are extended to a further characterization of data as representative functions and their inverses. However, the fullest characterization of the analytical properties of data results from the calculus that functionally relates the sums and differences by the inverse operations of *integration* and *differentiation*, respectively.

Modern calculus courses tend to characterize data more in terms of analytical expressions or equations than in the original graphical format that Newton and Leibniz, the co-inventors of calculus, accessed with rulers and ink quills. **Figure 1.2** gives some commonly considered textbook examples [e.g., **[2]**] to help illustrate the basic equivalence of analytical and graphical data integration and differentiation.

For each profile in **Figure 1.2**, the data are plotted as $f(t)$-values of the *dependent variable* along the *ordinate* or vertical axis for the corresponding $t$-values of the *independent variable* along the *abscissa* or horizontal axis. In the integrals, $dt$ is a vanishingly small interval or differential of $t$ that in the rest of this book is written without loss of generality as $\partial t$ or the partial differential of $t$ - i.e., $dt \equiv \partial t$ simplifies using the letter $d$ as an equation variable in the remaining elements of this book.

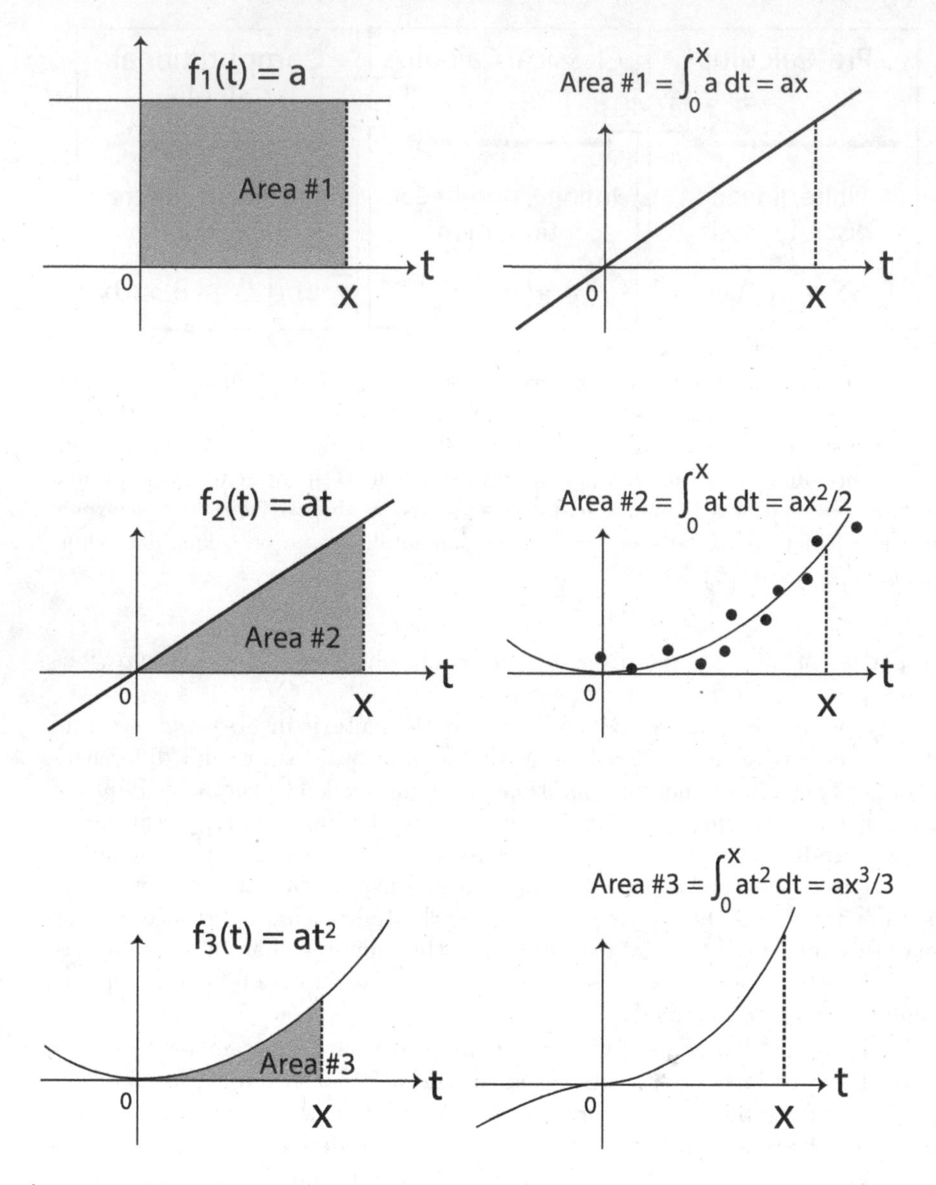


**Figure 1.2** The calculus relates the function to the integral of its derivative and vice versa [e.g., [**2**]].

The *First Fundamental Theorem of Calculus* [e.g., [**2**]] holds that a function [e.g., $f_2(t)$ in **Figure 1.2**] is the integral of its derivative [e.g., $f_2(t) = \int f_1(t)dt \equiv \int f_1(t)\partial t$] to within a constant. The *Second Fundamental Theorem of Calculus*, by contrast, holds that a function [e.g., $f_1(t)$] is the derivative of its integral [e.g., $f_1(t) = df_2(t)/dt \equiv \partial f_2(t)/\partial t$].

As an example, the top-left insert graphs the function $f_1(t)$ which is equal to the constant $a$ for all $t$-values. The top-right insert shows the integral of $f_1(t)$ for all $t$-values, where the incremental value of the integral at $t = x$ is the grey-shaded Area #1 under the curve $f_1(t)$ from $t = 0$ to $t = x$. This value can be determined both analytically by $f_2(t) = at|_0^x = a[x - 0]$ or graphically using a ruler to measure the area $[= a \times x]$ of the grey-shaded rectangle under $f_1(t)$.

The derivatives of a function also may be evaluated both analytically and numerically. Thus, the $t$-derivative of the middle-left insert of **Figure 1.2**, for instance, may be taken analytically by

$$f_2'(t) = \frac{\partial[f_2(t)]}{\partial t} = a = f_1(t), \tag{1.1}$$

In this case, the derivative $f_2'(t) = f_1(t)$ is the constant rate of change of the integral $f_2(t)$ with $t$.

More generally, the graphical evaluation of the derivative may be obtained by digitizing the $f(t)$-curve into $[i = 1, 2, ..., n]$ discrete observations and plotting the $[n - 2]$ local 3-data point differentials

$$\frac{\partial f(t_i)}{\partial t} = f'(t_i) \approx \frac{[f(t_{i+1}) - f(t_{i-1})]}{[t_{i+1} - t_{i-1}]} \tag{1.2}$$

against $t$. Note that each numerical 3-data point derivative is constrained by an observation except at the two end-points $t_1$ and $t_n$ where the 3-point derivatives are undefined. However, the fundamental theorems of calculus show that two consecutive integral values may be related to their respective derivatives by

$$f'(t_i) + f'(t_{i+1}) = \frac{2[f(t_{i+1}) - f(t_i)]}{\Delta t}, \tag{1.3}$$

where $\Delta t = [t_{i+1} - t_i]$. Thus, the derivatives at the two end-points $t_1$ and $t_n$ may be obtained from

$$f'(t_1) = \frac{2[f(t_2) - f(t_1)]}{\Delta t} - f'(t_2) \text{ and } f'(t_n) = \frac{2[f(t_n) - f(t_{n-1})]}{\Delta t} - f'(t_{n-1}). \tag{1.4}$$

Furthermore, these numerical procedures may be repeated on the $n$ numerical derivatives to estimate higher order derivatives which also satisfy the two fundamental theorems of calculus. In other words, the 3-point differentiation to order $p$ of an $n$-point signal involves $[n - (2p)]$ undefined boundary values that can be established fully with the two fundamental calculus theorems.

Like a function's derivative, its integral to within a constant can also be determined either analytically if the function is known or numerically from the data plot if the function is unknown. The more general graphical determination involves measuring the total area between the data and horizontal axis and plotting the numerical estimates with $t$ for the integral. For example, the function $f_3(t)$ in the middle-right insert of **Figure 1.2** is the integral of its derivative $f_2(t)$ in the top-right insert. Thus, at any $t$, the integral $f_3(t)$ numerically equals the shaded area under the $f_2(t)$-curve in the middle-left insert, which is one-half the area of the rectangle with sides $t$ and $at$, or

$$f(t) = a[\frac{t^2}{2}]. \tag{1.5}$$

For higher order functions [e.g., $f_3(t)$], the incremental area between two consecutive data points is the area $[TA]$ of the intervening trapezoid. This area is simply the area of the rectangle with top at the absolute magnitude minimum of $f(t_i)$ and $f(t_{i+1})$ plus the area of the triangle with base at the minimum, which simplifies to

$$TA = [\frac{f(t_i) + f(t_{i+1})}{2}]\Delta t. \tag{1.6}$$

The numerical integration value is clearly undefined at $t_1$, but can be estimated to within a constant from **Equation 1.3** by

$$f(t_1) = f(t_2) - [\frac{f'(t_1) + f'(t_2)}{2}]\Delta t. \tag{1.7}$$

Here, the *boundary value condition* [i.e., **Equation 1.3**] was obtained from combining numerical integration by rectangles and triangles [i.e., **Equation 1.6**] with numerical differentiation [i.e., **Equation 1.2**] via the two fundamental calculus theorems.

Thus, the two fundamental theorems of calculus guarantee that the mere act of plotting or mapping data fully reveals their analytical properties. However, measurements of $f(t)$ in practice are always based on a finite set of discrete point observations containing errors. Thus, the observations can never completely replicate the signal from which they originate, but only approximate a portion of it such as illustrated in the middle-right insert of **Figure 1.2**. However, to the degree to which $f(t)$ is represented by the discrete observations, the related derivative and intergral properties of $f(t)$ may also be evaluated, irrespective of the analytical form of the function. In other words, for establishing the analytical form of the function underlying the data, the investigator can obtain significant insights on the related differential and integral properties by simply processing the discrete observations for their respective numerical 3-point derivatives and cumulative 2-point trapezoidal areas.

To illustrate the great analytical power of data graphics, consider as ***Example 01.3a*** the discrete data listed and plotted in **Figure 1.3**. These 7 discrete data values simulate the late 16th century experimental observations of Galileo Galilei on the motion of a mass in the Earth's gravity field. Here, the dependent vertical axis records the effective free-fall distance $z(t_i)$ in feet [ft] traveled by any mass as a function of time $t_i$ in seconds [s] along the independent horizontal axis. These *travel-time* data were experimentally determined from measuring the time it took for different sized spherical masses

| i | 1 | 2 | 3 | 4 | 5 | 6 | 7 |
|---|---|---|---|---|---|---|---|
| ***t*** (s) | 0 | 1 | 2 | 3 | 4 | 5 | 6 |
| ***z*** (ft) | 0 | 16 | 64 | 144 | 256 | 400 | 576 |
| ***v*** (ft/s) | ? (0) | 32 | 64 | 96 | 128 | 160 | ? (192) |
| ***a*** (ft/$s^2$) | ? (32) | ? (32) | 32 | 32 | 32 | ? (32) | ? (32) |

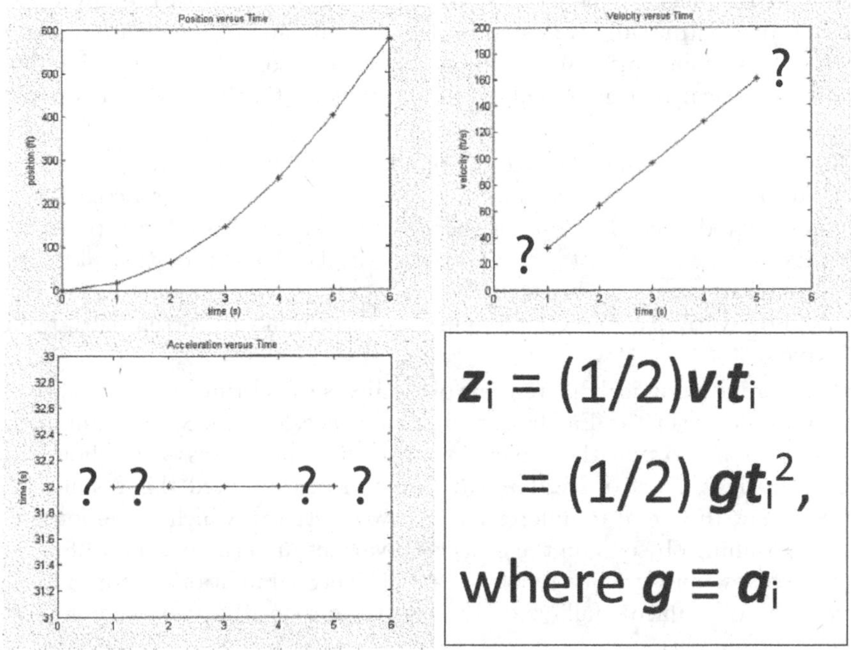

$$z_i = (1/2)v_i t_i$$
$$= (1/2)\, g t_i^2,$$
$$\text{where } g \equiv a_i$$

**Figure 1.3** Seven discrete data points simulating Galileo Galilei's observations of a mass's free fall [$z_i$(ft) in the top-left graph] in the Earth's gravity field [$g(ft/s^2)$] as a function of time [$t_i$(s)] uniformally sampled at one-second intervals. Other items listed in the table include each datum's index [$i$], and the related velocities [$v_i(ft/s)$ in the top-right graph] and accelerations [$a_i \equiv g$ in the bottom-left graph] obtained by numerical 3-data point differentiation via **Equation 1.2**. Question marks [?] denote the signal values left undefined by numerical differentiation. The ?-accompanying parentheses show these signal boundary values obtained using the two fundamental calculus theorems as described in the text.

or balls to roll down an inclined plane. Galileo and his assistants used rulers and their bodily pulse rates to measure the data coordinates of distance and time, respectively. The travel-time plots were remarkable because they were basically the same for all the balls, and thus independent of mass. These results, accordingly, disproved Aristole's intuitive claim of heavier objects falling faster than lighter ones that had prevailed for the previous 2,000 years.

This fallacy is evident numerically from the graphical differentials of the travel-time data that are the *velocities* $v(t_i)$ in ft/s listed in the table and plotted in the top-right graph of **Figure 1.3**. In general, for any $n$-point signal, the 3-point differentials in **Equation 1.2** are not defined for the two end-points at $t_1$ and $t_n$. However, the two boundary values are available from **Equation 1.4** that combines numerical differentiation and integration through the two fundamental calculus theorems. Accordingly, in the table, question marks [**?**] denote the undefined boundary velocities $v(t_1)$ and $v(t_7)$ with the appropriate values from **Equation 1.4** listed in the **?**-accompanying parentheses. Note also that these boundary value estimates are fully consistent with straightforward numerical and graphical extrapolations of the five defined 3-point differentials.

These results show that the velocity of a freely falling object increases with time $t$ at the constant rate $a$. The rate of velocity increase or *acceleration* can be estimated from the slope of the velocity line given by the ratio of its rise to its run [i.e., $a = (v(t_n) - v(t_1))/(t_n - t_1)$]. The acceleration also can be evaluated from any estimate $v(t_i)$ as $a = v(t_i)/t_i$, or by plotting up the numerical derivatives of the velocity curve as shown in the bottom-left graph of **Figure 1.3**.

Second order numerical differentials of Galileo's travel-time data reveal that the magnitude [$g$] of the Earth's gravitational acceleration is constant [i.e., $g \equiv a \approx 32$ ft/s$^2$] so that the travel-time data may be expressed as shown in the bottom-right insert of **Figure 1.3**. Newton incorporated this result into his second law [$\mathbf{F} = m\mathbf{a}$] to differentiate between *weight*, which is the force $\mathbf{F}$ acting on a falling object, and the object's invariant material property of *mass* [$m$]. Thus, Newton's gravity theory is a relatively straightforward graphical consequence of Galileo's falling body observations by the two fundamental theorems of calculus.

During his 3-year tenure as Professor of Mathematics at the University of Pisa, Galileo publicly demonstrated his results from Pisa's famous Leaning Tower that is pictured on this book's cover. He simultaneously dropped two balls of common diameters, but vastly different weights that simultaneously hit the ground before the gathered crowd. The August, 1971 video at **www.youtube.com/watch?v = 5C5$_\mathbf{d}$OEyAfk** shows a comparable travel-time experiment conducted on the Moon with a falcon's feather and a rock hammer by NASA's Apollo 15 mission commander, Dave Scott.

## 1.4 COMPUTING DEVICES

Like computing, the development of computing devices has a long history dominated by themes that persist to the present day. The history, which is outlined in **Tables 1.1** and **1.2**, is worth reviewing for insights on the forces that produced these inventions and continue to drive current and future computing requirements.

**Table 1.1**
**Pre-20th century computing devices [e.g., [88]; [83]].**

| Period | Device | Remarks |
|---|---|---|
| Prehistoric | Fingers, Toes, Sticks, Stones, Bones, etc. | Digital computing |
| 1100 BCE | Abacus | Digital computing with strung beads |
| 1500 ACE | Arabian Numbers | Adam Riese and others establish computing schools |
| 1603 | Logarithms | Invented by Jost Bürgi, watchmaker to Emperor Rudolf II |
| 1620 | Log Tables | Computed by Lord J. Napier |
| 1625 | Slide Rule | Invented by Wm. Oughtred |
| 1641 | Adding Machine | Developed by B. Pascal |
| 1650 | Double Scale Slide Rule | Developed by S. Partridge |
| | Calculating Clock | Developed by Wm. Schickard to perform Kepler's astronomical calculations |
| 1703 | Arithmetica Dyadica | Leibniz formulizes binary representation of numbers |
| 1800 | Mechanical Calculator | I.H. Müller rationalizes the efficiency of machine computation |
| 1820 | Commercial Calculator | C.X. Thomas manufactures calculators in Paris for the insurance industry |
| 1833 | Programmable Mechanical Calculator | C. Babbage of Cambridge University develops modern computing system |
| 1880 | Electronic Computation | Electrical circuits and relays duplicate mechanical computations |
| 1890 | Punched Cards | H. Hollerith uses punched cards for processing the U.S. Census |

Except for the 17th century analog slide rule, the computing aids were basically digital data processing tools. Indeed, the most popular computing tool of all time is the abacus that was invented over three millennia ago. This mechanical digital computer remains in wide use today among the majority of humans living outside North America and western Europe.

In the late 17th century, while working to adapt Pascal's calculator for multiplication and division, Leibniz formalized the *base-2* or *binary* number system. This binary system is the basis of modern electronic digital computing that began to be developed in the late 19th century. It represents numbers and their arithmetic operations as patterns of two different symbols - e.g., ones and zeroes, positive and negative poles on magnetic tape, open and closed electronic circuits, open and closed holes on punched paper tape and cards, etc.

However, the modern computing system dates back to the early 19th century with C. Babbage's invention of a programmable mechanical computer.

**Table 1.2**
**Post-19th century computing devices [e.g., [88]; [83]].**

| Period | Device | Remarks |
|---|---|---|
| 1930 | Magnetic/Photo/Electrical Transcription and Scanning | G. Tauschek develops several devices to read Hollerith's punched card code |
| 1938 | ABC-Electronic Digital Computer | Atanasoff and Berry [Iowa State College] develop first electronic digital computer |
| 1946 | ENIAC [Electronic Numerical Integrator and Computer] | Mauchly and Eckert [U. of Penn.] use the ABC concept to develop the first large-scale electronic digital computer which was programmed by external rewiring and switches |
| | Stored Program Concept | J. von Neumann [U. of Penn.] develops idea of placing numerically coded computer instructions in main computer storage |
| 1949 | EDSAC [Electronic Delay Storage Automatic Calculator] | M.V. Wilkes at Cambridge University builds first computer using the stored program concept written in machine language |
| 1951 | UNIVAC I | First commerically available computer at the U.S. Census Bureau |
| 1957 | FORTRAN [FORmula TRANslation] | IBM introduces high-level programming language that converts machine language into mathematical code |
| 1958 | Transistors | Replace vacuum tubes to usher in the second computer generation |
| 1975 | Microchips | Greatly down-size computers and enhance their capacities and capabilities |
| 1980 | Vector Computers | Supercomputers currently [ca. 2018] process over 200 quadrillion operations per second [ie., over 200 petaflops] |
| 2010 | Wireless | Electromagnetic signal broadcasting to operate computing devices |

It consisted of a *central processing unit* [CPU] made up of numerous, variably sized wheels that were manually cranked to perform the desired *input* and *output*, *arithmetic*, and *logical* or *Boolean operations*. Modern input operations process data from terminals, and card, magnetic tape, and disk readers, scanners, and other input devices. Output operations distribute data to terminals, storage files and disks, printers, plotters, and other output devices. Arithmetic operations are limited simply to taking data sums and differences that, of course, can be combined to completely describe the analytical properties of any dataset. Boolean operations, by contrast, compare data for values being *equal to*, *greater than*, or *less than* each other within working precision. These logical operations facilitate using variable data processing cycles [e.g., do-loops] in analysis.

In general, the modern computing system involves the same input, CPU, and output operations that Babbage originally developed for his system. However, the technologies are quite different with the early digital mechanical operations being supplanted by modern, faster, higher capacity, and more reliable digital electromagnetic operations. Currently, computing techonology is progressing with a doubling time of about 5 years or a 14% annual growth rate.

## 1.5 KEY CONCEPTS

1. Modern data analysis is based on the internationally recognized Arabic numerals $0, 1, 2, 3, 4, 5, 6, 7, 8,$ and $9$ which can be combined to represent any number no matter how large or small.
2. Data, the plural of datum, are discrete numbers with or without attached units of measurement.
3. Analog data are referenced against continuous quantities [e.g., lengths of a ruler, heights of mercury in a tube, slide rule calculations, etc.].
4. Digital data are referenced against discrete quantities [e.g., clock ticks, fingers, abacus beads, open and closed electronic circuits, etc.].
5. All data analysis is based on the use of simple arithmetic sums and differences to establish the inverse differentiation and integration properties of data, which are related by the two fundamental calculus theorems.
6. Specifically, the First Fundamental Theorem of Calculus states that the signal is the integral to within a constant of its derivative, whereas the Second Fundamental Theorem states that the signal is also the derivative of its integral.
7. Within working precision, the $p$th order derivatives of the $n$-point signal may be estimated from the rise-to-run ratios of successive 3-data point sequences. This process defines $[n - 2p]$ derivative values which are constrained by actual observations. The $p$ undefined derivatives at each end of the signal can be estimated from the boundary value problems established using the two fundamental calculus theorems.
8. Within working precision, the $p$th order integrals of the $n$-point signal may be estimated from summing the incremental trapezoidal areas under the signal between two consecutive data points. This process defines $[n - p]$ integral values that are fully constrained by the observations. The $p$ undefined integrals at the beginning of the signal can be estimated from the boundary value problems established using the two fundamental calculus theorems.

9. Working out a signal's numerical derivative and integral properties fundamentally constrains how environmental parameters may be quantitatively combined to account for the signal. A classical example of this approach is Newton's use of Galileo's free-falling body observations to show that the distance traveled by the body over time $t$ in the earth's gravity acceleration field $g$ is $[gt^2/2]$, and thus independent of its mass.
10. The development of computational devices is worthwhile considering for insightful examples of the elegant operational mechanics that underpin modern data analysis. Since pre-historic times, these devices have focused mostly on digital data and mathematics with the early mechanical operations being now largely supplanted by faster, higher capacity, and more reliable electronic operations.
11. The transformation from mechanical to electronic computations was facilitated greatly by Leibniz's late 17th century formulation of a binary numeral system and Babbage's early 19th century programmable mechanical computer. Babbage's computer was the forerunner of the modern computing system involving a central processing unit [CPU] capable of performing input, output, arithmetic, and logical or Boolean operations.

# 2 Data Attributes

## 2.1 OVERVIEW

*Non-deterministic data refer to numbers with or without attached measurement units that are too poorly understood to relate to environmental parameters. However, their study using graphical, dimensional, and statistical methods may convert them into deterministic data that estimate effective forward modeling or prediction parameters.*

*A data point or datum basically consists of the dependent and independent variable magnitudes and attached measurement units. In electronic computing, the data array or matrix of the dependent magnitudes in proper sequence is the fundamental format for analysis because the independent variable information does not require storage. Rather, it can be computed internally as needed using the array's origin and station interval.*

*With the advent of electronic computing, the matrix has become the dominant mode of modern data analysis. The matrix is a rectangular array of numbers or elements indexed by their row and column numbers. In this book, bold upper case letters mark matrices with two or more rows and columns. The symmetric matrix has symmetrically distributed elements about the diagonal. The transpose of a matrix involves interchanging the rows with the columns. The rows and columns of a matrix are often called its respective row and column vectors, which in this book are marked by bold lower case letters with and without the transpose symbol, respectively.*

*Matrices with the same number of rows and columns may be added and subtracted element-by-element. Multiplication of the matrix with k-columns by n-rows by another matrix with k-rows by m-columns consists of $(n \times m)$-elements where each is the summation of the respective row element-by-column element products. The n-order matrix inverse when multiplied either from the left or the right by the matrix gives the identity matrix with ones in the diagonal and zeros elsewhere. However, the matrix inverse exists only if the scalar value of its determinant is not zero. The above few matrix concepts driven by electronic computing offer considerable advantages in modern digital data analysis.*

*Data units that are the same can be added, subtracted, multiplied, and divided. For an equation to be valid, the measurement units on both sides of the equality sign must be the same - i.e., the equation must be dimensionally homogeneous.*

*Measurement units are based on length, mass, time, temperature, and other primary or base dimensions. Standardizing data makes data dimensionless and simplifies data processing. Modifying each datum by subtracting the data mean*

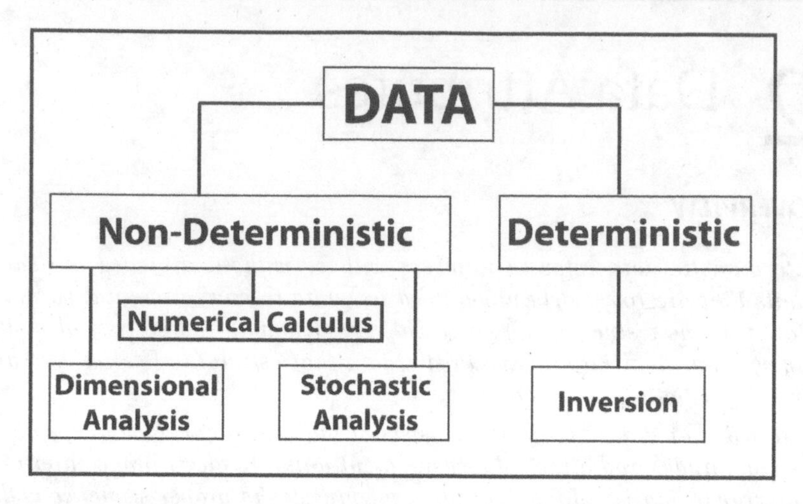


**Figure 2.1** Classification of data based on their analytical properties. Deterministic data may be related to parameters of a forward model by inversion, whereas non-deterministic data lack a quantifying forward model. Thus, to transform non-deterministic into deterministic data, effective forward models are developed using analytical procedures like numerical calculus, dimensional analysis, and statistics.

*and dividing through the standard deviation obtains standardized data with a zero mean and unit standard deviation.*

*Dimensional analysis provides insights on possible quantitative relationships between the data and environmental parameters reflected in the measurement units. Buckingham's $\pi$-theorem isolates [$n - m$] dimensionless groupings of n-parameters in m-primary or base dimensions for comparisons and insights on the functional relationships.*

*Dimensional analysis considerations are also useful for setting up scale models to study wind effects, mountain building, erosion and other environmental phenomena [i.e., prototypes] that are difficult to quantify. Effective scale modeling requires the model and prototype to maintain similitude or similarity in their geometric, kinematic, and dynamic parameters.*

*Dimensional analysis applied to Galileo's falling body measurements also obtains Newton's graphical calculus result, except for the proportionality constant [$\doteq 1/2$] that a scale modeling experiment with known time and distance parameters can provide.*

## 2.2 TYPES

Humans fundamentally operate like stamp collectors or taxonomists in that they are disposed to bin or sort and classify their observations of the environment. The resultant taxonomy typically provides insights for exploiting and communicating the observations. Accordingly, **Figure 2.1** presents a simple classification scheme for data in terms of their basic analytical properties.

Specifically, data range from *non-deterministic* to *deterministic* numbers with or without attached measurement units. Non-deterministic data cannot be described by analytical or mathematical expressions involving variables of natural processes, and hence must be explored using the *numerical calculus*, *dimensional analysis*, and *stochastic* [ie., *statistical*] *analysis* methods. The analysis of non-deterministic digital data is the principal focus of the first five chapters in **Part I** of this book.

Deterministic data, by contrast, are numbers with or without attached measurement units that can be related to equations based on variables of natural processes. The primary objective of analyzing non-deterministic data is to learn enough about them to *predict* or *forward model* their variations, and thus transform them into deterministic data. The forward model involves assumed or known variable values, as well as the unknown variable values that are determined from the data by *inversion*. Solving the inverse problem yields a complete set of variable values for analytically describing the differential and integral properties of data. The analysis of deterministic digital data is the primary focus of the last two chapters in **Part I** that considers fitting data to straight line models involving two unknowns. In **Part II**, this focus is generalized to fitting data to more complex models involving more than two unknowns.

## 2.3 FORMATS

As described in **Chapter 1.3** above, the $i$th data point or datum $[b_i, a_i]$ consists of two fundamental elements of information that include the dependent and independent variable magnitudes $[b_i]$ and $[a_i]$, respectively, along with their applicable measurement units. In electronic analysis, data incorporated as a collection or set of individual data points require maximum processing time and storage commitments to accommodate both fundamental elements. However, where the data are formatted as a uniformly sampled or gridded array in one, two, or more dimensions, only the data values or amplitudes $b_{i(j\cdots)}$ in their proper order or sequence need be considered. The independent variable infomation can be computed internally as needed from the data array using its *reference point* or *origin* and the *station interval*.

These computational efficiencies also hold in the mathematical manipulation of the data array or matrix. The algebra of matrices was pretty well established by the turn of the 20th century. The arduous nature of manually performing matrix operations has been supplanted by the widespread availability of linear equation solvers [e.g., MATLAB, Mathematica, Mathcad, etc.], which allow users to become quickly facile in manipulating matrices for analysis. Accordingly, with the advent of electronic computing, the matrix has become the dominant mode of modern data analysis.

### 2.3.1 ELEMENTARY MATRIX ALGEBRA

A matrix is a rectangular array of numbers such as given by

$$\mathbf{A} = \begin{pmatrix} a_{11} & a_{12} & a_{13} \\ a_{21} & a_{22} & a_{23} \\ a_{31} & a_{32} & a_{33} \\ a_{41} & a_{42} & a_{43} \end{pmatrix}. \tag{2.1}$$

The elements of $\mathbf{A}$ are $a_{ij}$, where the $i$-subscript denotes its row number ($\rightarrow$) and the $j$-subscript is its column number ($\downarrow$). Thus, the matrix $\mathbf{A}$ above has dimensions of 4 rows by 3 columns, or is of order $4 \times 3$ which also may be indicated by the notation $\mathbf{A}_{(4\times3)}$. The matrix is *even-ordered* or *square* if the number of rows equals the number of columns, otherwise it is *odd-ordered* or *non-square*.

The matrix $\mathbf{A}$ is *symmetric* if it is even-ordered and $a_{ij} = a_{ji}$, like in ***Example 02.3.1a*** with the $(3 \times 3)$ matrix

$$\mathbf{A} = \begin{pmatrix} 1 & 3 & 6 \\ 3 & 4 & 2 \\ 6 & 2 & 5 \end{pmatrix}. \tag{2.2}$$

The symmetric matrix of order $n$ can be packed into a singly dimensioned array of $p = n(n+1)/2$ elements. Thus, symmetric matrices are convenient to work with because they minimize storage requirements in computing.

The *transpose* of the $n \times m$ matrix $\mathbf{A}$ is the $m \times n$ matrix $\mathbf{A^t}$ with elements ${a^t}_{ij} = a_{ji}$. Thus, in taking the transpose of a matrix, the rows of the matrix become the columns of its transpose, or equivalently the columns of the matrix become the rows of its transpose. As ***Example 02.3.1b***, note that

$$\mathbf{A} = \begin{pmatrix} 1 & 2 & 3 \\ 4 & 5 & 6 \end{pmatrix} \quad \text{has the transpose} \quad \mathbf{A^t} = \begin{pmatrix} 1 & 4 \\ 2 & 5 \\ 3 & 6 \end{pmatrix}. \tag{2.3}$$

The *column vector* is made up of a single column filled with row elements that, in this book, is designated by a bold lower case letter - e.g.,

$$\mathbf{x} = (x_{1k} \quad x_{2k} \quad \ldots \quad x_{nk})^t \tag{2.4}$$

is a column vector with constant $k$. Its transpose, on the other hand, is the *row vector* made up of a single row filled with column elements - e.g.,

$$\mathbf{x^t} = (x_{k1} \quad x_{k2} \quad \ldots \quad x_{kn}) \tag{2.5}$$

is a row vector with constant row index $k$. The adopted notational convention for matrices and their vectors facilitates understanding the array's role in solving linear systems of simultaneous equations as developed in later sections of this book.

Matrix *addition* and *subtraction*, $\mathbf{C} = \mathbf{A} \pm \mathbf{D}$, involves the element-by-element operations defined by $c_{ij} = a_{ij} \pm d_{ij}$. For the operations to be defined, the matrices $\mathbf{A}$, $\mathbf{D}$, and $\mathbf{C}$ must all have the same dimensions. These operations are *associative* because $(\mathbf{A}\pm\mathbf{D})\pm\mathbf{B} = \mathbf{A}\pm(\mathbf{D}\pm\mathbf{B})$, and *commutative* because $\mathbf{A} \pm \mathbf{D} = \mathbf{D} \pm \mathbf{A}$.

Matrix *multiplication*, $\mathbf{C} = \mathbf{AD}$, involves the row-by-column multiplication of matrix elements given by $c_{ij} = \sum_{k=1}^{n} a_{ik}d_{kj}$. Matrix multiplication is possible only between *conformal* matrices where the dimensions of the product matrices satisfy $\mathbf{C}_{(m\times n)} = \mathbf{A}_{(m\times k)}\mathbf{D}_{(k\times n)}$. Clearly, $\mathbf{C} = \mathbf{AD}$ does not imply that $\mathbf{C} = \mathbf{DA}$, and hence matrix multiplication is not commutative. As ***Example 02.3.1c***, the products of the matrices in **Equation 2.3** are

$$\mathbf{AA}^{\mathrm{t}} = \begin{pmatrix} 14 & 32 \\ 32 & 77 \end{pmatrix} \quad \text{and} \quad \mathbf{A}^{\mathrm{t}}\mathbf{A} = \begin{pmatrix} 17 & 22 & 27 \\ 22 & 29 & 36 \\ 27 & 36 & 45 \end{pmatrix}. \tag{2.6}$$

Note that $\mathbf{A}^{\mathrm{t}}\mathbf{A}$ is a symmetric $(m \times m)$ matrix, whereas $\mathbf{AA}^{\mathrm{t}}$ is a symmetric $(n \times n)$ matrix. These product matrices are important because the matrix *inverse* is defined only for a square matrix.

For matrices $\mathbf{A}$ and $\mathbf{D}$ that are square in the same order, $\mathbf{D}$ is the inverse of $\mathbf{A}$ or $\mathbf{D} = \mathbf{A}^{-1}$ if $\mathbf{DA} = \mathbf{I}$, where

$$\mathbf{I} = \begin{pmatrix} 1 & 0 & \cdots & 0 \\ 0 & 1 & \cdots & 0 \\ \vdots & \vdots & \ddots & \vdots \\ 0 & 0 & \cdots & 1 \end{pmatrix} \tag{2.7}$$

is the identity matrix that has the special property $\mathbf{DI} = \mathbf{ID} = \mathbf{D}$. The inverse $\mathbf{A}^{-1}$ can be found by applying elementary row operations to $\mathbf{A}$ until it is transformed into the identity matrix $\mathbf{I}$. The row operations consist of adding and subtracting linear multiples of the rows from each other. The effects of these row operations on the corresponding elements of the identity matrix give the coefficients of the inverse $\mathbf{A}^{-1}$.

As ***Example 02.3.1d***, to obtain the inverse of the matrix

$$\mathbf{A} = \begin{pmatrix} 75.474 & 1.000 \\ 4.440 & 1.000 \end{pmatrix}, \tag{2.8}$$

elementary row operations are applied to transform $\mathbf{A}$ into the identity matrix $\mathbf{I}$ as shown in the following *row-equivalent* [$\sim$] systems

$$\begin{pmatrix} 75.474 & 1.000 & \vdots & 1.000 & 0.000 \\ 4.440 & 1.000 & \vdots & 0.000 & 1.000 \end{pmatrix} \sim$$

$$\begin{pmatrix} 1.000 & 1/75.474 & \vdots & 1/75.474 & 0.000 \\ 4.440 & 1.000 & \vdots & 0.000 & 1.000 \end{pmatrix} \sim$$

$$\begin{pmatrix} 1.000 & 1/75.474 & \vdots & 1/75.474 & 0.000 \\ 0.000 & 0.941 & \vdots & -0.059 & 1.000 \end{pmatrix} \sim$$

$$\begin{pmatrix} 1.000 & 1/75.474 & \vdots & 1/75.474 & 0.000 \\ 0.000 & 1.000 & \vdots & -0.063 & 1.063 \end{pmatrix} \sim$$

$$\begin{pmatrix} 1.000 & 0.000 & \vdots & 0.014 & -0.014 \\ 0.000 & 1.000 & \vdots & -0.063 & 1.063 \end{pmatrix} \ni$$

$$\mathbf{A}^{-1} = \begin{pmatrix} 0.014 & -0.014 \\ -0.063 & 1.063 \end{pmatrix}. \tag{2.9}$$

Here, the $a_{11}$ coefficient is transformed into the unity coefficient of the identity matrix by dividing 75.474 into the first row of the top system to obtain the second row-equivalent system. To convert the new $a_{21}$ coefficient into the zero coefficient of the identity matrix, the new first row is multiplied by $-4.440$ and added to the second row for the third row-equivalent system. The elementary row operations are continued until $\mathbf{A} \rightarrow \mathbf{I}$ with the coefficients of $\mathbf{A}^{-1}$ in the right-hand side of the final row-equivalent system - i.e., $\mathbf{I} \rightarrow \mathbf{A}^{-1}$. Of course, a test to see if the coefficients for the inverse were correctly determined is to verify that $\mathbf{A}^{-1}\mathbf{A} = \mathbf{A}\mathbf{A}^{-1} = \mathbf{I}$.

However, not every square matrix $\mathbf{A}$ has an inverse. Indeed, the inverse exists only if the *determinant* of $\mathbf{A}$ is not zero - i.e., $|\mathbf{A}| \neq 0$. Specifically, the determinant of the square matrix $\mathbf{A} = (a_{ij})$ of order $n$ is the scalar

$$|\mathbf{A}| = |\mathbf{a_{ij}}| = \begin{vmatrix} a_{11} & \cdots & a_{1n} \\ \vdots & \ddots & \vdots \\ a_{n1} & \cdots & a_{nn} \end{vmatrix} \tag{2.10}$$

given by the simple difference in the sums of the diagonal products. In particular, for order $n = 1$,

$$|\mathbf{A}_{(1\times1)}| = a_{11}, \tag{2.11}$$

whereas for $n = 2$,

$$|\mathbf{A}_{(2\times2)}| = a_{11}a_{22} - a_{12}a_{21}, \tag{2.12}$$

which is just the product of the two elements on the diagonal sloping down to the right minus the product of the elements on the diagonal sloping down to the left.

Note that **Equation 2.12** also is the product of the top first-column element [i.e., $a_{11}$] and the determinant of the submatrix or *minor* obtained by deleting the first row and column [i.e., $a_{22}$] minus the product of the top

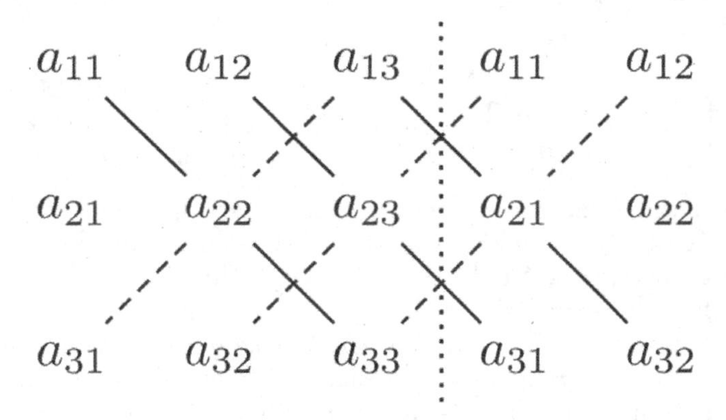


**Figure 2.2** By Sarrus' Rule, $|\mathbf{A}_{(3\times3)}|$ is simply the difference in the 6 diagonal 3-element products, which are added [+] for the diagonals sloping downwards to the right and subtracted [-] for those sloping downwards to the left [**Equation 2.13**].

second-column element [i.e., $a_{12}$] and the determinant of the minor from deleting the first row and second column [i.e., $a_{21}$]. Extending this process for the determinant of the order 3 matrix gives

$$|\mathbf{A}_{(3\times3)}| = a_{11}a_{22}a_{33}+a_{12}a_{23}a_{31}+a_{13}a_{21}a_{32}-a_{31}a_{22}a_{13}-a_{32}a_{23}a_{11}-a_{33}a_{21}a_{12}, \tag{2.13}$$

which is also consistent with the *Rule of Sarrus* in **Figure 2.2**. The example for $n = 4$ in **Figure 2.3** further highlights the fundamentally recursive nature of determinant calculations for orders $n > 1$. The determinant involves a sum with $n!$ terms, and thus can also be rather laborious to compute.

In general, a test for the existence of an inverse $\mathbf{A}^{-1}$ for $\mathbf{A}$ is to see if $|\mathbf{A}| \neq 0$. If the determinant is zero, then the matrix $\mathbf{A}$ is said to be *singular*. Matrix singularity occurs if a row or column is filled with zeroes or two or more rows or columns are linear multiples of each other. As ***Example 02.3.1e***, the

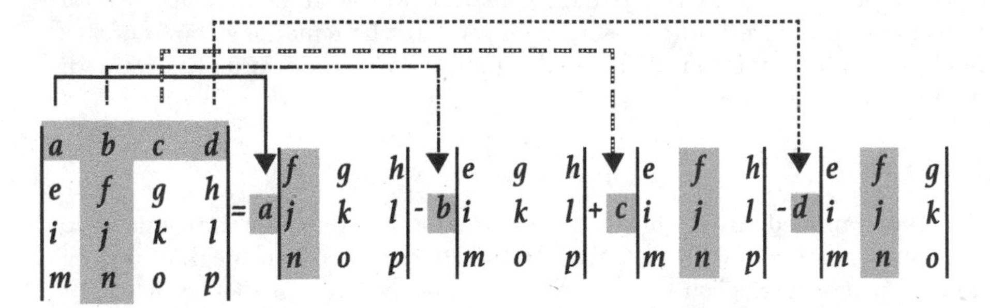


**Figure 2.3** Finding $|\mathbf{A}_{(4\times4)}| = \sum_{j=1}^{4}(-1)^{j-1}a_{1j} \times |\mathbf{M}_{1j}|$ recursively using determinants of the $(3\times3)$ minors $\mathbf{M}_{1j}$ formed by deleting the first row and $j$th column of $\mathbf{A}_{(4\times4)}$.

matrices given by

$$\mathbf{A1} = \begin{pmatrix} 1.0 & 1.0 & 0.0 \\ 3.0 & 7.0 & 0.0 \\ 1.7 & 2.7 & 0.0 \end{pmatrix} \quad \text{and} \quad \mathbf{A2} = \begin{pmatrix} 1.0 & 1.0 & 1.0 \\ 3.0 & 7.0 & 2.0 \\ 2.7 & 2.7 & 2.7 \end{pmatrix}$$

are both singular and have zero determinants. Specifically, $|\mathbf{A1}| = 0$ because the 3rd column is filled with zeroes so that each product element of the determinant is zero and thus $\mathbf{A1}^{-1}$ does not exist. Additionally, within working precision $|\mathbf{A2}| = 0$ because the 3rd row is the multiple of 2.7 times the 1st row. The rows 1 and 3 are *linearly dependent* or *co-linear* and can zero each other out with simple row operations so that $\mathbf{A2}^{-1}$ also does not exist.

With the determinant, conceptually simpler estimates of elements of the inverse $\mathbf{A}^{-1} = \mathbf{D}$ result from $d_{ij} = |\mathbf{D_{ij}}|/|\mathbf{A}|$. Here, the *augmented matrix* $\mathbf{D_{ij}}$ is obtained from $\mathbf{A}$ by replacing the $i$th column of $\mathbf{A}$ with the $j$th column of the identity matrix $\mathbf{I}$. As ***Example 02.3.1f***, for the matrix $\mathbf{A}$ in **Equation 2.8** with $|\mathbf{A}| = 71.034$, the elements of its inverse [i.e., **Equation 2.9**] are

$$d_{11} = (\frac{1.000}{71.034}) = 0.014, \quad d_{12} = (\frac{-4.440}{71.034}) = -0.063, \tag{2.14}$$
$$d_{21} = (\frac{-1.000}{71.034}) = -0.014, \text{ and } d_{22} = (\frac{75.474}{71.034}) = 1.063.$$

Although this approach for finding $\mathbf{A}^{-1}$ may be conceptually more straightforward than using the elementary row operations of **Equation 2.9**, it also becomes significantly more laborious to compute with increasing order $n$ because each determinant evaluation requires estimating $n!$ terms of the summation.

In summary, this section introduced the organization of the elements for a handful of widely used matrices in data analysis [e.g., the general, symmetric, transpose, vector, and identity matrix]. In addition, a comparable number of matrix operations [e.g., matrix addition/subtraction, multiplication, inverse, and determinant] were described that are readily implemented by electronic equation solvers [e.g., MATLAB, Mathematica, Mathcad, etc.]. These several concepts will be incorporated as appropriate into the remaining topics of this book to highlight the considerable advantages of the array format in modern digital data analysis.

## 2.4 UNITS

Originally applied to weights and measures, units were first introduced to facilitate commerce with each trading area having its own local system of units. The use of English pounds, shillings, and pence as monetary units, or degrees, minutes, and seconds to measure angles, was simply a convenience to avoid the use of fractions. However, this convenience bred a diversity of systems that became a severe handicap as trade expanded and scientists and engineers began to exchange data.

Towards the end of the 18th century, the problems were so acute that, with the exception of England and its colonies, the European countries began adopting the metric system based on the meter, kilogram, and second [ie., the MKS system of units or MKSu]. The metric system has the distinct advantage that its units and subunits are related as powers of ten. Thus, any quantity can be expressed as a multiple or a decimal fraction of a single unit of measurement. The metric system presently enjoys almost universal use among scientists, and in national and international trade and engineering [e.g., **[66]**].

In the US, the use of the metric system was legalized by the Congress in 1866. The "Mendenhall Order" of 1893 required all legal US units of measure be of metric units or exact numerical multiples of metric units. The order was issued by Superintendent T.C. Mendenhall of the US Bureau of Standard Weights and Measures. Mendenhall, a geodesist, also was the Superintendent of the US Coast and Geodetic Survey and the first Ohio State University professor of physical and mechanical sciences hired when the university was founded in 1870.

Despite the long official commitment of the US to the metric system, its assimilation into practice has been slow and incomplete with the English foot-pound-second units still largely dominant in engineering and commerce. This lack of commitment has not been inconsequential to US national interests. For example, NASA's loss of the 125 million USD Mars Climate Orbiter mission in late 1999 is attributed to a mix-up in the units of measurement. NASA's Jet Propulsion Laboratory in Pasadena, CA was applying orbital course corrections in metric units that the maker of the spacecraft, Lockheed Martin in CO, was supplying in English units. The resulting orbital errors caused the spacecraft to enter the Martian atmosphere too low and be either burned up or torn apart. A similar mix-up of units also may have contributed to the late-1999 failure of NASA's Mars Polar Lander mission.

The metric system is formally based on a length unit of the meter [m] defined by the 1875 Treaty of the Meter, a mass unit of the kilogram [kg], an atomic time unit of the second [s], an electric current unit of the ampere [A], a temperature unit of the degree-Kelvin [$^o$K], and a luminous intensity unit of the candela [cd]. In 1960, at the Eleventh General Conference on Weights and Measures convened in Paris, the metric system of units was adopted as the "International System of Units" and given the French name "Systéme international d'unités" which is abbreviated "SI" or "SIu" in all languages. The International System of Units has pretty much achieved global dominance and is recommended for all scientific, technical, practical, and teaching purposes [e.g., **[66]**].

### 2.4.1 DIMENSIONAL ANALYSIS

Data dimensions refer to the kind of units attached to every measurement. Like units can be added, subtracted, multiplied and divided. Furthermore, every valid equation exhibits *dimensional homogeneity* in that the units on both sides of the equality sign must be the same. Thus, dimensional homogeneity is an effective check on the veracity of any equation being considered for application or undergoing development. It also complements the numerical calculus as another rudimentary approach for obtaining insights on quantifying data in terms of applicable environmental parameters.

Dimensional analysis is based on the requirement for *homogeneous equations* to have consistent units. Fundamental dimensions of measurement include length $L$ in cm, ft, etc., mass $M$ in gm, slug, etc., and time $T$ in s, min, hr, day, etc. In mechanics, the fundamental dimensions also commonly include force $F$ such as from Newton's Second Law of Motion,

$$F = M \times a, \tag{2.15}$$

for mass $M$, and acceleration $a$ that is the velocity per unit time or distance per unit time per unit time. Thus, the dimensions of force are $F = MLT^{-2}$ and those of mass are $M = FL^{-1}T^2$ so that force may be expressed in the unit of mass and vice versa. Accordingly, a quantity like stress, which is the force per unit area, can be expressed in the dual units of $FL^{-2}$ and $ML^{-1}T^{-2}$. Other physical quantities of common interest in environmental studies are listed in **Table 2.1** along with their dimensional formulae.

Reducing the dimensional complexity of equations yields further insights on the involved variables and their relationships to each other. As ***Example 02.4.1a***, to make an equation like $x = y + \log(z)$ dimensionally homogeneous, the units of $x$, denoted by :($x$):, must equal the units :($y$): and :($\log(z)$): of $y$ and $\log(z)$, respectively - ie., :($x$): = :($y$): = :($\log(z)$):. This requirement is most easily satisfied if the variables $x, y$, and $\log(z)$ are *dimensionless* - ie., :($x$): = 1, :($y$): = 1, and :($\log(z)$): = 1.

In general, combining variables into dimensionless groups is useful for satisfying homogeneity requirements, searching out functional relationships in multi-parameter processes, and processing experimental data. For example, any dataset can be made dimensionless by dividing each value of the dataset by the dataset's standard deviation. If the mean of the dataset has also been subtracted from each value, the data are said to be *starndardized*. Dimensionless standardized data have a zero mean and unit standard deviation, and thus indeed permit quantitative comparisons of apples and oranges.

Plotting data from dimensionless groups of variables against each other can often yield insight into functional relationships for developing the theory of a process. A method for obtaining the minimum number of dimensionless groups of the variables in a process is given by the *Buckingham* $\pi$ *theorem* **[12]**. To apply the $\pi$ theorem, let $n$ be the number of *linearly independent* variables that include the $m$ base dimensions [e.g., $F, L, T$, etc.], and $\pi_i$ be the $i$th dimensionless group of variables. By the $\pi$ theorem, the number of dimensionless groups or $\pi$'s is $[n - m]$.

**Table 2.1**
**Common environmental quantities expressed in primary or base dimensions of $M, L, T$ and $F, L, T$. Dimensionless quantities are designated by the numeral 1.**

| QUANTITY | $M, L, T$ | $F, L, T$ |
|---|---|---|
| Length | $L$ | $L$ |
| Area | $L^2$ | $L^2$ |
| Volume | $L^3$ | $L^3$ |
| Time | $T$ | $T$ |
| Mass | $M$ | $FL^{-1}T^2$ |
| Velocity | $LT^{-1}$ | $LT^{-1}$ |
| Acceleration | $LT^{-2}$ | $LT^{-2}$ |
| Force, Load | $MLT^{-2}$ | $F$ |
| Mass Density | $ML^{-3}$ | $FL^{-4}T^2$ |
| Specific or Unit Weight | $ML^{-2}T^{-2}$ | $FL^{-3}$ |
| Angle | 1 | 1 |
| Angular Velocity | $T^{-1}$ | $T^{-1}$ |
| Angular Acceleration | $T^{-2}$ | $T^{-2}$ |
| Pressure or Stress | $ML^{-1}T^{-2}$ | $FL^{-2}$ |
| Work or Energy | $ML^2T^{-2}$ | $FL$ |
| Power | $ML^2T^{-3}$ | $FLT^{-1}$ |
| Momentum | $MLT^{-1}$ | $FT$ |
| Force Moment | $ML^2T^{-2}$ | $FL$ |
| Inertia Moment of Area | $L^4$ | $L^4$ |
| Inertia Moment of Mass | $ML^2$ | $FLT^2$ |
| Strain | 1 | 1 |
| Poisson's Ratio | 1 | 1 |
| Elasticity Modulus | $ML^{-1}T^{-2}$ | $FL^{-2}$ |
| Rigidity Modulus | $ML^{-1}T^{-2}$ | $FL^{-2}$ |
| Bulk Modulus | $ML^{-1}T^{-2}$ | $FL^{-2}$ |

The $\pi$ theorem's application requires the selection of linearly independent variables that cannot be expressed as a linear combination of the base dimensions of other variables. As ***Example 02.4.1b***, the circumference, radius and diameter all have the same length dimension $L$. Thus, they are linearly related quantities so that only one of them may be selected for the $\pi$ theorem analysis of any process involving all three of the quantities.

To form the $\pi_{[n-m]}$-equations, $[n - m]$ of the variables are selected and each one multiplied against the exponential products of the other $[n-m-1]$-variables. The dimensions for each variable are inserted into each $\pi$-equation and the exponential indices are solved for assuming that the sum of the indices of the base dimensions [eg., $F, L, T$, etc.] must equal zero for the $\pi$-equation to be dimensionless.

**Table 2.2**
**Physical variables of potential significance for using the rebound height of a dropped hammer to estimate the compression strength of rocks.**

| QUANTITY | Symbol | Units |
|---|---|---|
| **1)** Rock's Uniaxial Compression Strength | $Q$ | $FL^{-2}$ |
| **2)** Rock's Deformation Modulus | $E$ | $FL^{-2}$ |
| **3)** Rock's Viscosity | $n$ | $FL^{-2}T$ |
| **4)** Rock's Density | $\rho$ | $FL^{-4}T^2$ |
| **5)** Hammer's Mass | $M$ | $FL^{-1}T^2$ |
| **6)** Hammer's Drop Height | $H_o$ | $L$ |
| **7)** Hammer's Rebound Height | $H_r$ | $L$ |
| **8)** Gravitational Acceleration | $g$ | $LT^{-2}$ |

As ***Example 02.4.1c***, consider a research project seeking to estimate the compression strength of a rock from the rebound height of a drop hammer. Establishing such an *in-situ* capability may yield a cheaper and more effective test than recovering and shipping core samples of the rock to a laboratory for strength testing. A review of possible variables and base dimensions involved here might suggest the list in **Table 2.2**.

By the $\pi$ theorem, accordingly, the variables from **Table 2.2** can be combined into $[n-m] = [8-3] = 5$ dimensionless groups or $\pi$-equations, which include

$$\begin{aligned}
\pi_1 = \quad Q^{x_1} H_o{}^{y_1} g^{z_1} M &= \quad [FL^{-2}]^{x_1}[L]^{y_1}[LT^{-2}]^{z_1} FL^{-1}T^2, \\
\pi_2 = \quad Q^{x_2} H_o{}^{y_2} g^{z_2} E &= \quad [FL^{-2}]^{x_2}[L]^{y_2}[LT^{-2}]^{z_2} FL^{-2}, \\
\pi_3 = \quad Q^{x_3} H_o{}^{y_3} g^{z_3} \rho &= \quad [FL^{-2}]^{x_3}[L]^{y_3}[LT^{-2}]^{z_3} FL^{-4}T^2, \\
\pi_4 = \quad Q^{x_4} H_o{}^{y_4} g^{z_4} H_r &= \quad [FL^{-2}]^{x_4}[L]^{y_4}[LT^{-2}]^{z_4} L, \text{ and} \\
\pi_5 = \quad Q^{x_5} H_o{}^{y_5} g^{z_5} n &= \quad [FL^{-2}]^{x_5}[L]^{y_5}[LT^{-2}]^{z_5} FL^{-2}T.
\end{aligned} \tag{2.16}$$

Because each $\pi$ group must be dimensionless, the sum of the exponents for each base dimension also must be zero. Thus, for each $\pi$-factor there are three equations to solve for the appropriate magnitudes of the exponents. Specifically, the exponents for the $\pi_1$ group are obtained from

$$\begin{aligned}
F: &\quad x_1 + 1 = 0 && \ni\ x_1 = -1, \\
L: &\quad -2x_1 + y_1 + z_1 - 1 = 0 && \\
&\quad \ni\ y_1 + z_1 = -1 && \text{since } x_1 = -1, \text{ and} \\
T: &\quad -2z_1 + 2 = 0 && \ni\ z_1 = 1 \text{ and } y_1 = -2.
\end{aligned} \tag{2.17}$$

Thus,

$$\pi_1 = Q^{-1} H_o^{-2} g M = Mg/QH_o^2. \tag{2.18}$$

Note, however, that the above equations for the exponents of the primary dimensions also form a linear system of simultaneous equations given by

$$\mathbf{A}_{\pi_1}\mathbf{x}_{\pi_1} = \mathbf{b}_{\pi_1}, \tag{2.19}$$

where by the definition of matrix multiplication

$$\mathbf{A}_{\pi_1} = \begin{pmatrix} 1 & 0 & 0 \\ -2 & 1 & 1 \\ 0 & 0 & -2 \end{pmatrix} \text{ and } \mathbf{x}_{\pi_1} = \begin{pmatrix} x_{11} = x_1 \\ x_{21} = y_1 \\ x_{31} = z_1 \end{pmatrix}, \text{ with } \mathbf{b}_{\pi_1} = \begin{pmatrix} -1 \\ 1 \\ -2 \end{pmatrix}. \tag{2.20}$$

Now, because the determinant $|\mathbf{A}_{\pi_1}| = -2 \neq 0$, the variables $Q, H_o$, and $g$ in the respective $x_1-, y_1-$, and $z_1-$exponents are linearly independent and the inverse $\mathbf{A}_{\pi_1}^{-1}$ exists. Accordingly, the solution is

$$\mathbf{x}_{\pi_1} = \mathbf{A}_{\pi_1}^{-1}\mathbf{b}_{\pi_1} = \begin{pmatrix} x_{11} = x_1 = -1 \\ x_{21} = y_1 = -2 \\ x_{31} = z_1 = \phantom{-}1 \end{pmatrix}, \tag{2.21}$$

which is the same as that obtained previously in **Equation 2.17.**

Similarly for the $\pi_2$ group, the exponents are derived from

$$\begin{aligned} F: &\quad x_2 + 1 = 0 && \ni\ x_2 = -1, \\ L: &\quad -2x_2 + y_2 + z_2 - 2 = 0 && \\ &\quad \ni\ y_2 + z_2 = 0 && \text{since } x_2 = -1, \text{ and} \\ T: &\quad -2z_2 = 0 && \ni\ z_2 = 0 \text{ and } y_2 = 0, \end{aligned} \tag{2.22}$$

which is the same solution that the equivalent matrix form of the above linear equation system also obtains with $\mathbf{A}_{\pi_2} = \mathbf{A}_{\pi_1}$ and $\mathbf{b}_{\pi_2} = [-1, 2, 0]^t$. Accordingly, the second dimensionless grouping of variables is

$$\pi_2 = Q^{-1}E = E/Q. \tag{2.23}$$

Likewise, for the $\pi_3$ group, the exponents are obtained from

$$\begin{aligned} F: &\quad x_3 + 1 = 0 && \ni\ x_3 = -1, \\ L: &\quad -2x_3 + y_3 + z_3 - 4 = 0 && \\ &\quad \ni\ y_3 + z_3 = 2 && \text{since } x_3 = -1, \text{ and} \\ T: &\quad -2z_3 + 2 = 0 && \ni\ z_3 = 1 \text{ and } y_3 = 1, \end{aligned} \tag{2.24}$$

or from $\mathbf{A}_{\pi_3} = \mathbf{A}_{\pi_1}$ with $\mathbf{b}_{\pi_3} = [-1, 4, -2]^t$, so that

$$\pi_3 = Q^{-1}H_o g\rho = H_o g\rho/Q. \tag{2.25}$$

For the $\pi_4$ group, on the other hand, the exponents are found from

$$\begin{aligned} F: &\quad x_4 = 0 && \ni\ x_4 = 0, \\ L: &\quad 2x_4 + y_4 + z_4 + 1 = 0 && \\ &\quad \ni\ y_4 + z_4 = -1 && \text{since } x_4 = 0, \text{ and} \\ T: &\quad -2z_4 = 0 && \ni\ z_4 = 0 \text{ and } y_4 = -1, \end{aligned} \tag{2.26}$$

or from $\mathbf{A}_{\pi_4} = \mathbf{A}_{\pi_1}$ with $\mathbf{b}_{\pi_4} = [0, -1, 0]^t$, so that

$$\pi_4 = Q^0 H_o^{-1} g^0 H_r^1 = H_r / H_o. \tag{2.27}$$

Finally, the exponents for the $\pi_5$ group are derived from

$$\begin{array}{llll} F: & x_5 + 1 = 0 & \ni \; x_5 = -1, & \\ L: & -2x_5 + y_5 + z_5 - 2 = 0 & & (2.28) \\ & \ni \; y_5 + z_5 = 0 & \text{since } x_5 = -1, \text{ and} & \\ T: & -2z_5 + 1 = 0 & \ni \; z_5 = 1/2 \text{ and } y_5 = -1/2. & \end{array}$$

or from $\mathbf{A}_{\pi_5} = \mathbf{A}_{\pi_1}$ with $\mathbf{b}_{\pi_5} = [-1, 2, -1]^t$, and thus

$$\pi_5 = Q^{-1} H_o^{-1/2} g^{1/2} n = [n/Q]\sqrt{g/H_o}. \tag{2.29}$$

As the $\pi$-factors are presumed to be related, each one can be written as an equation of one or more of the remaining $\pi$-factors - e.g.,

$$\pi_1 = f[\pi_2, \pi_3, \pi_4, \pi_5] \;\ni\; [\frac{Mg}{QH_o^2}] = f[(\frac{E}{Q}), (\frac{g\rho H_o}{Q}), (\frac{H_r}{H_o}), (\frac{n}{Q}\sqrt{\frac{g}{H_o}})]. \tag{2.30}$$

If a variable occurs in only one $\pi$-factor, the solution is direct. With respect to the objective of relating the rock's compression strength $Q$ to the hammer's rebound height $H_r$, the relationship between the groups $\pi_3$ and $\pi_4$ seems most promising. Equating these two factors suggests that $Q$ may be proportional to $(g\rho)(H_o^2/H_r)$, or perhaps more simply to $K(H_o^2/H_r)$, where taking $g\rho$ as the constant $K$ acknowledges that variations in $g$ and $\rho$ may be negligible in practice. However, the suggested relationship must be tested with experimental data, as it could be completely wrong if the dimensional analysis has missed a significant variable [e.g., temperature] of the process.

Solving for a variable that occurs in more than one of the dimensionless groups requires a trial-and-error approach. The experimental data from the project would be grouped into the various $\pi$-factors, and curves obtained by plotting one $\pi$-factor against another analyzed for possible functional relationships [e.g., **[71, 90]**].

### 2.4.2 SIMILITUDE

Environmental phenomena that are difficult to quantify are commonly studied using scale modeling experiments. For example, tornado effects can be simulated with wind machines, mountain building processes studied using clay tables, earthquake and landslide effects evaluated using shake tables, and river, lake and ocean shoreline erosion modeled using water and sand tables, etc.

In general, dimensional analysis considerations are necessary for developing effective scale models of natural phenomena. In these studies, the *prototype* refers to the natural phenomenon under study, whereas the *model* relates to the scaled simulation of the prototype. Effective scale modeling experiments require the model and prototype to maintain *geometric similarity*, *kinematic similarity*, and *dynamic similarity*.

Geometric similarity involves scaling all pertinent geometric dimensions so that their ratios are equal in the model and prototype. Kinematic similarity equalizes the ratios of temporal elements, velocities, and accelerations between model and prototype. Dynamic similarity, on the other hand, involves scaling all forces so that their ratios are equal between model and prototype.

Similitude is established where the above three similarity conditions are fulfilled. However, in practice, it is often difficult, if not impossible, to simultaneously satisfy all the requirements for similitude. An alternate approach to satisfying similitude conditions is to maintain the significant $\pi$-factors equal between model and prototype.

As ***Example 02.4.2a***, suppose setting up a model to examine the deflection, $d$, of a simple beam under a concentrated load, $P$. Let the beam with elasticity modulus $E$ and thickness or depth $D$ span the distance or length $B$. Then two of the dimensionless parameters might be

$$\pi_1 = d/B \text{ and } \pi_2 = E(D^2/P). \tag{2.31}$$

If $\pi_1$ is the same for the model and prototype, then geometric similarity is maintained, whereas dynamic similarity is obtained if $\pi_2$ is constant.

Suppose the investigator now wants to test the material of the prototype beam directly in the scale model, where the elasticity modulus $E$ would not be scaled down. Thus, if the geometric scale is (1/50), the load $P$ would have to be scaled up $[50^2 = 2,500]$ times to keep $\pi_2$ constant. Accordingly, structural scale modeling often requires a centrifuge to produce the large loading increase necessary to offset the geometric scale factor and fulfill similitude.

Another approach to managing similitude is to consider the scaling factors individually that are required for the modeling. In maintaining geometric similarity, the length scale factor is

$$K_L = L_m/L_p \tag{2.32}$$

for model length $L_m$ and prototype length $L_p$. Additionally, because areas and volumes vary as $L^2$ and $L^3$, respectively, the respective area and volume scale factors are

$$K_A = K_L^2 \text{ and } K_{vol} = K_L^3. \tag{2.33}$$

For kinematic similarity, on the other hand, where velocity $V = \partial x/\partial t \ni \partial x = V\partial t$, the scale factors may be written as

$$\begin{aligned} K_L &= (\partial x)_m/(\partial x)_p &&= K_V K_t &&\rightarrow \text{geometric scale factor,} \\ K_V &= V_m/V_p &&&&\rightarrow \text{velocity scale factor,} \\ K_t &= (\partial t)_m/(\partial t)_p &&&&\rightarrow \text{time scale factor, and} \\ K_a &= K_L/K_t^2 &&= K_V^2/K_L &&\rightarrow \text{acceleration scale factor.} \end{aligned} \tag{2.34}$$

Furthermore, for dynamic similarity, the force scale factor from Newton's Second Law [**Equation 2.15**] is

$$K_F = K_M K_a. \tag{2.35}$$

As ***Example 02.4.2b***, consider the problem of extrapolating the crater dimensions [eg., crater radii] from small test explosions in the ground to those from large scale explosions in the same ground. This type of testing is typically done to estimate the amount of explosive needed to excavate a desired volume of ground.

From **Equations 2.34** and **2.35**, the force scale factor may also be written as

$$K_F = K_M K_V^2 / K_L. \tag{2.36}$$

Now, the strain, $\epsilon$, and ground's particle velocity, $V$, and the propagation velocity of the shock wave, $C$, are related by

$$\epsilon = V/C. \tag{2.37}$$

Thus, the scale factor for strain is

$$K_\epsilon = K_V / K_C, \tag{2.38}$$

so that the force scale factor may be written as

$$K_F = K_M (K_\epsilon K_C)^2 / K_L. \tag{2.39}$$

Because the force, $F$, pressure, $P$, and area, $A$, are related by $F = P \times A$, it follows that

$$K_F = K_M (K_\epsilon K_C)^2 / K_C = K_P K_L^2 \tag{2.40}$$

or

$$K_L^3 = K_M (K_\epsilon K_C)^2 / K_P. \tag{2.41}$$

In both the model and prototype, the stresses and strains at the boundaries of the broken ground are equal so that $K_P = 1 = K_\epsilon$. In addition, the same ground also propagates the shock in the model and prototype so that $K_C = 1$. Thus,

$$K_L = K_M^{1/3}. \tag{2.42}$$

As ***Example 02.4.2c***, suppose 100 gm of an explosive produces a 5-m crater in the ground. Thus, by **Equation 2.42**, the amount of the explosive needed to excavate a 50-m crater is 100 kgm, where $K_M = M_m/M_p \ni [5/50]^3 = [1/10^3] = 100/M_p$. In other words, the scale factor for extrapolating crater dimensions is the cube root of the mass of the explosives used, which is a well known ground blasting and excavation result in the geotechnical community [e.g., [**16**]].

**Figure 2.4** High-speed photos showing the time history of the 'ball of fire' from the Trinity atomic bomb test at Alamagordo, NM in 1945. [Adapted from [**91**].]

A famous example [***Example 02.4.2d***] of the power and elegance of dimensional analysis is a study of high-speed photographs of the Trinity atomic bomb test at Alamagordo, NM in 1945 [**91**]. The photographs [e.g., **Figure 2.4**] were available in de-classified movies of nuclear tests and also published by *Life Magazine* in 1947 when the yield of the bomb in the test was still strictly classified. However, to the surprise of the American intelligence community, a good estimate of the bomb's yield was publically available to anyone viewing the photos and proficient in dimensional analysis - including Soviet scientists.

To see this, note that a nuclear explosion involves an almost instantaneous release of the bomb's energy or yield $Y$ at an effective point region of space. The initial air pressure may be ignored because of the immensely greater pressure of the spherical shock wave produced by the explosion. Other variables of apparent significance include the radius $R$ of the shock wave, the yield $Y$, time $t$, and the initial air density $\rho_o$ with respective dimensions $:[R]: = L$, $:[Y]: = ML^2T^{-2}$, $:[t]: = T$, and $:[\rho_o]: = ML^{-3}$. By the $\pi$ theorem, there is only a single dimensionless factor for the four variables in three base units or dimensions.

Specifically, the dimensionless group that should be constant during the shockwave's expansion is

$$\begin{aligned}\pi_1 &= Y^{x_1}\rho_o{}^{y_1}t^{z_1}R &&= [ML^2T^{-2}]^{x_1}[ML^{-3}]^{y_1}[T]^{z_1}L \\ & &&= M^{x_1+y_1}L^{2x_1-3y_1+1}T^{-2x_1+z_1},\end{aligned} \quad (2.43)$$

with the solution $x_1 = [-1/5]$, $y_1 = [1/5]$, and $z_1 = [-2/5]$, so that

$$\pi_1 = [Y^{-1/5}\rho_o{}^{1/5}t^{-2/5}]R = K, \quad (2.44)$$

where $K$ is an undetermined constant. The yield, accordingly, may be obtained from

$$Y = K^{-1/5}[\rho_o t^{-2}]R^5 = \frac{\rho_o R^5}{K^5 t^2}. \quad (2.45)$$

The US Army provided de-classified photos of the test with spatial scales and time stamps to effectively estimate $R$ and $t$, respectively. Taylor determined the constant of proportionality as $K \approx 1$ by scaling the results of a numerical model of a shockwave, but its value could also be determined from scale model experiments of the nuclear test using lower energy explosives [e.g., dynamite]. Thus, an accurate estimate of around 25 kilotons for the yield of the Trinity atom bomb test was obtained from the de-classified photos and dimensional analysis.

### 2.4.3 SUMMARY

Following the mid 17th-century discovery of calculus for using the graphical attributes of data to recover their analytical properties, the basic concepts of similarity and dimensional analysis were being worked out near the beginning of the 19th-century. By the early 20th-century, the power of dimensional homogeneity for quantifying relationships between variables of a process was well established [e.g., **[12, 9]**]. This approach finds wide application in earth science and engineering for revealing relationships between variables of poorly understood processes, as well as difficult-to-quantify processes that must be studied by scale model experiments [e.g., **[49, 81, 5, 90]**].

Both data exploration approaches yield consistent results for any valid equation of a process. As ***Example 2.4.3a***, consider **Equation 1.5** for the fall of a body in a vacuum that the calculus obtained from the graphical analysis of experimental data [e.g., in middle-right insert of **Figure 1.2**]. In Galileo's time, apparently significant variables included the mass $m$, as well as the distance fallen $f(t)$, acceleration of gravity $g$, and time $t$ with respective dimensions $:[m]: = M$, $:[f(t)]: = L$, $:[g]: = LT^{-2}$, and $:[t]: = T$. By the $\pi$ theorem then, only a single dimensionless factor of the four variables in the three base units or dimensions is possible.

The dimensionless group, accordingly, is

$$\begin{aligned}\pi_1 &= m^{x_1}g^{y_1}t^{z_1}f(t) &&= [M]^{x_1}[LT^{-2}]^{y_1}[T]^{z_1}L \\ & &&= M^{x_1}L^{y_1+1}T^{-2y_1+z_1},\end{aligned} \quad (2.46)$$

with the solution $x_1 = [0]$, $y_1 = [-1]$, and $z_1 = [-2]$, so that

$$\pi_1 = [m^0 g^{-1} t^{-2}] f(t) = K, \tag{2.47}$$

where $K$ is an undetermined constant. However, plugging in the results from an observation [i.e., an experimental datum] will show that $K = [1/2]$. Thus, dimensional analysis indicates that the distance a body falls is indeed given by **Equation 1.5** and also independent of its mass.

In principle, any valid equation of a process can be derived from exploring the related data graphics and/or measurement units using the respective calculus and dimensional analysis. The next chapter describes a third approach to establishing a valid equation based on the affiliated data errors and their propagation.

## 2.5 KEY CONCEPTS

1. Data are numbers with or without attached measurement units that are classified non-deterministic if they cannot be quantitatively forward modeled, and deterministic where they can be related to a modeling equation based on environmental parameters. In general, environmental parameters may be related via numerical, statistical, and dimensional analyses to non-deterministic data, and by inversion to deterministic data.
2. The datum consists of dependent and independent variable magnitudes and attached measurement units. The dominant format for electronic analysis is the data array where only the dependent magnitudes in proper order or sequence are needed with the independent variables computed internally as necessary from the array's origin and station interval.
3. Several array or matrix concepts dominate electronic computing including the general matrix with two or more rows and columns, the symmetric matrix with upper and lower triangular elements that are symmetric across the diagonal, the matrix transpose with interchanged rows and columns, the $1D$ row or column vector, and the identity matrix with ones in the diagonal and zeros elsewhere.
4. In this book, bold upper case letters signify matrices with two or more rows and columns, whereas bold lower case letters with and without the transpose symbol respectively mark row and column vectors.
5. The roughly handful of matrix operations that dominate electronic computing include adding and subtracting two matrices with common numbers of rows and columns, multiplying a matrix with $k$-columns by a matrix with $k$-rows, taking the inverse of a matrix where the numbers of rows or columns are equal, which is possible only if the scalar value of its determinant is not zero.

6. Like measurement units can be added, subtracted, multiplied, and divided. Any valid equation is dimensionally homogeneous with the same measurement units on both sides of the equality sign.
7. Measurement units are based on length, mass, time, temperature, and other primary or base dimensions.
8. Standardizing data makes data dimensionless to simplify data processing. Standardized data have a zero mean and unit standard deviation, and result by subtracting the data's mean from each datum and dividing the difference by the data's standard deviation.
9. Dimensional analysis can help to quantify relationships between data and their measurement units. By Buckingham's $\pi$-theorem, there are $[n - m]$ dimensionless groupings of $n$-parameters in $m$-primary or base dimensions that can be isolated and compared for the functional relationships.
10. Dimensional analysis also is useful for simulating difficult-to-quantify environmental phenomena with scale models of effectively similar geometric, kinematic, and dynamic properties.
11. Newton's gravity theory also can be obtained from the dimensional analysis of Galileo's free-falling body observations to within the proportionality constant $[= 1/2]$ that, in turn, can be determined from any scale modeling experiment with known time and distance parameters.

# 3 Error Analysis

## 3.1 OVERVIEW

*All measurements and calculations contain uncertainties [e.g., systematic, random, calculation, and sampling errors] that must be properly estimated and propagated to obtain effective conclusions in data analysis. Indeed, the analysis of errors is a third fundamental approach for quantifying data variations in terms of environmental parameters.*

*Measures of error include accuracy, which refers to how close an observation is to the true value either known or presumed, and precision that refers to an observation's repeatability. The difference between a precise and true observation is the observation's bias.*

*Absolute precision refers to the observation ± the error given in the same measurement units as the observation, whereas relative precision indicates the uncertainty as a fraction or percentage of the observation's magnitude.*

*Absolute error is the absolute magnitude of the difference between the true and estimated values, whereas relative error is the absolute error divided by the estimated value given in either a fraction or percentage.*

*The minimum number of digits required to report a value without loss of accuracy is the number of significant figures, which varies according to the arithmetic [i.e., addition, subtraction, multiplication, division] and rounding off operations involved.*

*For a function involving $m$ measured parameters with related uncertainties, the propagated error of the function is given using Taylor's theorem. To first order accordingly, the function's uncertainty is the sum of $m$ terms with each term given by the product of a parameter's uncertainty times the function's change with respect to that parameter's change.*

*Dividing the propagated error by the function yields the function's propagated relative error.*

## 3.2 ERROR TYPES

No observation or measurement can be made without error. Thus, to make effective conclusions based on observations, it is necessary to estimate errors and know how they propagate in data analysis. A principal application of statistics is to quantify data uncertainties that broadly include systematic, random, calculation, and sampling errors.

*Systematic errors* result from faulty measurement equipment, observer biases [e.g., reading error, etc.] and other repeatable, but incorrect results introduced during the data collection process. These errors are difficult to evaluate statistically, and commonly require alternate measurement equipment, observers, and other modifications of the data collection process to resolve.

*Random errors* result from inexact measurements that yield random fluctuations of the observed values about the true values. Statistics can be used to determine if the fluctuations are random, but these errors can also be assumed in practice. As ***Example 03.2a***, consider spatial heat flow mapping, where the repeat heat flow measurements can be statistically processed to establish the random error component of the observation's dependent element. However, if the station's location was determined from a map with applicable uncertainties, then a probable random error component must be inferred and propagated for the observation's independent element - i.e., the coordinates of the station.

*Calculation errors* are due to inexact theory for data reduction, truncation of calculations, imprecise knowledge or truncation of values of the constants used in the data processing, and other computational uncertainties. These errors are commonly addressed using both statistical and non-statistical error propagation arguments.

*Sampling errors* due to improper sampling may be either systematic or random. These errors cannot be easily estimated from a dataset, but they can be very important, leading to severely contrasting results. To recognize them in the field, there is no substitute to additional mapping. In computer simulations [e.g., gridding], the sampling procedures may be varied and their outputs compared. Typically, an effective sampling parameter is indicated where the output statistics begin to stabilize as the sampling parameter is decreased [e.g., station interval] or increased [e.g., number of observations].

## 3.3 ACCURACY AND PRECISION

Popular dictionaries consider accuracy and precision as synonyms, which is contrary to how the terms are used in science [**Figure 3.1**]. The *accuracy* of an observation is a measure of how close it comes to the *true value*, even though the true value commonly may not be known in practice.

*Precision*, on the other hand, is a measure of how repeatable or reproducible an observation is without reference to the true value. The difference between a precise observation and the true value is a measure of the observation's *bias* [e.g., see upper-right insert in **Figure 3.1**].

*Absolute precision* indicates the magnitude of the uncertainty in the same units as the result - e.g., [10.0$\pm$0.1] m. *Relative precision*, by contrast, indicates the uncertainty as a fraction or percentage of the value of the result - e.g., 10.0 m $\pm 1\%$.

## 3.4 ABSOLUTE AND RELATIVE ERRORS

*Absolute error* [Aerror] is defined as the absolute magnitude of the difference between the true and estimated values - i.e., $Aerror \equiv |true - estimate|$. Normally, only the estimated value may be known in practice, but good insight on the process may allow the evaluation of the other two quantities in the above equation.

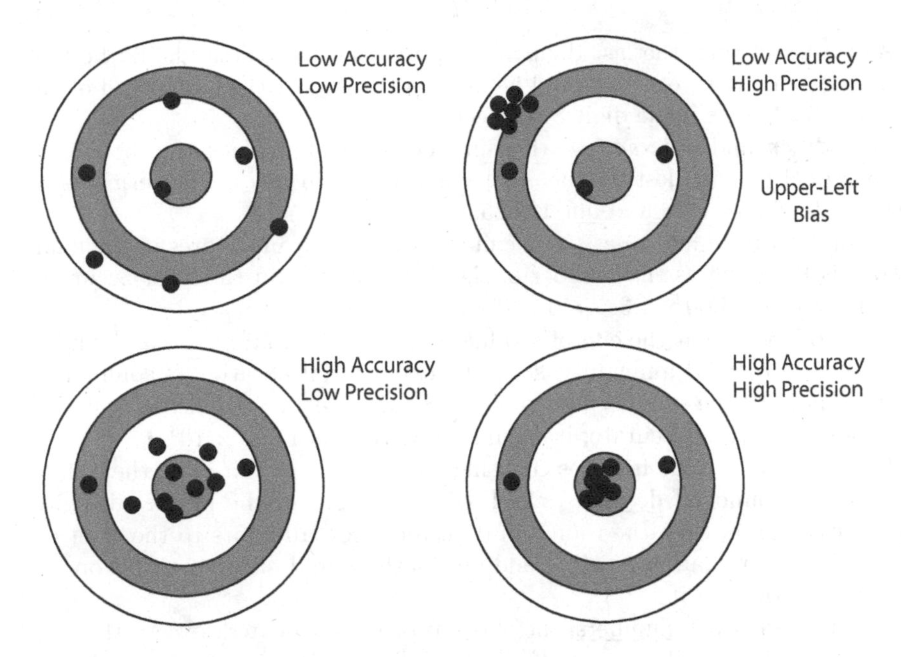


**Figure 3.1** Concepts of accuracy, precision, and bias used in science.

*Relative error* [Rerror] refers to the ratio of the absolute error divided by the estimated value given in either a fraction or percentage - i.e., $Rerror \equiv [Aerror/estimate]$.

## 3.5 SIGNIFICANT FIGURES

The minimum number of digits required to report a value without loss of accuracy is the number of significant figures. The left-most non-zero digit is the *most significant* digit - e.g., the most significant figure in the number 72,340 is 7.

If there is no decimal point, the right-most non-zero digit is the *least significant* - e.g., the least significant figure in the above example is 4. If there is a decimal point, the right-most digit is the least significant, even if it is zero - e.g., the least significant figure in 72,340.00100 is the right-most zero.

All digits between the the least and most significant digits are counted as *significant figures.* As an example, suppose the radius of a first-order earth is calculated and reported in km as

$$6 \times 10^3; \quad 6,400; \quad 6,300; \quad 6,370; \quad 6,371.; \quad \text{and} \quad 6371.54896321. \tag{3.1}$$

The correct value would depend on the estimates of the errors of the parameters in the calculation, but clearly it would not be the first three values or the last value given above.

As a general rule, the last digit in a reported measurement should be the first uncertain digit - e.g., it would be incorrect to report that the speed of an object is 5.284 m/s if the digit 8 is uncertain.

In *addition* and *subtraction*, the number of decimal places in the result should equal the smallest number of decimal places in any of the terms used - e.g., $123 + 5.35 = 128$ and not 128.35.

In *multiplication* and *division*, the number of significant figures in the final answer is the same as the number of significant figures in the least accurate term used - e.g., $3.60 \times 5.387 = 19.39$ and not 19.3932.

Reporting a zero at the end of a value without a decimal point can be misinterpreted. To avoid ambiguity, scientific exponential notation is commonly used to indicate the number of significant figures - e.g., to indicate that the last digit in 1,010 is accurate, it should be written as $1.010 \times 10^3$.

*Rounding off* a value involves choosing a digit to round off - i.e., the round off digit. To round off decimals, check the digit right of the round off digit. If it is less than 5, keep the round off digit and drop all digits to the right of it. If it is greater than or equal 5, add one to the round off digit and drop all digits to the right of it.

To round off whole numbers, the above process is repeated, except that all digits to the right of the round off digit, which were previously dropped, are now set to zero. As examples, consider the value 72,340.8939 that rounded off to the nearest thousandth place is 72,340.894, or to the nearest ones place is 72,341, or to the nearest thousands place is 72,000.

## 3.6 ERROR PROPAGATION

Using inexact data in calculations causes errors in the results of the analysis. Insight on how data uncertainties or errors propagate in the calculations is available by expanding the function in terms of each variable plus its uncertainty [e.g., **[2, 6]**].

As ***Example 03.6a***, consider estimating the uncertainty $\Delta V$ in the box's volume

$$V = L \times W \times H \tag{3.2}$$

from length $[L]$, width $[W]$, and height $[H]$ measurements with respective uncertainties $\Delta L = [L_o - L], \Delta W = [W_o - W]$, and $\Delta H = [H_o - H]$ about their true values $L_o, W_o$, and $H_o$. The volume error can be obtained from

$$\begin{aligned} V + \Delta V &= [L + \Delta L] \times [W + \Delta W] \times [H + \Delta H] \\ &= LWH + W\Delta LH + L\Delta WH + \Delta L\Delta WH + LW\Delta H \\ &\quad + W\Delta L\Delta H + L\Delta W\Delta H + \Delta L\Delta W\Delta H. \end{aligned} \tag{3.3}$$

As long as the errors $\Delta L, \Delta W$, and $\Delta H$ are small, the terms involving products of two or more of these uncertainties can be ignored as suggested in **Figure 3.2**. Thus, upon subtracting the volume $V$, the volume error can be

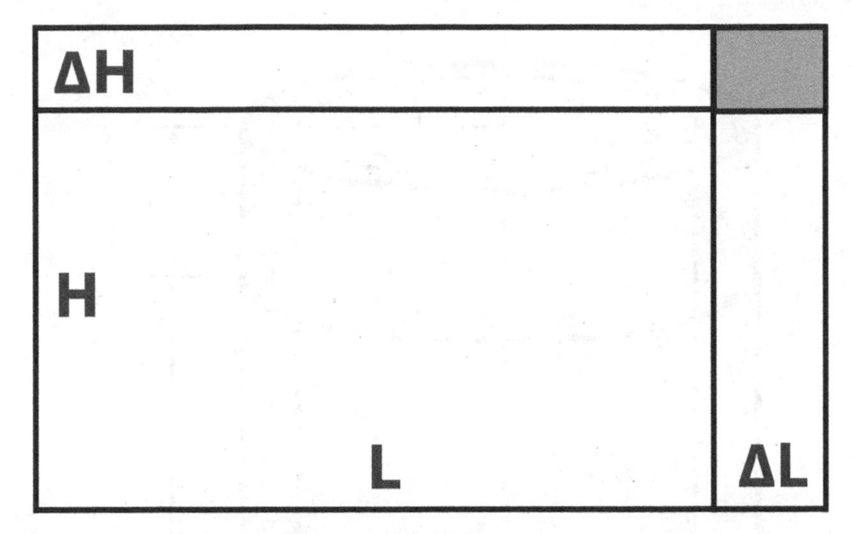


**Figure 3.2** Estimating the area of a rectangle $A = L \times H$ with associated errors $\Delta L$ and $\Delta H$ in the length $L$ and height $H$, respectively. The shaded area in the upper right corner is the second order area $[\Delta H \times \Delta L]$ ignored in evaluating the first order error in the area $\Delta A \approx \Delta L \frac{\partial A}{\partial L} + \Delta H \frac{\partial A}{\partial H}$ by Taylor's theorem.

estimated from

$$\Delta V \approx \Delta L \frac{\partial V}{\partial L} + \Delta W \frac{\partial V}{\partial W} + \Delta H \frac{\partial V}{\partial H}. \tag{3.4}$$

This error also is commonly and more elegantly written as the relative or fractional error

$$\frac{\Delta V}{V} \approx \frac{\Delta L}{L} + \frac{\Delta W}{W} + \frac{\Delta H}{H}. \tag{3.5}$$

As ***Example 03.6b***, consider the error $\Delta V$ associated with estimating a cylinder's volume

$$V = \pi r^2 h \tag{3.6}$$

with uncertainties $\Delta r$ and $\Delta h$ in the radius $r$ and height $h$ measurements as shown in **Figure 3.3**.

Replacing each variable in **Equation 3.6** with the variable plus its uncertainty obtains

$$V + \Delta V = \pi[r + \Delta r]^2 \times [h + \Delta h], \tag{3.7}$$

so that by taking the product and subtracting $V$ from both sides yields

$$\Delta V = \pi[r^2 \Delta h + 2rh\Delta r + \Delta r^2 h + 2r\Delta r\Delta h + \Delta r^2 \Delta h]. \tag{3.8}$$

Now, if $\Delta r$ and $\Delta h$ are small compared to $r$ and $h$, the terms involving $\Delta r^2$ and $\Delta r\Delta h$ will be negligible with the volume error approximation of

$$\Delta V \approx \pi[r^2 \Delta h + 2rh\Delta r] = \Delta h \frac{\partial V}{\partial h} + \Delta r \frac{\partial V}{\partial r}. \tag{3.9}$$

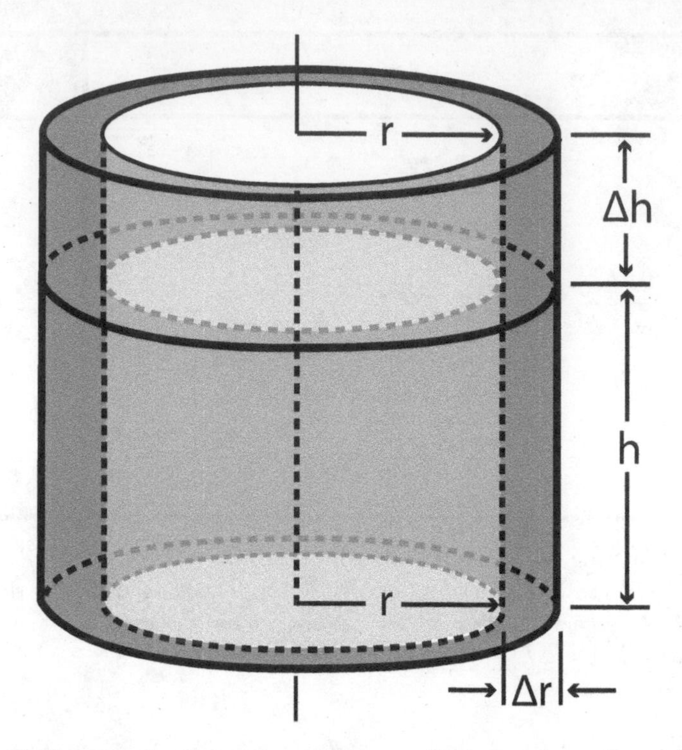


**Figure 3.3** Estimating the uncertainty $\Delta V$ in a cylinder's volume $V = \pi r^2 h$, where the radius $r$ and height $h$ measurements have respective errors $\Delta r$ and $\Delta h$.

Furthermore, dividing the above equation by $V$ yields the fractional error

$$\frac{\Delta V}{V} \approx \frac{\pi[r^2\Delta h + 2rh\Delta r]}{\pi r^2 h} = \frac{\Delta h}{h} + \frac{2\Delta r}{r}. \tag{3.10}$$

The above two algebraic examples of error propagation may be generalized via *Taylor's theorem*, which obtains an infinite power series representation of any function with continuous derivatives. In particular, suppose the infinite power series representation of $f(x)$ is

$$f(x) = f(x_o + h) = A_o + B_o h + C_o h^2 + D_o h^3 + \cdots, \tag{3.11}$$

where $h = [x - x_o]$ is the difference between an arbitrary $x$-value and a specific $x_o$-value. Then, for the series representation to be true for all $h$, it must hold for $h = 0$ so that $A_o = f(x_o)$.

Furthermore, for the series to hold for all $h$, its first derivative must equal the first derivative of $f(x)$ or $f'(x)$. Thus,

$$f'(x) = f'(x_o + h) = B_o h + 2C_o h + 3D_o h^2 + \cdots, \tag{3.12}$$

which is true for all $h = 0$ so that $B_o = f'(x)$.

Differentiating the series representation repeatedly, and then making $h = 0$ and equating the results to successive differentials of $f(x)$ at $x = x_o$ yields $C_o = f''(x_o)/2 = (\frac{1}{2!})\frac{\partial^2 f(x_o)}{\partial x^2}$, $D_o = F'''(x_o)/6 = (\frac{1}{3!})\frac{\partial^3 f(x_o)}{\partial x^3}, \ldots$, and $N_o = (\frac{1}{n!})\frac{\partial^n f(x_o)}{\partial x^n}$. Using this approach, Taylor's theorem becomes

$$\begin{aligned} f(x) = f(x_o + h) &= f(x_o) + h(\tfrac{\partial f(x_o)}{\partial x}) + \tfrac{h^2}{2!}(\tfrac{\partial^2 f(x_o)}{\partial x^2}) + \tfrac{h^3}{3!}(\tfrac{\partial^3 f(x_o)}{\partial x^3}) + \\ &+ \tfrac{h^4}{4!}(\tfrac{\partial^4 f(x_o)}{\partial x^4}) + \cdots + \tfrac{h^n}{n!}(\tfrac{\partial^n f(x_o)}{\partial x^n}). \end{aligned} \tag{3.13}$$

As ***Example 03.6c***, consider the Taylor's series expansion for $\sin(x)$ using **Equation 3.13**. Accordingly, the successive derivatives of $\sin(x)$ at $x = 0$ are $f(x_o) = \sin(0) = 0$; $f'(x_o) = \frac{\partial \sin(0)}{\partial x} = \cos(0) = 1$; $f''(x_o) = \frac{\partial^2 sin(0)}{\partial x^2} = -\sin(0) = 0$; $f'''(x_o) = \frac{\partial^3 sin(0)}{\partial x^3} = -\cos(0) = -1$; $f''''(x_o) = \frac{\partial^4 sin(0)}{\partial x^4} = \sin(0) = 0; \ldots$. Substituting the above values into **Equation 3.13** and putting 0 for $x_o$ and $x$ for $h$ yields

$$\begin{aligned} f(x) = \sin(x) &= 0 + x(1) + \tfrac{x^2}{2!}(0) + \tfrac{x^3}{3!}(-1) + \tfrac{x^4}{4!}(0) + \cdots, \text{ or} \\ &= \quad x - \tfrac{x^3}{6} + \tfrac{x^5}{120} - \tfrac{x^7}{5{,}040} + \cdots. \end{aligned} \tag{3.14}$$

The electronic evaluation of any function, be it by pocket calculators through super computers, is always based on the series representation of the function that is available from Taylor's theorem. In practice, only a finite number of terms can be included in the series representation. However, the remainder or error in using the truncated series can be estimated [e.g., **[2]**]. In addition, usually only the first order differences $h = [x - x_o]^1 = \Delta x$ are considered because $x$ is typically well enough estimated that $[\Delta x/x] << 1$ so that $\Delta x^n$ for all $n \geq 2$ can be neglected.

Taylor's theorem also generalizes algebraic error propagation for any arbitrary function $f$ that depends on observable parameters $a, b, c, \ldots$ [e.g., **Equations 3.4, 3.5, 3.9** and **3.10**]. Specifically, for the function $f[a, b, c, \ldots]$ with associated errors $\Delta a, \Delta b, \Delta c, \ldots$, the first-order error is

$$\Delta f \approx \Delta a\frac{\partial f}{\partial a} + \Delta b\frac{\partial f}{\partial b} + \Delta c\frac{\partial f}{\partial c} + \cdots, \tag{3.15}$$

with the fractional error

$$\frac{\Delta f}{f} \approx (\frac{\Delta a}{f})\frac{\partial f}{\partial a} + (\frac{\Delta b}{f})\frac{\partial f}{\partial b} + (\frac{\Delta c}{f})\frac{\partial f}{\partial c} + \cdots. \tag{3.16}$$

The above error estimates are quite general and do not draw on any statistical concept in characterizing error. However, since the beginning of the 20th century, it has become relatively commonplace to report errors in terms of the mean, variance, and other statistical properties of data. The next chapter, accordingly, invokes Taylor series expansions of these statistical concepts to obtain straightforward statistical error expressions.

## 3.7 KEY CONCEPTS

1. All measurements and calculations contain errors that must be evaluated and propagated for effective data analysis. Statistics is a principal tool for quantifying errors.
2. Systematic errors result from repeatable but incorrect results introduced during the data collection process. They are difficult to evaluate statistically.
3. Random errors result from inexact measurements that yield random fluctuations about the true value. Statistics can determine if the fluctuations are random.
4. Calculation errors result from the truncation of calculations, data reduction errors, and other computational uncertainties. Both statistical and non-statistical assessments are used to address these errors.
5. Sampling errors due to improper sampling may be either systematic or random, and thus may be addressed by non-statistical and statistical assessments, respectively.
6. Accuracy refers to how close a measurement comes to the true value, either known or presumed. Precision refers to the observation's repeatability, where absolute precision expresses the observation $\pm$ the error in the same measurement units as the observation, and relative precision gives the error as a fraction or percentage of the observation's magnitude. The observation's bias is the difference between its precise and true values.
7. The number of significant figures of an observation is the number of digits that report the value without loss of accuracy. The number of significant figures in addition, subtraction, multiplication, and division is always the number of significant figures in the least accurate term involved. A digit can be rounded up by adding one to it only if the digit to its right is greater than or equal to 5.
8. Taylor's theorem is the basis for propagating data uncertainties or errors in any calculation. Thus, for a function of $m$ measured parameters with affiliated uncertainties, the function's first order uncertainty is given by the sum of $m$ terms with each the product of a parameter's uncertainty times the variation of the function with the parameter. Dividing this uncertainty by the function gives the function's fractional error.

# 4 Statistics

## 4.1 OVERVIEW

*Statistics measure data errors or uncertainties in the context of the data mean and variance or power as respectively defined by their average value and the average of the squared deviations of the data about the mean. The square root of the variance defines the data's standard deviation or energy. For small numbers of data [i.e., $n \leq 15$], biased variance and standard deviation estimates are avoided by replacing n with the degrees of freedom $\nu = n - 1$ to account for using a data-derived parameter [i.e., the mean] in obtaining the estimates.*

*Application of Taylor's theorem obtains the first order propagation rules for the variance and mean of a function with parameters containing measurement errors. Specifically, the variance propagation rule holds that the function's variance is the sum of each parameter's measurement variance multiplied by the squared variation of the function with the parameter. Applying the variance propagation rule to the definition of the data mean shows that the standard deviation of the mean [i.e., the standard error] is the data's standard deviation divided by the root squared number of data.*

*A useful summary of a dataset is given by its statistical distribution that the data histogram approximates. The principal parameters used to describe the distribution include the data mean and variance. The total area of the distribution constitutes the total probability or 100%, whereas the probability of finding the dependent variable value within an interval of the independent variable is given by the area under the curve over the interval [i.e., the bin area of* **Equation 1.6***].*

*The binomial distribution describes the probability of observing a combination of $x$ events from a total of n discrete events [$x \leq n$] given that the probability of observing any one of the events is $p = 1/n$. It governs processes with a fixed number of independent trials or runs where each has only two possible outcomes - e.g., tossing a coin a fixed number of times to see how many heads [or tails] occur, rolling a die for a particular number from 1 to 6, lotto games, etc.*

*Perhaps the most widely used distribution in data analysis is the bell-shaped distribution based on Gauss' normal law of errors. The Gaussian or normal distribution is centered on the mean where roughly 68.3%, 95.5%, and 99.7% of the data values are respectively contained within ±1, ±2, and ±3 standard deviations of the mean. By the central limits theorem, well-sampled data [i.e., numbering $n \geq 25$] have means that follow the normal distribution. Accordingly, standard error-based statistical confidence limits may be obtained for these means to more fully constrain their applications. Also, the overlap in these error bars or confidence limits tests if two or more means are* *sufficiently close to each other that the corresponding datasets may be drawn from a common population. For smaller datasets [i.e., $n < 25$] with approximately normal or bell-shaped distributions, the weaker Student t-test is applied to establish the error bars and inference testing.*

*For normally distributed data, statistical error bars or confidence intervals on the data variance may be established using the chi-squared [$\chi^2$] distribution. In addition, hypothesis testing of the variance can be done using Fisher's f distribution that governs the variance ratios of datasets extracted randomly from a normal population. Fisher's f distribution also is useful for breaking down the total variance between two or more datasets into within-sample, among-sample, and other meaningful variance sources via analysis of variance [ANOVA].*

*To help establish the statistical distribution that data may follow, a goodness of fit test is available based on the $\chi^2$ distribution. This test breaks up the data distribution into 8 or more probability intervals and compares the observed numbers of data in each interval against those expected from the presumed distribution.*

*Non-parametric or distribution-free tests yield statistical inferences that avoid assuming a specific probability distribution for the data. These tests develop inferences by ranking or otherwise characterizing data within a sample into parameters for comparison with all possible rankings or parameters of a distribution that is independent of the data distribution's shape or form.*

## 4.2 BASIC PARAMETERS

Humans are worrywarts who incessantly doubt data collected by themselves and anyone else. Typically, once a measurement has been made, doubt about its validity sets in which leads to taking more measurements. Of course, none of these measurements are the same because every measurement in principle contains error. However, humans have evolved to accept the notion of the *mean* or *average* of the measurements as the most representative value of the observations. Furthermore, they have evolved to take the notion of the variance or its square root, the standard deviation, as an effective measure of the scatter or dispersion of the measurements about their most likely mean value.

An important application of statistics is to estimate representative values of a *population* from a *sample* or subset of the population. In practice, the parent population is an idealized notion that can only be partly measured or observed using sample datasets.

### 4.2.1 POPULATION MEAN, VARIANCE, AND STANDARD DEVIATION

For any population of $x$-values, the population *mean* or *expectation* is its average defined as

$$\mu_x \equiv \lim_{n\to\infty} (\frac{1}{n} \sum x_i) \equiv < x > . \tag{4.1}$$

The population *variance*, on the other hand, is a measure of the deviation of the population from its mean as defined by

$$\sigma_x^2 \equiv \lim_{n\to\infty} (\frac{1}{n}\sum[x_i - \mu_x]^2) \equiv < [x_i - \mu_x]^2 >, \quad (4.2)$$

where the population *standard deviation* is defined by

$$\sigma_x \equiv \sqrt{\sigma_x^2}. \quad (4.3)$$

If the data are widely distributed about the mean, $\sigma_x$ is large, and it is small where the values of the population are nearly equal to $\mu_x$. Additionally, statistical variance and standard deviation respectively equate with a signal's power and energy. The units of the mean and standard deviation also are in the units of $x$ - i.e., $:[\mu_x]: = :[\sigma_x]: = :[x]:$ - and the variance is in the units of $x^2$ - i.e., $:[\sigma_x^2]: = :[x^2]:$.

### 4.2.2 SAMPLE MEAN, VARIANCE, AND STANDARD DEVIATION

For an observable finite sample rather than the unobservable infinite parent population, the sample mean or expectation value may be obtained from

$$\bar{x} \equiv (\frac{1}{n})\sum_{i=1}^{n} x_i \approx \mu_x. \quad (4.4)$$

This is a desirable approximation of $\mu_x$ because the sum of its squared deviations is minimized.

To see this, expand

$$\sum[x_i - \bar{x}]^2 = \sum x_i^2 - 2\bar{x}\sum x_i + n\bar{x}^2 \quad (4.5)$$

and vary $\bar{x}$ until the expression is minimum - i.e.,

$$\begin{aligned} \frac{\partial}{\partial x}\sum[x_i - \bar{x}]^2 &= -2\sum x_i + 2n\bar{x} \quad (4.6)\\ &= -2\sum x_i + 2n(1/n)\sum x_i = 0. \end{aligned}$$

The process developed in **Equations 4.5** and **4.6** is the basis of the least squares method, which is further examined in **Chapter** 6.2.

Computing the sample variance via **Equation 4.2** is complicated by the bias that results when $n$ is small. For example, the sample mean of a dataset containing a single observation is just the observation itself, but the variance is invariably biased to zero no matter what the observation's value is.

To avoid biased estimation as $n \to 0$, the *degrees of freedom* $[\nu]$ of the dataset are invoked, which is the number of observations in excess of those needed to determine the parameters in the equation [e.g., **[20]**]. The degrees of freedom for estimating the sample variance is $\nu = n-1$ because the sample mean $\bar{x}$ is needed. Therefore, the sample variance is defined by

$$s_x^2 \equiv (\frac{1}{n-1})\sum[x_i - \bar{x}]^2 \approx \sigma_x^2. \tag{4.7}$$

Clearly, when $n = 1$, the sample variance is indeterminate as would be expected. Accordingly, **Equation 4.7** is taken as an unbiased estimator of the population variance $\sigma_x^2$.

In addition, the sample standard deviation is taken by

$$s_x = \sqrt{s_x^2} \approx \sigma_x = \sqrt{\sigma_x^2}. \tag{4.8}$$

Using the unbiased estimator is good policy in any application, although the biases are relatively marginal for data numbers $n \geq 15$.

### 4.2.3 VARIANCE/COVARIANCE PROPAGATION

Quantifying propagation of error in the above basic statistical concepts involves developing their first-order Taylor series expansions [e.g., **[6]**]. In particular, consider $X = f(u, v, \ldots]$ where $x_i = f[u_i, v_i, \ldots]$ in the context of data variance [e.g., **Equation 4.2**]. Expanding $[x_i - \bar{x}]$ in terms of the observed parameters $[u_i - \bar{u}]$, $[v_i - \bar{v}]$, etc. yields the first-order series

$$x_i - \bar{x} \approx [u_i - \bar{u}]\frac{\partial X}{\partial u} + [v_i - \bar{v}]\frac{\partial X}{\partial v} + \cdots. \tag{4.9}$$

Combining the above result with **Equation 4.2** gives

$$\begin{aligned}\sigma_X^2 \approx & \lim_{n\to\infty} \tfrac{1}{n}\sum[(u_i - \bar{u})\tfrac{\partial X}{\partial u} + (v_i - \bar{v})\tfrac{\partial X}{\partial v} + \cdots]^2 \\ \approx & \lim_{n\to\infty} \tfrac{1}{n}\sum[(u_i - \bar{u})^2\tfrac{\partial^2 X}{\partial u^2} + (v_i - \bar{v})^2\tfrac{\partial^2 X}{\partial v^2} + \\ & +2(u_i - \bar{u})(v_i - \bar{v})(\tfrac{\partial X}{\partial u})(\tfrac{\partial X}{\partial v}) + \cdots].\end{aligned} \tag{4.10}$$

Now the variances in $u$ and $v$ are respectively

$$\sigma_u^2 = \lim_{n\to\infty}\frac{1}{n}\sum[u_i - \bar{u}]^2 \text{ and } \sigma_v^2 = \lim_{n\to\infty}\frac{1}{n}\sum[v_i - \bar{v}]^2, \tag{4.11}$$

whereas the *covariance* between $u$ and $v$ is defined by

$$\sigma_{uv}^2 \equiv \lim_{n\to\infty}\frac{1}{n}\sum[u_i - \bar{u}] \times [v_i - \bar{v}]. \tag{4.12}$$

Thus, **Equation 4.10** may be rewritten as the *variance/covariance propagation rule*

$$\sigma_X^2 \approx \sigma_u^2(\frac{\partial X}{\partial u})^2 + \sigma_v^2(\frac{\partial X}{\partial v})^2 + 2\sigma_{uv}^2(\frac{\partial X}{\partial u})(\frac{\partial X}{\partial v}) + \cdots, \tag{4.13}$$

which simplifies to the *variance propagation rule*

$$\sigma_X^2 \approx \sigma_u^2(\frac{\partial X}{\partial u})^2 + \sigma_v^2(\frac{\partial X}{\partial v})^2 + \cdots \tag{4.14}$$

if $u$ and $v$ are uncorrelated so that $\sigma_{uv}^2 = 0$.

The above propagation rules are widely used to transform simple numerical differences or uncertainties of data into statistical errors. Suppose, for example, that $X$ is a function of the addition and/or subtraction of variables $u$ and $v$ - i.e., $X = au \pm bv$, where $a$ and $b$ are constants. Then, the variance in $X$ is

$$\sigma_X^2 \approx a^2\sigma_u^2 + b^2\sigma_v^2 \pm 2ab\sigma_{uv}^2, \tag{4.15}$$

which is the *addition/subtraction* variance propagation rule.

For *multiplication* where $X = \pm auv$, the variance rule is

$$\sigma_X^2 \approx a^2v^2\sigma_u^2 + a^2u^2\sigma_v^2 + 2a^2uv\sigma_{uv}^2, \tag{4.16}$$

with the simplified relative variance

$$\frac{\sigma_X^2}{X^2} \approx \frac{\sigma_u^2}{u^2} + \frac{\sigma_v^2}{v^2} + 2\frac{\sigma_{uv}^2}{uv}. \tag{4.17}$$

For *division*, on the other hand, with $X = \pm au/v$, the relative variance is

$$\frac{\sigma_X^2}{X^2} \approx \frac{\sigma_u^2}{u^2} + \frac{\sigma_v^2}{v^2} - 2\frac{\sigma_{uv}^2}{uv}. \tag{4.18}$$

Examples of variables raised to *powers* include $X = au^{\pm b}$ for which

$$\frac{\sigma_X^2}{X^2} \approx \frac{b^2\sigma_u^2}{u^2}. \tag{4.19}$$

As ***Example 04.2.3a***, consider the area of a circle $A = \pi r^2$, where the radius $r$ is determined as 10 cm with a 3 mm standard deviation. Thus, the area is $A = 100\pi$ cm$^2$ with $\sigma_A = 6\pi$ cm$^2$.

*Exponential* examples include $X = ae^{\pm bu}$ for which

$$\frac{\sigma_X^2}{X^2} \approx b^2\sigma_u^2, \tag{4.20}$$

and $X = a^{\pm bu}$ for which the relative standard deviation is

$$\frac{\sigma_X}{X} \approx [b\ln(a)]\sigma_u. \tag{4.21}$$

The *logarithmic* function $X = a[\ln(\pm bu)]$, on the other hand, has the variance

$$\sigma_X^2 \approx \frac{a^2\sigma_u^2}{u^2}. \tag{4.22}$$

The variance propagation rule applied to the sample mean [**Equation 4.4**] yields

$$[\sigma_{\bar{x}}^2 \approx (\frac{1}{n})^2 n s_x^2 = \frac{s_x^2}{n}] \approx [\sigma_{\mu_x}^2 \approx \frac{\sigma_x^2}{n}]. \quad (4.23)$$

Taking the square root of the above result obtains the *standard error of the mean*

$$[\sigma_{\bar{x}} \approx \frac{s_x}{\sqrt{n}}] \approx [\sigma_{\mu_x} \approx \frac{\sigma_x}{\sqrt{n}}], \quad (4.24)$$

which in practice is often simply referred to as the *standard error*.

In general, the error $\Delta x$ reported for a measured value $x$ may take on several possible interpretations. For example, if the true value $x_o$ is known, then the uncertainty in the measured $x$ may be expressed in terms of the *true error* $\Delta x = [x_o - x]$ - i.e., the observation is reported as $x \pm \Delta x$. On the other hand, if the use of $\Delta x$ requires the greatest safety factor, $\Delta x$ can be set as the *maximum error* possible so that the true value has the highest likelihood of being in the interval $[x - \Delta x] \leq x \leq [x + \Delta x]$.

A further option common in modern practice assumes that the errors are distributed randomly about the true value, and thus follow the *normal* or *Gaussian distribution* described in **Chapter 4.4**. In this case the error is taken as the standard deviation of the observations - i.e., $\Delta x = \sigma_x$.

Invoking the Gaussian distribution of errors permits the development of statistical confidence intervals or error bars for the data and their applications. As ***Example 04.2.3b***, reporting $x$ with an uncertainty of $\sigma_x$ implies that it is in the interval $[x - \sigma_x] \leq x \leq [x + \sigma_x]$ with a 68% probability - i.e., there is a 68% probability that the true value $x_o$ lies between $[x - \sigma_x]$ and $[x + \sigma_x]$. The probability can be increased to 95%, however, by reporting $x$ in the interval $[x \pm 2\sigma_x]$. Further insights on statistical error distributions and their applications result from the probability concepts outlined below.

### 4.2.4 PROBABILITY AND DISTRIBUTIONS

The key focus of statistics is to determine the probable outcomes from the possible outcomes. This objective is facilitated by the notion of *probability* $[P]$, which is the chance of an event happening as defined by the ratio of the number of the event's occurrences $[\Sigma_o]$ divided by the total number of occurrences of all the possible events $[\Sigma_t]$ - i.e.,

$$P \equiv [\Sigma_o]/[\Sigma_t] \quad (4.25)$$

expressed as either a fraction or percent.

As ***Example 04.2.4a***, a coin toss offers only two equally likely events (i.e., either a head or a tail) so that the chance of the head coming up is $P = 0.5$ or 50%. In ***Example 04.2.4b***, by contrast, the chance of drawing an ace from a full pack of 52 cards is [4/52], or 4 : 52, or 0.077, or 7.7%, whereas the chance of drawing a second ace is now [3/51] = 5.9%. However, the chance of a single draw of two cards which are aces is one to the number of two-card pairs in the deck - or [1/1326] ≈ 0.08%.

If the chance for an event to happen is $P_1$ and that of a second event is $P_2$, and so on, then the chance that $n$-events will happen in $n$-trials is

$$\prod_{i=1}^{n} P_i = P_1 \times P_2 \times \cdots \times P_n \tag{4.26}$$

Thus, the chance for two heads to come up in successive tosses of a coin is $P = 25\%$, because only 4 possibilities or equally likely outcomes are possible. The outcomes include 1) 2 heads, 2) 1 head and 1 tail, 3) 1 tail and 1 head, and 4) 2 tails, so that on average 2 heads will occur in each four trials of the double coin throw.

In general, the probability distibution is an idealization of a histogram of rectangular bins with the vertical lengths counting out how many data items are found within the horizontal length intervals of the independent variable. It is basically a frequency distribution with the number of data items on the ordinate per interval of bin width on the abscissa. Thus, the area of the bin is a measure of the probability of finding an item associated with a value of the independent variable in the interval spanned by the bin's width [e.g., **Figure 4.1**].

Histograms are convenient diagrams for charcterizing and describing data distributions. The raw frequency histogram in **Figure 4.1**, for example, graphically illustrates the distribution of sediment sizes from the accompanying table. Other equivalent characterizations of this distribution are shown in **Figure 4.2** including the raw distribution in the top graph obtained by connecting the center top of each bin with a straight line segment. This graph further details the shape of the distribution and integrates or sums to the total number of data $n = 306$.

The middle graph shows the differential distribution obtained from dividing each raw frequency value by $n$ for the relative frequency. This graph integrates or sums to 1 or 100% as shown in the bottom graph of the cumulative frequency distribution.

Parameters commonly reported for a distribution include its mean or average value which is the most likely value of the distribution. For non-symmetric distributions, other attributes may be applied. The *median*, for example, is the value which splits the distribution into two equal halves - i.e., 50% of the values lie above the median and 50% below. The *mode*, on the other hand, is the value that has the greatest frequency of occurrence. **Figure 4.3** contrasts the meanings of the mean, median, and mode for describing asymmetrical [top] and symmetrical [bottom] data distributions

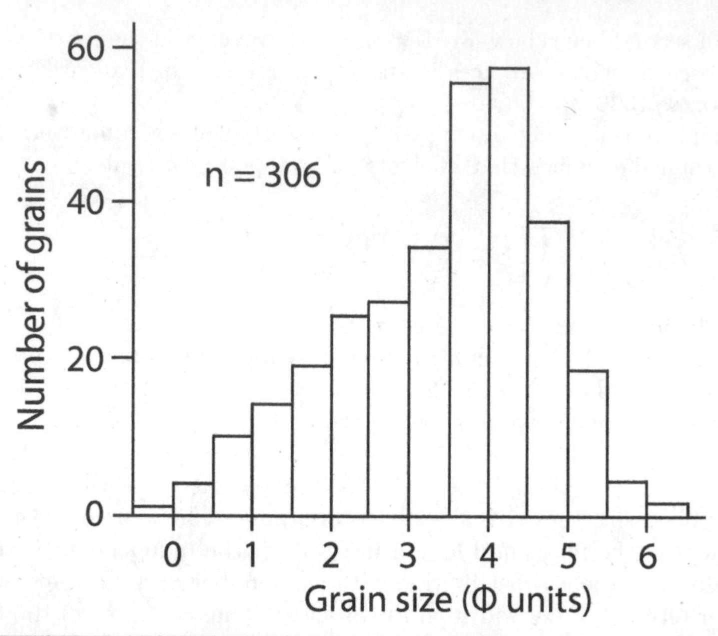


| Interval (Φ units) | Frequency (number of grains) | Relative frequency | Cumulative frequency |
|---|---|---|---|
| -0.51-0.0 | 1 | 0.003 | 0.003 |
| 0.01-0.5 | 4 | 0.013 | 0.016 |
| 0.51-1.0 | 10 | 0.033 | 0.049 |
| 1.01-1.5 | 14 | 0.046 | 0.095 |
| 1.51-2.0 | 19 | 0.062 | 0.157 |
| 2.01-2.5 | 25 | 0.082 | 0.239 |
| 2.51-3.0 | 27 | 0.088 | 0.327 |
| 3.01-3.5 | 34 | 0.111 | 0.438 |
| 3.51-4.0 | 55 | 0.180 | 0.618 |
| 4.01-4.5 | 57 | 0.186 | 0.804 |
| 4.51-5.0 | 37 | 0.121 | 0.925 |
| 5.01-5.5 | 18 | 0.059 | 0.984 |
| 5.51-6.0 | 4 | 0.013 | 0.997 |
| 6.01-6.5 | 1 | 0.003 | 1.000 |

**Figure 4.1** Histogram of sediment grain sizes based on the data in the table binned in $\Phi$ units, which are in negative base 2 logarithms of the size in mm - e.g., $0\Phi = 1$ mm, $1\Phi = 0.5$ mm, $2\Phi = 0.25$ mm, etc.

## 4.3 BINOMIAL DISTRIBUTION

In general, if the chance of a discrete event happening in a particular trial is $P$, then the chance of it not happening in any trial is $[1 - P]$, because the total probability of all possible outcomes must sum up to 1 or 100%. Thus,

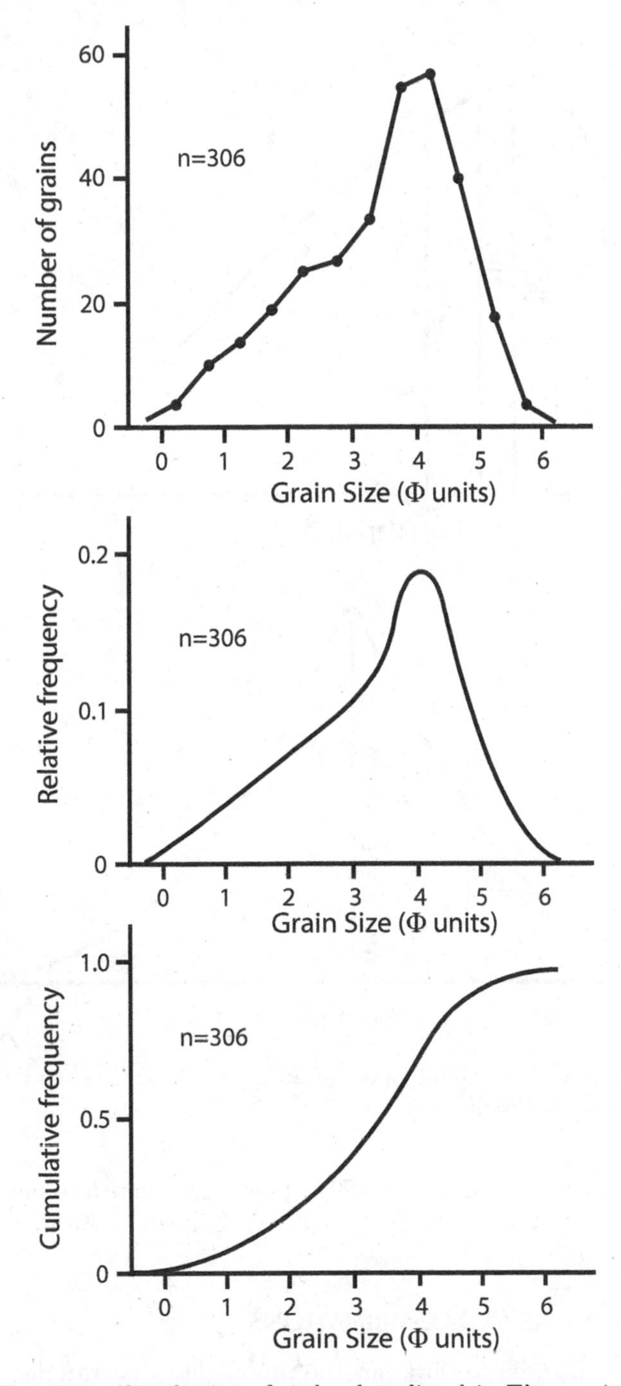


**Figure 4.2** Frequency distributions for the data listed in **Figure 4.1** include the frequency polygon [top], relative or differential frequency distribution [middle], and cumulative or integral frequency distribution [bottom].

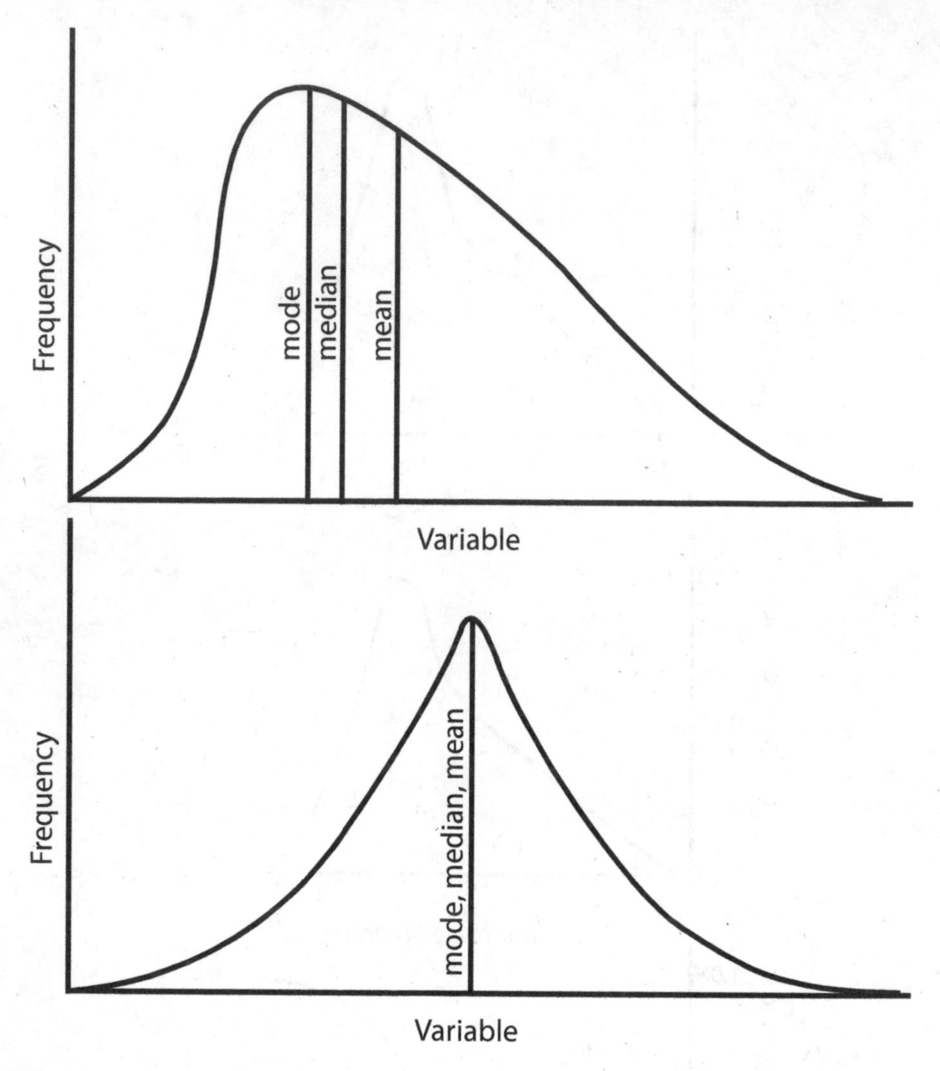


**Figure 4.3** The mean, median, and mode for an asymmetrical [top] and a symmetrical [bottom] frequency distribution.

the chance that this particular event will happen every time in $n$ independent trials is $\prod_1^n P = P^n$ and the chance of it not happening in all $n$ trials is $\prod_1^n [1 - P] = [1 - P]^n$.

### 4.3.1 PERMUTATIONS AND COMBINATIONS

These results can be used to work out the probabilities for various combinations of possible outcomes in a process. As ***Example 04.3.1a***, suppose that on average one of every ten rock specimens collected in a field area yields a trilobite fossil. Then the chance of collecting 4 successive rock samples containing no trilobites is $[0.9]^4 = 65.61\%$, as opposed to the $[0.1]^4 = 0.01\%$ chance of finding trilobites in all 4 samples.

Now, the chance of getting just one trilobite in 4 successive samples involves working out all possible outcomes or ways of collecting a single trilobite. For example, the chance of getting the fossil the first time with the three other rock samples containing no trilobite is $[0.1][0.9]^3 = 7.29\%$. However, there are three other possible sequences that involve finding a trilobite in either the second, third, or fourth sample with the rest barren of the fossil. The sum of these chances is the total probability or chance of observing 1 trilobite in $[1+3] = 4$ rock samples given by

$$\begin{aligned} P_1[3] = & \quad [0.1][0.9]^3 + [0.9][0.1][0.9]^2 + [0.9]^2[0.1][0.9] + [0.9]^3[0.1] \\ = & \quad 4 \times [0.1][0.9]^3 = 29.16\% \end{aligned} \tag{4.27}$$

Similarly, the chance of getting 2 trilobite fossils in 4 successive rock samples is

$$\begin{aligned} P_2[2] = & \quad [0.1]([0.1][0.9]^2 + [0.9][0.1][0.9] + [0.9]^2[0.1]) + [0.9][0.1]([0.1][0.9] + \\ & +[0.9][0.1]) + [0.9]^2[0.1]^2 = 6 \times [0.1]^2[0.9]^2 = 4.86\% \end{aligned} \tag{4.28}$$

Finally, the chance of getting 3 fossils in 4 rock samples is

$$\begin{aligned} P_3[1] = & \quad [0.1]^3[0.9] + [0.1]^2[0.9][0.1] + [0.1][0.9][0.1]^2 + [0.9][0.1]^3 \\ = & \quad 4 \times [0.1]^3[0.9] = 0.36\% \end{aligned} \tag{4.29}$$

Note that the sum of these five chances is unity, as it should be - i.e., the total probability for all possible outcomes of getting a trilobite in 4 successive samples is

$$0.6561 + 0.2916 + 0.0486 + 0.0036 + 0.0001 = 1 \text{ or } 100\%. \tag{4.30}$$

The above example may be generalized to give the chance of just $x$ outcomes in $n$ independent trials, when the chance of an outcome in each trial is $P$. This probability, accordingly, is

$$\begin{aligned} P_x[n-x] = & \quad \frac{n[n-1][n-2]\cdots[n-x+1]}{x!}[P]^x[1-P]^{n-x} \\ = & \quad \frac{n!}{x![n-x]!}[P]^x[1-P]^{n-x} \\ = & \quad \binom{n}{x}[P]^x[1-P]^{n-x}, \end{aligned} \tag{4.31}$$

where the $n$-factorial, for example, is $n! = 1 \times 2 \times \cdots \times n$ and $0! = 1$.

Recasting the above trilobite example in terms of **Equation 4.31** for $n = 4$ trials or rock samples with a $P = 0.1$ chance per sample of recovering a trilobite fossil shows the chances of obtaining $x = 4, 3, 2, 1$ or $0$ fossils as

$$\begin{aligned} P_4[0] = & \quad [1][0.9]^0[0.1]^4 = & 0.0001 \\ P_3[1] = & \quad [4][0.9][0.1]^3 = & 0.0036 \\ P_2[2] = & \quad [6][0.9]^2[0.1]^2 = & 0.0486 \\ P_1[3] = & \quad [4][0.9]^3[0.1] = & 0.2916 \\ P_0[4] = & \quad [1][0.9]^4[0.1]^0 = & \underline{0.6561} \\ \ni \;= & \quad \textstyle\sum_0^4 P_i \;\; = & 1.0000. \end{aligned} \tag{4.32}$$

Accordingly, combining the chances $P_1$ through $P_4$ shows that there is a 34.4% chance of finding a trilobite fossil in a collection of 4 rock samples as opposed to the 65.6% chance of not finding a fossil.

In addition, note from **Equation 4.31** that the total number of different arrangements or permutations of $n$ things taken $x$ at a time is given by the *binomial coefficient*

$$\binom{n}{x} = \frac{n!}{x![n-x]!} \tag{4.33}$$

$\forall\ x \leq n$. The coefficient, which is zero when $x > n$, is useful for evaluating how many different outcomes can occur given that their number is fixed.

As ***Example 04.3.1b***, note that **Equation 4.33** governs state and national lotteries where people pay to pick a winning subset of numbers from a fixed set of numbers. Until recently, the state of Ohio's lotto game involved picking six numbers from the sequence of integers running from 1 through 46. Thus, the chance of picking the winning ticket with $x = 6$ and $n = 46$ in **Equation 4.33** is one in $9,366,819$.

It is no problem to write computer code to identify each of the somewhat more than 9 million combinations of the six numbers in the sequence. Thus, when the payout exceeds the number of combinations, a profit is guaranteed if the purchase of the one-dollar tickets for all of the combinations can be carried out. The lotto payout of February 11, 1998, for example, was about thirty million dollars, which would have realized a tidy profit of little more than twenty million dollars using the **Equation 4.33**-based strategy. Nowadays, however, this profit would be reduced to sixteen million dollars because the selection sequence in the Ohio lottery has been extended to the first $n = 49$ non-zero integers.

### 4.3.2 PROBABILITY INTERPRETATION

Given the probability $p = 1/n$ of observing the outcome $x$ in any one trial, the probability $P_B[x, n, p]$ for observing $x$ of the $n$ discrete items is

$$P_B[x, n, p] = \binom{n}{x} p^x q^{n-x}, \tag{4.34}$$

where $q = [1-p]$. This equation is called the binomial distribution because it is closely related to the *Binomial Theorem* given by

$$\begin{aligned}[p+q]^n = & \sum_{x=0}^{n}[\binom{n}{x}p^x q^{n-x}] \ \forall \ p^2 > q^2 \text{ and real } n \\ = & \ q^n + nq^{n-1}p + \tfrac{n[n-1]}{2!}q^{n-2}p^2 + \tfrac{n[n-1][n-2]}{3!}q^{n-3}p^3 + \cdots .\end{aligned} \quad (4.35)$$

The binomial theorem or expansion describes the algebraic expansion of powers of a binomial, which is the sum of two terms or monomials. It is widely used to simplify the algebraic expression into an approximation involving usually just the first order term of the expansion, although sometimes one or more of the higher order terms may also be included.

The defining properties of the binomial distribution are its mean and variance. Specifically, the mean or average of $P_B$ is given by

$$\mu_B = \sum_{x=0}^{n}[x\frac{n!}{x![n-x]!}p^x q^{n-x}] = np. \quad (4.36)$$

Thus, the number $x$ of outcomes observed for an experiment performed $n$ times will average to $\bar{x} \rightarrow \mu_B = np$. In tossing coins, for example, half the coins on average land heads up.

The variance of $P_B$, on the other hand, is obtained from

$$\sigma_B^2 = \sum_{x=0}^{n}[(x-\mu_B)^2\frac{n!}{x!(n-x)!}p^x q^{n-x}] = npq. \quad (4.37)$$

If $p = q = 0.5$, then $P_B[x,n,p]$ is symmetric about $\mu_B$ with variance $\sigma_B^2 = \mu_B/2$ [e.g., **Figure 4.4**]. As ***Example 04.3.2a***, suppose $n = 10$ coins are tossed into the air $N = 100$ times where the number $x_i$ of heads is observed for each toss $i$. Then the probability function governing the distribution of $x_i$ is $P_B[x,n,p] = P_B[x_i,10,1/2]$, which is not affected by $N$. **Figure 4.4** shows the distribution with $\mu_B = 10[1/2] = 5$, and $\sigma_B = \sqrt{10[1/2][1/2]} = \sqrt{2.5} = 1.58 \approx 1.6$. $P_B[x_i,10,0.5]$ is symmetric about its mean $\mu_B$ and the magnitudes of the distribution at the discrete points sum to 1 or 100%. Only the discrete points have physical significance, but the smooth curve through them helps to illustrate the distribution's behavior over the $x = x_i$ outcomes.

Where $p \neq q$, on the other hand, $P_B[x,n,p]$ is asymmetric with a smaller $\sigma_B^2$ [e.g., **Figure 4.5**]. As ***Example 04.3.2b***, consider rolling $n = 10$ dice, where the probability of throwing one die with 1 up is $p = 1/6$. Thus, the probability of $x$ of them landing with the 1 up is given by

$$P_B[x,10,1/6] = \frac{10!}{x![10-x]!}[\frac{1}{6}]^x[\frac{5}{6}]^{10-x}, \quad (4.38)$$

which is shown in **Figure 4.5** with mean $\mu_B = 10/6 \approx 1.67$ and standard deviation $\sigma_B = \sqrt{10[1/6][5/6]} = [5/6]\sqrt{2} = 1.18 \approx 1.2$. Here, $P_B[x,10,1/6]$ is

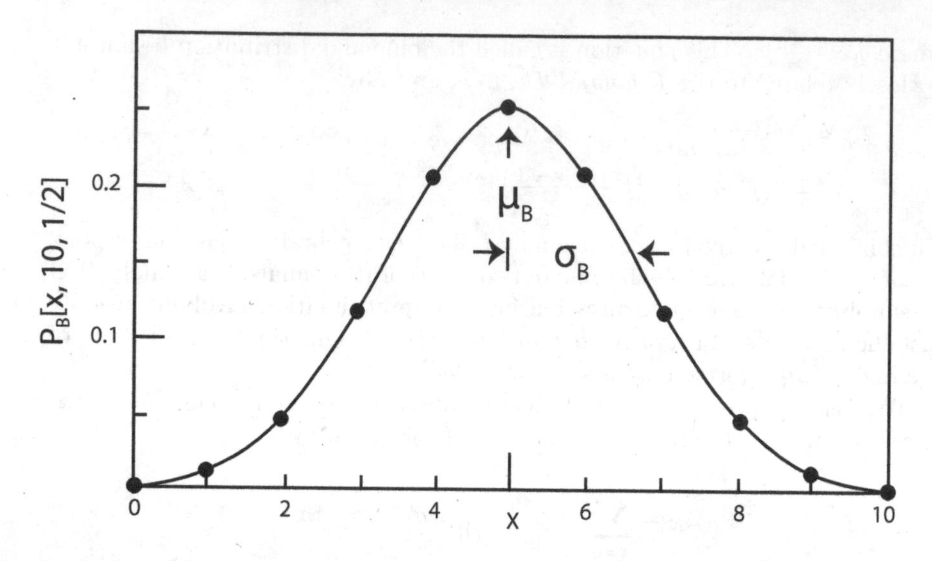


**Figure 4.4** Binomial distribution $P_B[x, n, p]$ for $n = 10, p = 1/2$, and $\sigma_B = 1.6$.

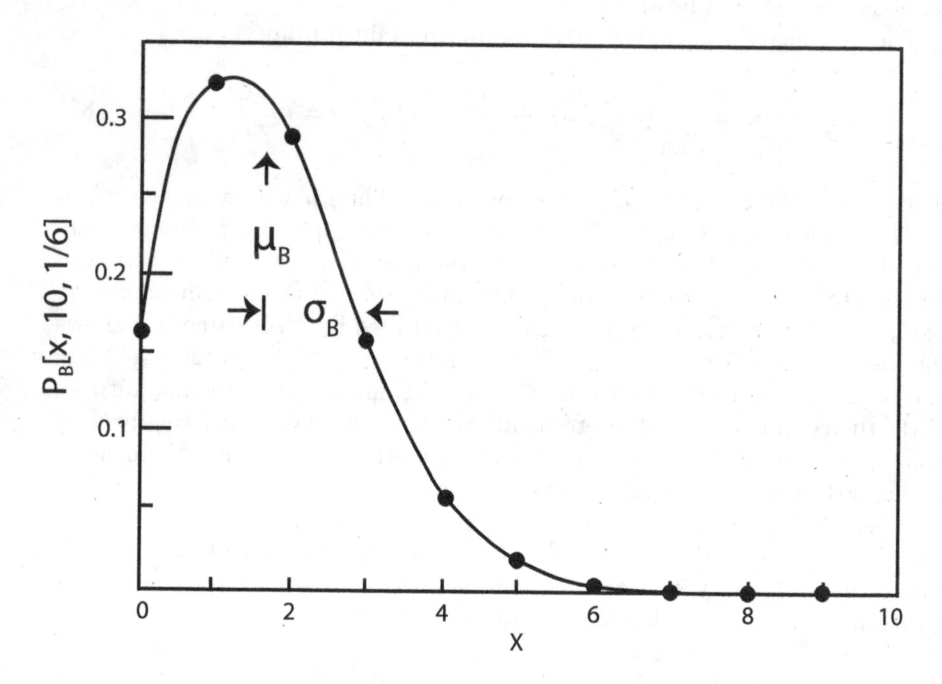


**Figure 4.5** Binomial distribution $P_B[x, n, p]$ for $n = 10, p = 1/6$, and $\sigma_B = 1.2$.

not symmetric about $\mu_B$ or any other point. The most probable value is for $x = 1$, but the peak of the smooth curve occurs for a slightly larger value of $x$. Also, note that the standard deviation is smaller than the one in the previous example of **Figure 4.4**, even though the number of items $n$ is the same.

As ***Example 04.3.2c***, consider the multiple choice exam of 50 questions each with 5 possible answers of which only one is correct. If the answers were picked randomly, then with $n = 50$ and $p = 0.20$, the average score would be $\mu_B = 10$ correct answers with a standard deviation of $\sigma_B = 8$. Accordingly, the scores for those students picking answers purely by chance would be expected to range from 36% to 4% at the 95% confidence level. Clearly, one would have to score at least above 36% to avoid professing more than statistical ignorance of the material covered by the exam.

## 4.4 GAUSSIAN OR NORMAL DISTRIBUTION

Perhaps the most widely used probability distribution in science and engineering is based on Gauss' normal law of errors. This distribution quantifies randomness in data in terms their mean $[\mu_x]$ and standard deviation $[\sigma_x]$ according to

$$P_G[x] = \frac{1}{\sigma_x\sqrt{2\pi}} \exp[-(\frac{1}{2})(\frac{x-\mu_x}{\sigma_x})^2] \ \forall \ -\infty < x < \infty. \tag{4.39}$$

The probability density function is centered on the data mean or *expectation* with the width of the bell-shaped curve proportional to the standard deviation of the data [e.g., **Figure 4.6**]. Standardizing the data [e.g., **Chapter** 2.4.1] yields the *standard normal distribution* with the $\mu_x = 0$ and $\sigma_x = 1$ in **Figure 4.6** marked by the 4-cross line segments.

Integrating the normal distribution gives the cumulative normal probability density distribution [e.g., **Figure 4.7**]. The integral function maintains the central inflection point on the mean with slope that is proportional to the standard deviation.

### 4.4.1 PROBABILITY INTERPRETATION

In general, the area under the normal curve within the interval $[a, b]$ gives the probability of observing $x$ within the interval - i.e., the probability that $x$ will fall somewhere in the interval $a \leq x \leq b$ is

$$Prob_G[x \in (a, b)] = \int_a^b P_G[x]dx. \tag{4.40}$$

However, there is a probability limit to resolving a specific $x$-value in the distribution because as the interval is decreased the probability for observing $x$ vanishes.

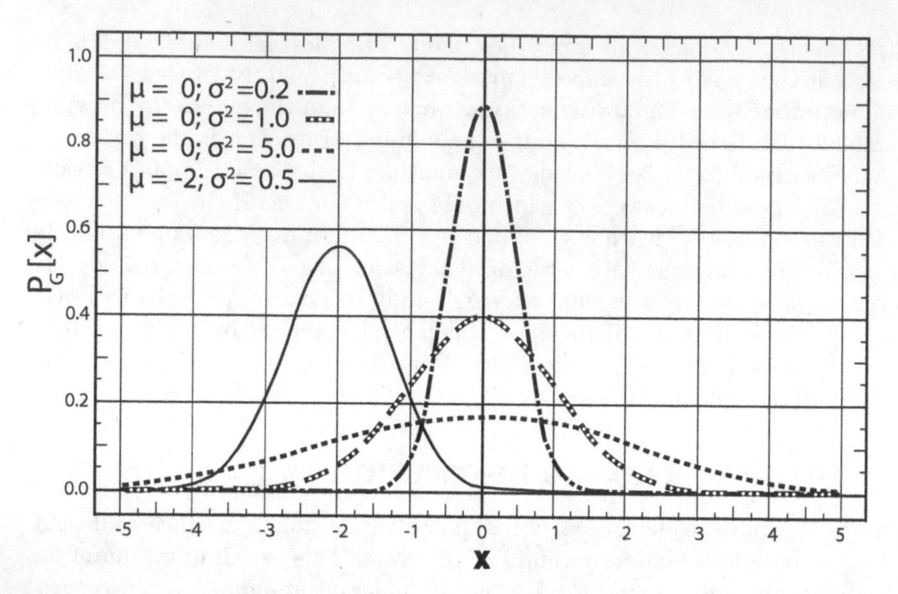


**Figure 4.6** Examples of the Gaussian or normal probability density distribution or function. The curve for $\mu = 0$ and $\sigma^2 = 1.0$ shows the standard normal distribution. **Figure 4.7** shows the related cumulative or integral probability functions.

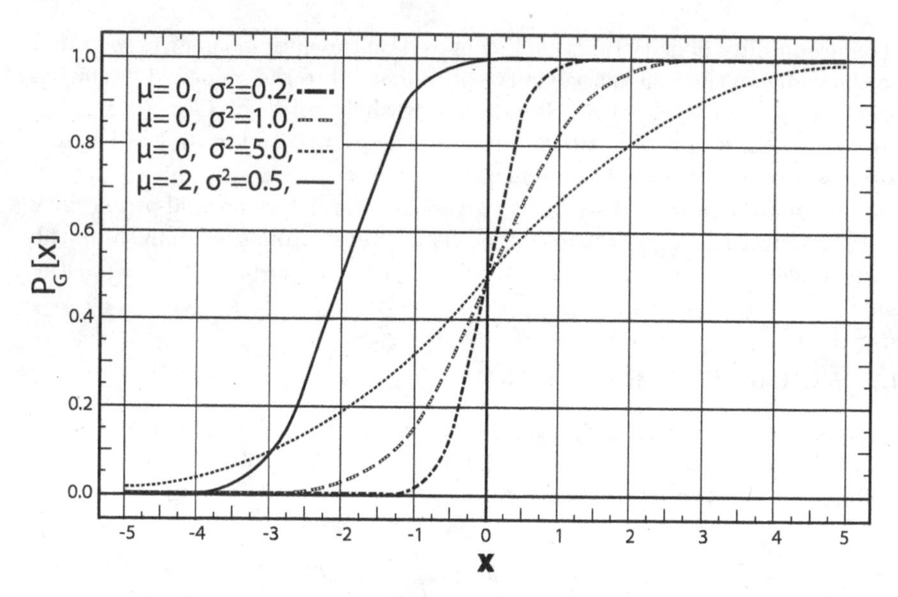


**Figure 4.7** Cumulative or integral probability functions for the normal probability density functions of **Figure 4.6**.

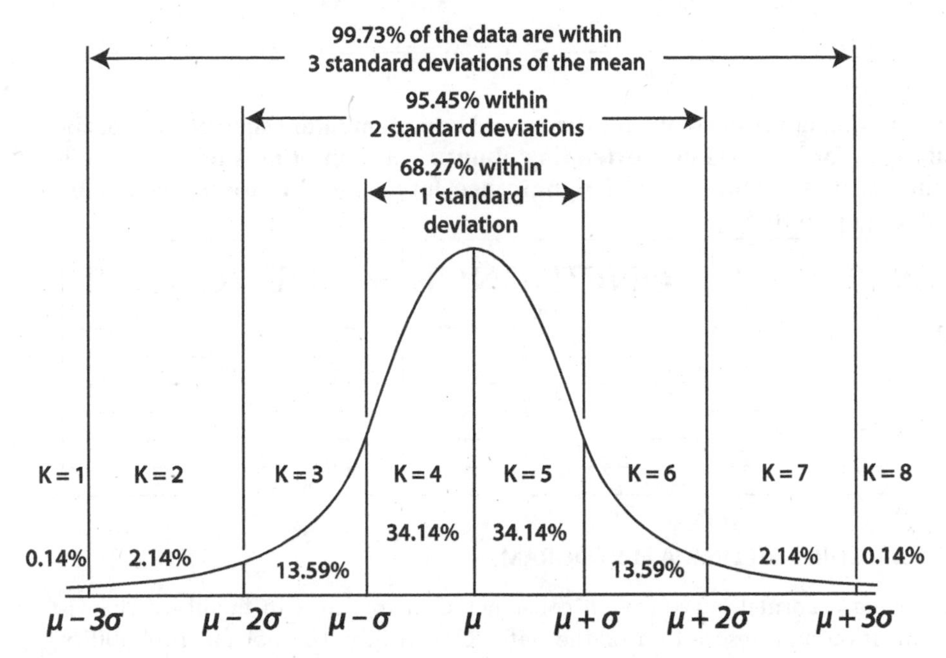


**Figure 4.8** The Gaussian or normal probability density function centered on the data mean $\mu_x \approx \bar{x}$ and subdivided into eight $[K]$ probability intervals in integer multiples of the standard deviation $\sigma_x \approx s_x$. Thus, the chance for a value of the distribution to fall within the range $[\mu_x + 1\sigma_x]$ of the $K = 5$ interval is about 34%, whereas the chance is roughly doubled to find the value within the $[\mu_x \pm 1\sigma_x]$ range that encompasses the $K = 4$ and 5 intervals. The chances increase roughly to 95% and over 99% for the value to fall within two and three standard deviations of the mean, respectively.

The normal law is usefully applied in a great many ways because the areas under the curve are easily established in terms of the distribution's mean $\mu_x$ and standard deviation $\sigma_x$ as shown in **Figure 4.8**. Thus, for example, there is roughly a 68% probability for $x$ to fall within the interval $\mu_x \pm \sigma_x$ and a 32% chance for it to fall outside this interval.

In general, the probability for $x$ to be contained within the interval is commonly referred to as the *confidence level*, whereas the *significance level* refers to the complementary probability of the distribution outside of the interval. The confidence and significance levels form a closed system in that together they account for 100% of the distribution. **Table 4.1** tabulates additional details concerning the confidence and significance levels of the normal distribution. Thus, fewer than 5 measurements in 100 will normally differ from $\mu_x$ by more that $2\sigma_x$, and fewer than 3 in $1,000$ measurements will differ from the mean by $3\sigma_x$. Values observed that are less than $\mu_x - 3\sigma_x$ or greater than $\mu_x + 3\sigma_x$ are sometimes considered *outliers* of the distribution reflecting possible data mistakes or other errors because the probability of obtaining them is less than about 0.14%.

**Table 4.1**
**Tabulation of the normal confidence and complementary significance probabilities for intervals of $x$ extending about $\mu_x$ in $k$-multiples of $\sigma_x$. For any interval, the confidence and significance levels form a closed system by adding up to 100%.**

| INTERVAL $[\mu_x \pm k\sigma_x]$ | CONFIDENCE Level | SIGNIFICANCE Level |
|---|---|---|
| $\pm 1\sigma_x$ | 68.27% | 31.73% |
| $\pm 1.96\sigma_x$ | 95.00% | 5.00 |
| $\pm 2\sigma_x$ | 95.45% | 4.55% |
| $\pm 2.576\sigma_x$ | 99.00% | 1.00% |
| $\pm 3\sigma_x$ | 99.73% | 0.27% |

### 4.4.2 CURVE FITTING A HISTOGRAM

Drawing a normal curve over a constructed histogram can facilitate the visualization and description of the data. Generically, the normal probability density $P_G[x]$ represents the Gaussian probability per unit area of the distribution. Thus, the distribution $P_G[x]$ must be drawn so that the area under the normal curve is the same as the area of the histogram, which is

$$c \times N, \tag{4.41}$$

where $c$ is the class interval or bin width and $N$ is the total frequency or number of data points.

Now, the maximum ordinate on the normal curve is

$$Y_{max} = \frac{cN}{s_x\sqrt{2\pi}} = P_G[\bar{x}], \tag{4.42}$$

with the other ordinates established from **Table 4.2** in terms of the height ratios relative to $Y_{max}$. Interpolating a smooth curve through the points at these ordinates yields an effective estimate of the related normal curve.

The example in **Figure 4.9** illustrates the procedure for fitting a normal distribution to a histogram of beach sand moisture measurements with $\bar{x} =$ 3.16%, $c = 1\%$, $N = 127$, and $s_x = 1.14\%$. Along the base of the histogram, the upper abscissa gives the original percentage classes, whereas the lower one is the standardized abscissa with its origin at the sample mean $\bar{x}$ and scaled in units of $s_x = 1.14\%$.

Hence, $Y_{max} = 44.4\%$ with the other ordinates obtained at $0.5s_x$ intervals using the relative heights from the last column in **Table 4.2**. At $\pm 1s_x$, for example, the ordinate is [44.4 × 0.6065 = 26.93]%. The curve is smoothly sketched through the points to be positive and stops just short of the base line to reflect the finiteness of the sample distribution.

**Table 4.2**
**Ordinates of the normal curve in sample standard deviation units about the sample mean.**

| $s_x$ Units About $\bar{x}$ | $P_G[0, 1]$ Ordinates | Ordinates Relative to $Y_{max}$ |
|---|---|---|
| $\pm 0.00 s_x$ | 0.3990 | 1.0000 |
| $\pm 0.50 s_x$ | 0.3520 | 0.8822 |
| $\pm 1.00 s_x$ | 0.2420 | 0.6065 |
| $\pm 1.50 s_x$ | 0.1300 | 0.3258 |
| $\pm 2.00 s_x$ | 0.0540 | 0.1353 |
| $\pm 2.50 s_x$ | 0.175 | 0.0439 |
| $\pm 3.00 s_x$ | 0.0044 | 0.0110 |
| $\pm 4.00 s_x$ | 0.0001 | 0.0003 |

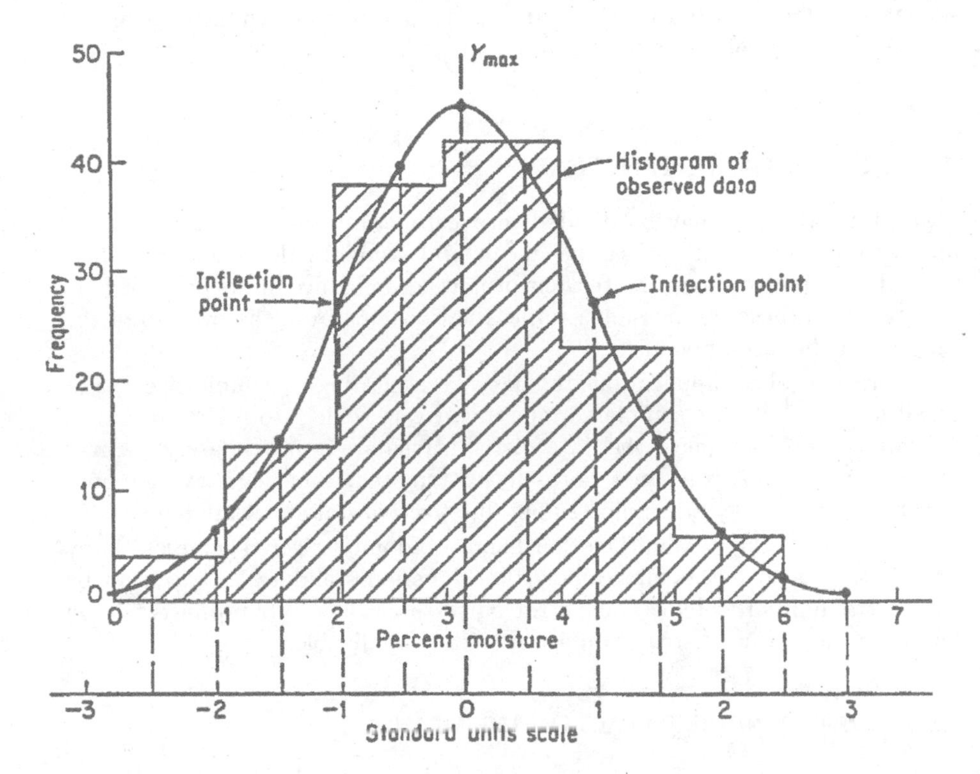


**Figure 4.9** Fitting a normal curve to a sample histogram [56].

### 4.4.3 CENTRAL LIMITS THEOREM

The most used statistical parameters in environmental science and engineering are the sample mean $\bar{x}$ and standard deviation $s_x$. If a population with unknown distribution is repeatedly sampled randomly, the distribution of the sample means $\bar{x}$ will follow an approximately normal distribution with mean $\mu_x$ and variance $\sigma_x^2/n$ provided the sample size is large - i.e., $\mu_x \approx \bar{x}$ and $\sigma_\mu \approx \sigma_x/\sqrt{n}$ for $n \geq 25 - 30$.

This profound result is a consequence of the *Central Limits Theorem* that says if $\bar{x}$ is the mean of a random sample of size $n$ taken from any population distribution with mean $\mu_x$ and variance $\sigma_x^2$, then the limiting form of the distribution of $Z = [\bar{x} - \mu_x]/[\sigma_x/\sqrt{n}]$ as $n \to \infty$ is the standardized normal distribution $P_G[z, 0, 1]$. This theorem provides the business model for all casino and gambling concerns - as long as they operate games that favor the house, the house is guaranteed to win over the long haul. Players on average are losers whose losses grow as they continue to play.

In general, the normal approximation for $\bar{x}$ will be good if $n \geq 25 - 30$ regardless of the shape of the population [**Figure 4.10**]. If $n < 25 - 30$, the approximation is good only if the data distribution is not too different from a normal distribution. On the other hand, if the data are known to be normally distributed, samples of $\bar{x}$ follow a normal distribution exactly, no matter how small $n$ is.

## 4.5 STATISTICAL INFERENCE

Statistical inference may be divided into *estimation* and *hypothesis testing* applications. They use the sample mean and standard deviation and their related distributions to estimate their probable confidence limits or error bars and test hypotheses on possible sample relationships from the overlap of the relevant confidence intervals.

In practice, these applications break down according to sample size $n$. For small $n \leq 25 - 30$, the sample variances can vary widely to make statistical inference unreliable unless the data distributions are approximately normal. Where small datasets are not normally distributed, somewhat weaker non-parametric tests may be applied to obtain effective statistical inferences.

For large $n \geq 25 - 30$, on the other hand, randomly selected samples of any distribution yield means that follow the Gaussian standard distribution by the central limits theorem. Thus, error bar estimation and hypothesis testing based on the normal distribution are routinely applicable.

### 4.5.1 CONFIDENCE INTERVALS ON THE MEAN

For a ***large number*** [i.e., $n \geq 25 - 30$] of samples randomly drawn from a parent distribution of arbitrary shape, the $q$% confidence [or $(100 - q = \alpha)$%

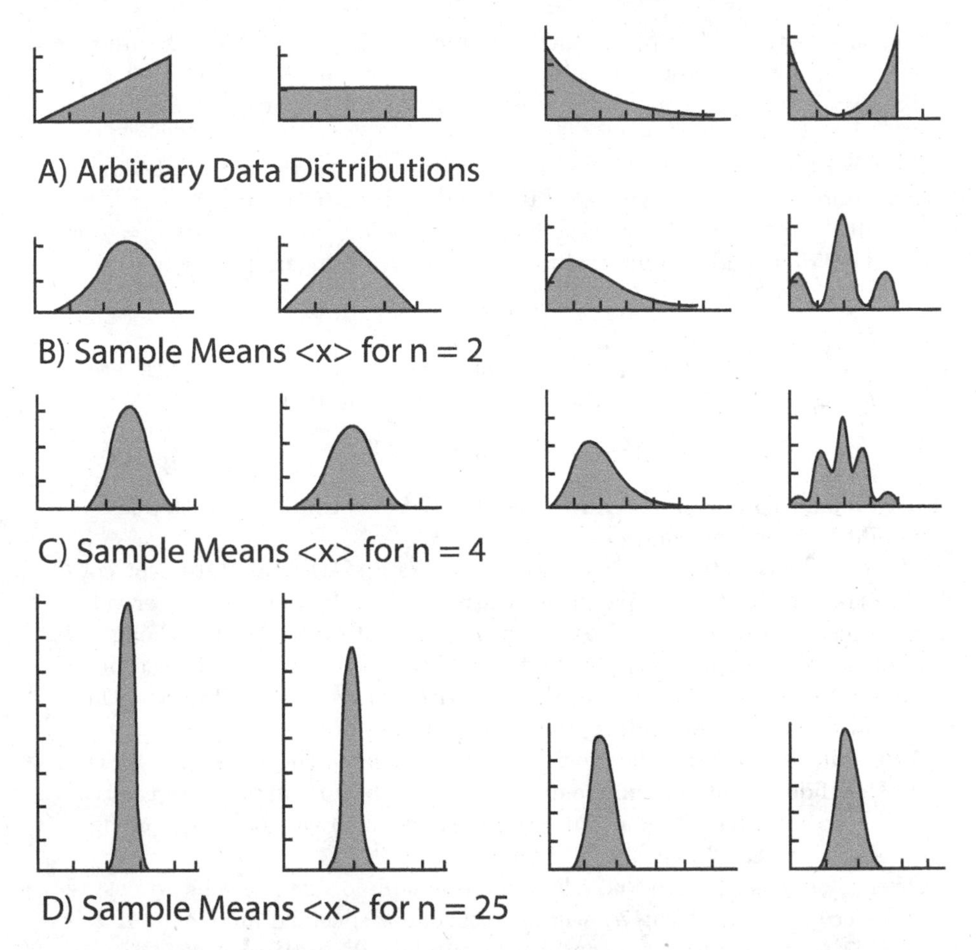


**Figure 4.10** The central limits theorem states that distributions of sample means $\bar{x} =< x >$ trend to the normal distribution as $n$ increases. Examples from arbitrary data distributions [**A**] include the $\bar{x}$-distributions from random samples of size $n = 2$ [**B**], $n = 4$ [**C**], and $n = 25$ [**D**]. [Adapted from [**20**] after [**58**].]

significance] interval for the population mean $\mu_x$ is given by

$$\bar{x} - k[\frac{s_x}{\sqrt{n}}] \leq \mu_x \leq \bar{x} + k[\frac{s_x}{\sqrt{n}}], \tag{4.43}$$

where $k$ is the number of standard deviation units about the mean that encompasses the desired probability for the confidence interval. For example, from **Table 4.1**, $k = 1.96$ for the $q = 95\%$ confidence [or $\alpha = 5\%$ significance] level, whereas for the 99% confidence [or $\alpha = 1\%$ significance] level, $k = 2.576$. Relevant $k$-values for other probabilities are commonly available in detailed tabulations [e.g., [**20**]] or from equation-solving software [e.g., MATLAB, Mathematica, Mathcad, etc.].

For this test, the sample mean $\bar{x}$ is calculated, the number of samples $n$ is known, and the approximation $\sigma_x \approx s_x$ is assumed. Accordingly, the probability $q$ that $\mu_x$ is contained in the interval about $\bar{x}$ is given by **Equation 4.43**.

As ***Example 04.5.1a***, suppose the density mean and standard deviation for 36 samples of a rock type are calculated to be 2.6 and 0.3 gm/cm$^3$, respectively. The sample size is large [i.e., $n \geq 25 - 30$] so that using $k = 1.96$ yields the 95% confidence interval for the rock type's mean density as

$$2.50 \leq \mu_x \leq 2.70 \text{ gm/cm}^3. \tag{4.44}$$

Intervals of greater certainty for containing the true mean $\mu_x$ require larger $k$-values - e.g., $k = 2.58$ yields the 99% confidence interval

$$2.47 \leq \mu_x \leq 2.73 \text{ gm/cm}^3. \tag{4.45}$$

Clearly, a longer interval or error bar is required to estimate $\mu_x$ with a higher probability or certainty and vice versa.

The confidence interval can be used to make a statistical statement concerning the error of the sample mean [**Figure 4.11**]. If $\mu_x$ is the center value of the interval, then $\bar{x} =< x >$ estimates $\mu_x$ without error. Mostly, though, $\bar{x}$ will not be exactly equal to $\mu_x$ so that some error is involved in the estimate. However, we can be $q$% confident that the size of this error will be less than $k[\sigma_x/\sqrt{n}]$ for the corresponding standard deviation unit $k$.

As ***Example 04.5.1b***, the previous example ***4.5.1a*** suggested that we can be 95% confident that the sample mean $\bar{x} = 2.6$ gm/cm$^3$ differs from the true mean $\mu_x$ by less than 0.1 gm/cm$^3$. However, we also can be 99% confident that this difference is less than 0.13 gm/cm$^3$.

Often, it is desirable to know how large a sample size must be to ensure that the error in estimating $\mu_x$ will be less than a specified amount $e$. If $\bar{x}$ is used as an estimate of $\mu_x$, we can be $q$% confident [**Equation 4.43**] that the error will be less than the amount $e$ [**Figure 4.11**] when the sample size is

$$n \geq [k\frac{\sigma_x}{e}]^2. \tag{4.46}$$

As ***Example 04.5.1c***, for the previous example ***4.5.1b***, determine how large the sample size must be if we want to be 95% confident that our estimate

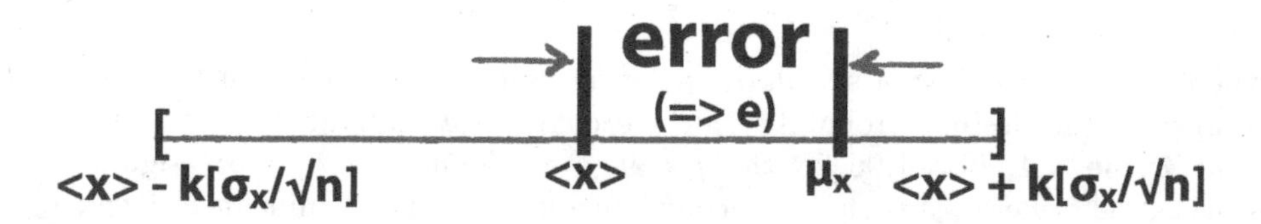


**Figure 4.11** Error $e$ of the sample mean $\bar{x} =< x >$ relative to the population mean $\mu_x$ in terms of the units $k$ of the $n$ samples' standard deviation $s_x \approx \sigma_x$.

of $\mu_x$ is off by less than 0.05 gm/cm$^3$. The preliminary sample size is $n = 36$ with $\sigma_x \approx s_x = 0.3$ gm/cm$^3$, so that by **Equation 4.46**, the required number of samples is $n \geq 138.3$. In other words, we can be 95% confident that a random sample size of 139 will provide an estimate of $\bar{x}$ that differs from $\mu_x$ by less than 0.05 gm/cm$^3$.

Note the tradeoff between precision and *reliability* or the probability of getting a good result. Confidence intervals are numerical measures of this tradeoff, where smaller confidence intervals reflect greater precision with the estimate $\bar{x}$ closer to $\mu_x$. Larger confidence intervals, on the other hand, highlight greater reliability with greater probability that $\bar{x}$ is significantly close to $\mu_x$.

For a ***small number*** [i.e., $n < 25 - 30$] of samples randomly drawn from an approximately bell-shaped distribution, the confidence interval may be established using *Student's t Test for* $\mu_x$. This test was introduced in 1908 by W.S. Gosset under the pen name "Student" because the Guinness brewery in Dublin, Ireland, where he worked disallowed publication of research by its staff. The $t$ distribution [**Figure 4.12**] is a combination of the standard normal distribution and the chi-squared [$\chi^2$] distribution [see next section-below], which is identical to the normal distribution when the sample size is infinite.

The $t$ distribution is appropriate for estimating the mean of a population when $\sigma_x$ is unknown and it is not possible to obtain a large sample size due to cost, or accessibility, or other limiting factors on data collection. However, as long as the sample distribution is approximately bell-shaped or normal, confidence intervals on $\mu_x$ may be computed by

$$\bar{x} - [\frac{s_x}{\sqrt{n}}]t_{q(n-1)} \leq \mu_x \leq \bar{x} + [\frac{s_x}{\sqrt{n}}]t_{q(n-1)}. \tag{4.47}$$

Here, $t_{q(n-1)}$ is the value of the $t$ distribution corresponding to the $q$ confidence [or $\alpha = (1 - q)$ significance] level for degrees of freedom $\nu = [n - 1]$ that are reduced because the $\chi^2$ part of the $t$ distribution depends on estimating $s_x$. In **Figure 4.12**, for example, the value of the $t$ distribution at the $q = 95\%$ confidence or $\alpha = 5\%$ significance level for $n = 6$ observations is $t_{q(\nu)} = 2.015$.

As ***Example 04.5.1d***, consider porosity measurements on 10 samples of Tensleep Sandstone from the Bighorn Basin [WY] with $\bar{x} = 21.3\%$ and $s_x = 5.52\%$ [**20**]. Then, by **Equation 4.47**, the 90% confidence interval on $\mu_x$ is $[21.3 \pm 2.4]\%$. This result also may be reported as a relative error of about 11.3% on the measurement of the mean for these samples.

Confidence intervals are convenient measures of the areas or probabilities associated with data histograms. Thus, the degree of overlap of confidence intervals provides a statistical test on whether the datasets are significantly different from each other and other hypotheses.

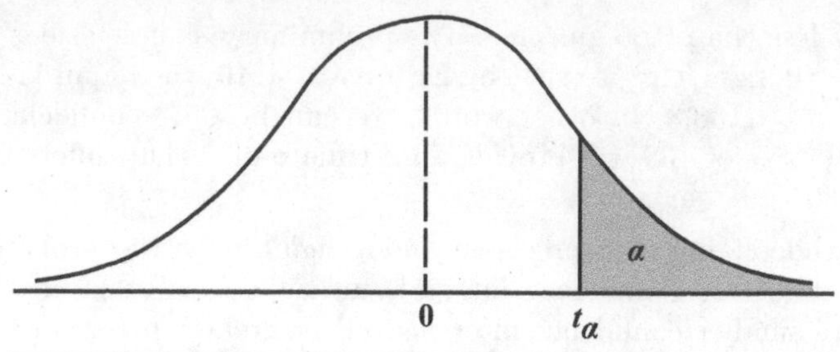


| ν | α | | | | |
|---|---|---|---|---|---|
| | 0.10 | 0.05 | 0.025 | 0.01 | 0.005 |
| 1 | 3.078 | 6.314 | 12.706 | 31.821 | 63.657 |
| 2 | 1.886 | 2.920 | 4.303 | 6.965 | 9.925 |
| 3 | 1.638 | 2.353 | 3.182 | 4.541 | 5.841 |
| 4 | 1.533 | 2.132 | 2.776 | 3.747 | 4.604 |
| 4 | 1.476 | 2.015 | 2.571 | 3.365 | 4.032 |
| 6 | 1.440 | 1.943 | 2.447 | 3.143 | 3.707 |
| 7 | 1.415 | 1.895 | 2.365 | 2.998 | 3.499 |
| 8 | 1.397 | 1.860 | 2.306 | 2.896 | 3.355 |
| 9 | 1.383 | 1.833 | 2.262 | 2.821 | 3.250 |
| 10 | 1.372 | 1.812 | 2.228 | 2.764 | 3.169 |
| 11 | 1.363 | 1.796 | 2.201 | 2.718 | 3.106 |
| 12 | 1.356 | 1.782 | 2.179 | 2.681 | 3.055 |
| 13 | 1.350 | 1.771 | 2.160 | 2.650 | 3.012 |
| 14 | 1.345 | 1.761 | 2.145 | 2.624 | 2.977 |
| 15 | 1.341 | 1.753 | 2.131 | 2.602 | 2.947 |
| 16 | 1.337 | 1.746 | 2.120 | 2.583 | 2.921 |
| 17 | 1.333 | 1.740 | 2.110 | 2.567 | 2.898 |
| 18 | 1.330 | 1.734 | 2.101 | 2.552 | 2.878 |
| 19 | 1.328 | 1.729 | 2.093 | 2.539 | 2.861 |
| 20 | 1.325 | 1.725 | 2.086 | 2.528 | 2.845 |
| 21 | 1.323 | 1.721 | 2.080 | 2.518 | 2.831 |
| 22 | 1.321 | 1.717 | 2.074 | 2.508 | 2.819 |
| 23 | 1.319 | 1.714 | 2.069 | 2.500 | 2.807 |
| 24 | 1.318 | 1.711 | 2.064 | 2.492 | 2.797 |
| 25 | 1.316 | 1.708 | 2.060 | 2.485 | 2.787 |
| 26 | 1.315 | 1.706 | 2.056 | 2.479 | 2.779 |
| 27 | 1.314 | 1.703 | 2.052 | 2.473 | 2.771 |
| 28 | 1.313 | 1.702 | 2.048 | 2.467 | 2.763 |
| 29 | 1.311 | 1.699 | 2.045 | 2.462 | 2.756 |
| ∞ | 1.282 | 1.645 | 1.960 | 2.326 | 2.576 |

**Figure 4.12** Critical values of the $t$ distribution tabulated according to degrees of freedom $\nu$ and significance levels $\alpha$. The tabulated values are commonly given by statistics texts [e.g., **[99]**] and electronic equation solvers [e.g., MATLAB, Mathematica].

### 4.5.2 HYPOTHESIS TESTS OF THE MEAN

Suppose a ***large number*** of random samples is collected to test the hypothesis that this sample with mean $\bar{x}$ is part of a population of known $\mu_x$ and $\sigma_x$. By the Central Limits Theorem, $\bar{x}$ follows a Gaussian distribution so that $\bar{x}$ can be compared to $\mu_x$ on the standard normal distribution by computing the statistic

$$Z_G = \frac{\bar{x} - \mu_x}{\sigma_x/\sqrt{n}}. \tag{4.48}$$

For example, at the significance level $\alpha = 0.05$ or 5% [or the confidence level $q = 1 - \alpha = 0.95$ or 95%], if either $Z_G > 1.96$ or $Z_G < 1.96$, or equivalently $|Z_G| > 1.96$, then the sample population does not fit the normal population within the 5% significance level. Thus, the hypothesis is *rejected* - i.e., $Z_G$ falls in the region outside the 95% confidence interval.

On the other hand, if $Z_G$ falls within the 95% confidence interval [i.e., $-1.96 < Z_G < 1.96$], then there is no reason to reject the hypothesis that the sample is part of the population. It cannot be claimed, however, that the sample is drawn from the population because the sample mean $\bar{x}$ is significantly close to $\mu_x$. It may only be said that based on the $Z_G$-test of the mean, there is no reason to believe that the sample is not part of the population at the $q$% confidence level. This weaker indirect statement is characteristic of all hypothesis tests.

In general, the hypothesis that is formulated for rejection is called the *null hypothesis* $[H_0]$ as opposed to the alternate *failure-to-reject hypothesis* $[H_1]$. Hypothesis tests tend to be either *one-tailed* in that the null hypothesis $H_0 : \bar{x} = \mu_x$, for example, is tested against the alternate $H_1 : \bar{x} > \mu_x$ [or $\bar{x} < \mu_x$], or *two-tailed* where $H_0 : \bar{x} = \mu_x$ is tested against the 2-sided alternative $H_1 : \bar{x} \neq \mu_x$.

Rejecting $H_0$ when it is true is called a *type I error*, whereas a *type II error* is accepting $H_0$ when it is false. The *$\alpha$-significance level* of the test is the probability of committing a type I error.

As ***Example 04.5.2a***, consider a geological engineer's claim that a highly weathered and friable rock unit has a tensile strength of 15 lbs with a standard deviation of 0.5 lbs. In particular, test the hypothesis that $\bar{x} = \mu_x = 15$ lbs against the alternative that $\bar{x} \neq \mu_x$ at the 1% significance level for a random sample of 50 specimens which upon laboratory testing was found to have a mean tensile strength of $\bar{x} = 14.8$ lbs.

For $\alpha = 0.01$ (or $q = 99\%$), the critical regions are $< -Z'_G = -2.58$ and $> Z'_G = 2.58$ so that **Equation 4.48** with $\bar{x} = 14.8$ lbs and $n = 50$ gives

$$Z_G = -2.828 \; < \; -Z'_G. \tag{4.49}$$

Thus, the hypothesis $H_0 : \bar{x} = \mu_x = 15$ lbs is rejected - i.e., $|Z_G| > |Z'_G|$. Furthermore, it may be concluded that the average tensile strength of the population from which the sample was drawn is in fact less than 15 lbs.

For hypothesis testing a ***small number*** of randomly selected samples from an approximately bell-shaped distribution, the statistic computed from **Equation 4.48** is compared against the $t$-distribution rather than the normal distribution. Specifically,

$$Z_t = \frac{\bar{x} - \mu_x}{\sigma_x / \sqrt{n}}. \tag{4.50}$$

where $\bar{x}$ and $s_x$ refer to the $n$ samples and $\mu_x$ is the hypothetical population mean against which $\bar{x}$ is being tested.

The interpretation involves comparing the test value $Z_t$ to the critical value $Z'_t = t_{\alpha(n-1)}$ of the distribution [e.g., **Figure 4.12**] for a desired $\alpha$-significance level and degrees of freedom $\nu = [n - 1]$. Accordingly, the null hypothesis is rejected if $|Z_t| > |Z'_t|$, whereas there is no reason to reject the hypothesis if $|Z_t| < |Z'_t|$ - i.e., there is no reason to believe that the sample is not part of the population.

As ***Example 04.5.2b***, suppose a subterrene [i.e., a tunnel boring machine (TBM) that melts rock with nuclear heating] excavates a given volume of rock in an average of 50 hours [hrs] with a standard deviation of 10 hrs. An expensive modification of the TBM uses forced air to clean the excavation heading during boring operations. A random sample of 12 experiments with the new procedure yields an approximately normal distribution about an average time of 42 hrs with a standard deviation of 11.9 hrs. Test the hypothesis that the population mean is now less than 50 hrs at significance levels of 5% and 1%.

For degrees of freedom $\nu = 11$, **Figure 4.12** gives $|Z_t| = |-2.33| > |Z'_t| = |-1.796|$ at the $\alpha = 5\%$ level, whereas at the $\alpha = 1\%$ level it shows $|Z_t| < |Z'_t| = |-2.718|$. Accordingly, the null hypothesis that the mean excavation time for the new procedure is 50 hrs [i.e, $H_0 : \bar{x} = \mu_x = 50$ hrs] is rejected at the 5% signifance level, but not at the 1% level. These results might be interpreted to suggest that the new mean is likely to be less than 50 hrs, but the difference may be too small to warrant the high cost of the forced air modification.

As ***Example 04.5.2c***, reconsider the porosity measurements in ***Example 04.5.1d*** on 10 samples of Tensleep Sandstone from Wyoming's Bighorn Basin with $\bar{x} = 21.3\%$ and $s_x = 5.52\%$. Test the hypothesis that the sample mean $\bar{x} = 21.3\%$ is less than the hypothetical mean $\mu_x = 18\%$ at the $\alpha = 5\%$ significance level. More specifically, test the null hypothesis $H_0 : \bar{x} = 21.3\% \leq \mu_x = 18\%$ against the alternative $H_1 : \bar{x} > \mu_x$ at the $\alpha = 5\%$ significance level. Accordingly, by **Equation 4.50**, $|Z_t| = 1.89 > |Z'_t| = t_{0.05(9)} = 1.83$ so that the null hypothesis can be rejected at the 95% confidence level.

The means $< x1 >$ and $< x2 >$ from two samples can also be compared against each other using the modification of **Equation 4.50** given by

$$Z_{t(x1,x2)} = \frac{< x1 > - < x2 >}{s_{(x1,x2)} \sqrt{(1/n1) + (1/n2)}}, \tag{4.51}$$

where $s_{(x1,x2)}$ is the *pooled standard deviation* based on the pooled variance

$$s^2_{(x1,x2)} = \frac{[n1-1]s^2_{x1} + [n2-1]s^2_{x2}}{[n1+n2-2]}. \tag{4.52}$$

The test value $Z_{t(x1,x2)}$ is compared to the critical value $Z'_{t(x1,x2)} = t_{\alpha(n1+n2-2)}$ for a desired $\alpha$-significance level, and the degrees of freedom $\nu = [n1 + n2 - 2]$ that account for incorporating both sample variances into the test value estimate.

As ***Example 04.5.2d***, compare the porosity mean $< x1 >= 21.3\%$ from the previous ***Example 04.5.2c*** with the mean $< x2 >= 18.9\%$ from a second batch of 10 porosity measurements of the Tensleep Sandstone taken now from Wyoming's Wind River Basin with $s_{x2} = 4.82\%$ [**20**]. In other words, test the hypothesis $H_0 :< x1 >=< x2 >$ that the mean of the first sample's population is the same as the mean of the second sample's population against the alternative $H_1 :< x1 > \neq < x2 >$. By **Equation 4.51**, the test value $|Z_{t(x1,x2)}| = 1.04$ is less than the critical value $|Z'_{t(x1,x2)}| = |t_{0.01(18)}| = 2.55$ at the 99% confidence level. Thus, the null hypothesis cannot be rejected - i.e., the test at the 1% significance level does not support the notion that the two samples came from populations with different means.

The assumptions underlying this test include the random collection of samples, which either are large in number [i.e., $n \geq 25 - 30$] or otherwise drawn from an approximately bell-shaped distribution. A further assumption is that the samples have equal variances. The next four subsections consider ways to check the variance and normality properties of datasets.

### 4.5.3 CONFIDENCE INTERVALS ON THE VARIANCE

An interval estimate on the population variance $\sigma^2_x$ can be established using the statistic

$$X^2 = [n-1]s^2_x/\sigma^2_x, \tag{4.53}$$

where $s^2_x$ is an estimator of $\sigma^2_x$. The statistic $X^2$ has a chi-squared $[\chi^2]$ distribution [**Figure 4.13**] with [n - 1] degrees of freedom when $n$ samples are chosen from a normal distribution. The probability for a value $X^2$ in the interval $[\chi^2_{(1-\alpha/2)}, \chi^2_{(\alpha/2)}]$ is

$$Pr[\chi^2_{(1-\alpha/2)} \leq X^2 \leq \chi^2_{(\alpha/2)}] = 1 - \alpha. \tag{4.54}$$

Accordingly, the $q\%$ or $[1-\alpha]100\%$ confidence interval or error bar for $\sigma^2_x$ of a normal population is

$$\frac{[n-1]s^2_x}{\chi^2_{(\alpha/2)}} \leq \sigma^2_x \leq \frac{[n-1]s^2_x}{\chi^2_{(1-\alpha/2)}}, \tag{4.55}$$

where $s^2_x$ is the variance of a random sample of size $n$, and $\chi^2_{(\alpha/2)}$ and $\chi^2_{(1-\alpha/2)}$ are the values of the $\chi^2$ distribution with $[n-1]$ degrees of freedom leaving areas $[\alpha/2]$ and $[1-\alpha/2]$, respectively, to the *right* of the distribution [e.g., **Figure 4.13**].

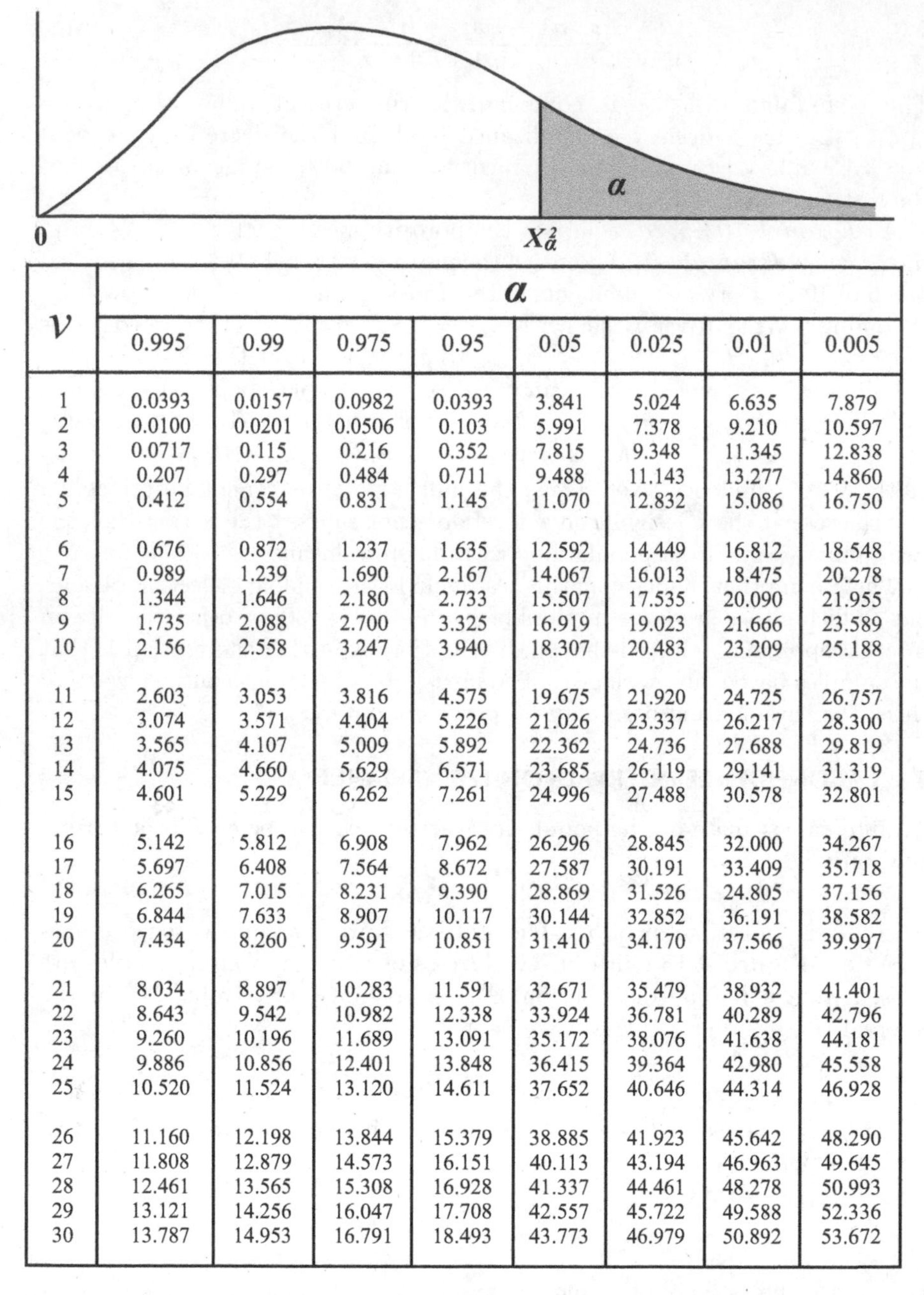


| ν | α | | | | | | | |
|---|---|---|---|---|---|---|---|---|
| | 0.995 | 0.99 | 0.975 | 0.95 | 0.05 | 0.025 | 0.01 | 0.005 |
| 1 | 0.0393 | 0.0157 | 0.0982 | 0.0393 | 3.841 | 5.024 | 6.635 | 7.879 |
| 2 | 0.0100 | 0.0201 | 0.0506 | 0.103 | 5.991 | 7.378 | 9.210 | 10.597 |
| 3 | 0.0717 | 0.115 | 0.216 | 0.352 | 7.815 | 9.348 | 11.345 | 12.838 |
| 4 | 0.207 | 0.297 | 0.484 | 0.711 | 9.488 | 11.143 | 13.277 | 14.860 |
| 5 | 0.412 | 0.554 | 0.831 | 1.145 | 11.070 | 12.832 | 15.086 | 16.750 |
| 6 | 0.676 | 0.872 | 1.237 | 1.635 | 12.592 | 14.449 | 16.812 | 18.548 |
| 7 | 0.989 | 1.239 | 1.690 | 2.167 | 14.067 | 16.013 | 18.475 | 20.278 |
| 8 | 1.344 | 1.646 | 2.180 | 2.733 | 15.507 | 17.535 | 20.090 | 21.955 |
| 9 | 1.735 | 2.088 | 2.700 | 3.325 | 16.919 | 19.023 | 21.666 | 23.589 |
| 10 | 2.156 | 2.558 | 3.247 | 3.940 | 18.307 | 20.483 | 23.209 | 25.188 |
| 11 | 2.603 | 3.053 | 3.816 | 4.575 | 19.675 | 21.920 | 24.725 | 26.757 |
| 12 | 3.074 | 3.571 | 4.404 | 5.226 | 21.026 | 23.337 | 26.217 | 28.300 |
| 13 | 3.565 | 4.107 | 5.009 | 5.892 | 22.362 | 24.736 | 27.688 | 29.819 |
| 14 | 4.075 | 4.660 | 5.629 | 6.571 | 23.685 | 26.119 | 29.141 | 31.319 |
| 15 | 4.601 | 5.229 | 6.262 | 7.261 | 24.996 | 27.488 | 30.578 | 32.801 |
| 16 | 5.142 | 5.812 | 6.908 | 7.962 | 26.296 | 28.845 | 32.000 | 34.267 |
| 17 | 5.697 | 6.408 | 7.564 | 8.672 | 27.587 | 30.191 | 33.409 | 35.718 |
| 18 | 6.265 | 7.015 | 8.231 | 9.390 | 28.869 | 31.526 | 24.805 | 37.156 |
| 19 | 6.844 | 7.633 | 8.907 | 10.117 | 30.144 | 32.852 | 36.191 | 38.582 |
| 20 | 7.434 | 8.260 | 9.591 | 10.851 | 31.410 | 34.170 | 37.566 | 39.997 |
| 21 | 8.034 | 8.897 | 10.283 | 11.591 | 32.671 | 35.479 | 38.932 | 41.401 |
| 22 | 8.643 | 9.542 | 10.982 | 12.338 | 33.924 | 36.781 | 40.289 | 42.796 |
| 23 | 9.260 | 10.196 | 11.689 | 13.091 | 35.172 | 38.076 | 41.638 | 44.181 |
| 24 | 9.886 | 10.856 | 12.401 | 13.848 | 36.415 | 39.364 | 42.980 | 45.558 |
| 25 | 10.520 | 11.524 | 13.120 | 14.611 | 37.652 | 40.646 | 44.314 | 46.928 |
| 26 | 11.160 | 12.198 | 13.844 | 15.379 | 38.885 | 41.923 | 45.642 | 48.290 |
| 27 | 11.808 | 12.879 | 14.573 | 16.151 | 40.113 | 43.194 | 46.963 | 49.645 |
| 28 | 12.461 | 13.565 | 15.308 | 16.928 | 41.337 | 44.461 | 48.278 | 50.993 |
| 29 | 13.121 | 14.256 | 16.047 | 17.708 | 42.557 | 45.722 | 49.588 | 52.336 |
| 30 | 13.787 | 14.953 | 16.791 | 18.493 | 43.773 | 46.979 | 50.892 | 53.672 |

**Figure 4.13** Critical values of the $\chi^2$-distribution tabulated according to degrees of freedom $\nu$ and significance levels $\alpha$ [e.g., **[99]**, MATLAB, Mathematica].

As ***Example 04.5.3a***, consider the 10 unit volume samples of sand grains of a particular geologic formation with the following weights in gm: 16.4 16.1, 15.8, 17.0, 16.1, 15.9, 15.8, 16.9, 15.2, and 16.0. Find the 95% confidence interval for the variance of these unit volume sand weights.

A computationally efficient estimate of $s_x^2$ that avoids first computing $\bar{x}$ directly is given by

$$s_x^2 = \frac{[(n\sum_{i=1}^{n} x_i^2) - (\sum_{i=1}^{n} x_i)^2]}{[n(n-1)]} = 0.286 \text{ gm}^2. \tag{4.56}$$

Choosing $\alpha = 0.005$ with $\nu = 9$ yields the critical values $\chi^2_{0.025} = 19.023$ and $\chi^2_{0.975} = 2.700$ from **Figure 4.13**. Accordingly, the 95% confidence interval for $\sigma_x^2$ is

$$[0.135 \leq 0.286 \leq 0.953] \text{ gm}^2. \tag{4.57}$$

## 4.5.4 HYPOTHESIS TESTS OF THE VARIANCE

Tests of the equality of variances are based on Fisher's $f$ distribution [**Figure 4.14**] that theoretically governs the distribution of all possible variance ratios for pairs of datasets randomly collected from a normal population. In **Figure 4.14**, for example, the distribution value for $\alpha = 0.05$, $\nu_1 = 2$, and $\nu_2 = 12$ is $f_{(0.05,2,12)} = 3.89$, whereas $f_{(0.05,12,2)} = 19.41$ in **Figure 4.15**.

The hypothesis test for $\sigma_x^2$ is defined by the statistic

$$F = \frac{s_{x1}^2}{s_{x2}^2} \;\forall\; s_{x1}^2 > s_{x2}^2 \tag{4.58}$$

for two samples with variances $s_{x1}^2$ and $s_{x2}^2$ of respective sizes $n_1$ and $n_2$ and degrees of freedom $\nu_1 = [n_1 - 1]$ and $\nu_2 = [n_2 - 1]$. The calculated $F$ value is compared to the critical value $F' = f_{(\alpha,\nu_1,\nu_2)}$ from the $f$ distribution. If $F > F'$, then the hypothesis that the two samples are from the same population is rejected. However, if $F < F'$, then on the basis of the $F$-test, there is no reason to reject the hypothesis.

As ***Example 04.5.4a***, consider the two sample variances $s_{x1}^2 = 5.52\%$ and $s_{x2}^2 = 4.82\%$ on the porosity measurements of the Tensleep Sandstone from examples ***4.5.1d*** and ***4.5.2d***, respectively. Using the $F$-ratio [**Equation 4.58**] at the 5% significance level, test the hypothesis $H_0 : s_{x1}^2 = s_{x2}^2$ against the alternate $H_1 : s_{x1}^2 \neq s_{x2}^2$. Accordingly, $F = 1.14 < F' = f_{(0.05,9,9)} = 3.18$ so that the $F$-test does not support the notion that the variances are different. However, accepting this result involves the risk of rejecting $H_0$ when it is correct once in every twenty tests.

Another test of the null hypothesis considers the $p$-probability for obtaining an estimated $F$ equal to or greater than the critical $F$' when the null hypothesis is correct. As ***Example 04.5.4b***, the top panel of **Figure 4.16** shows

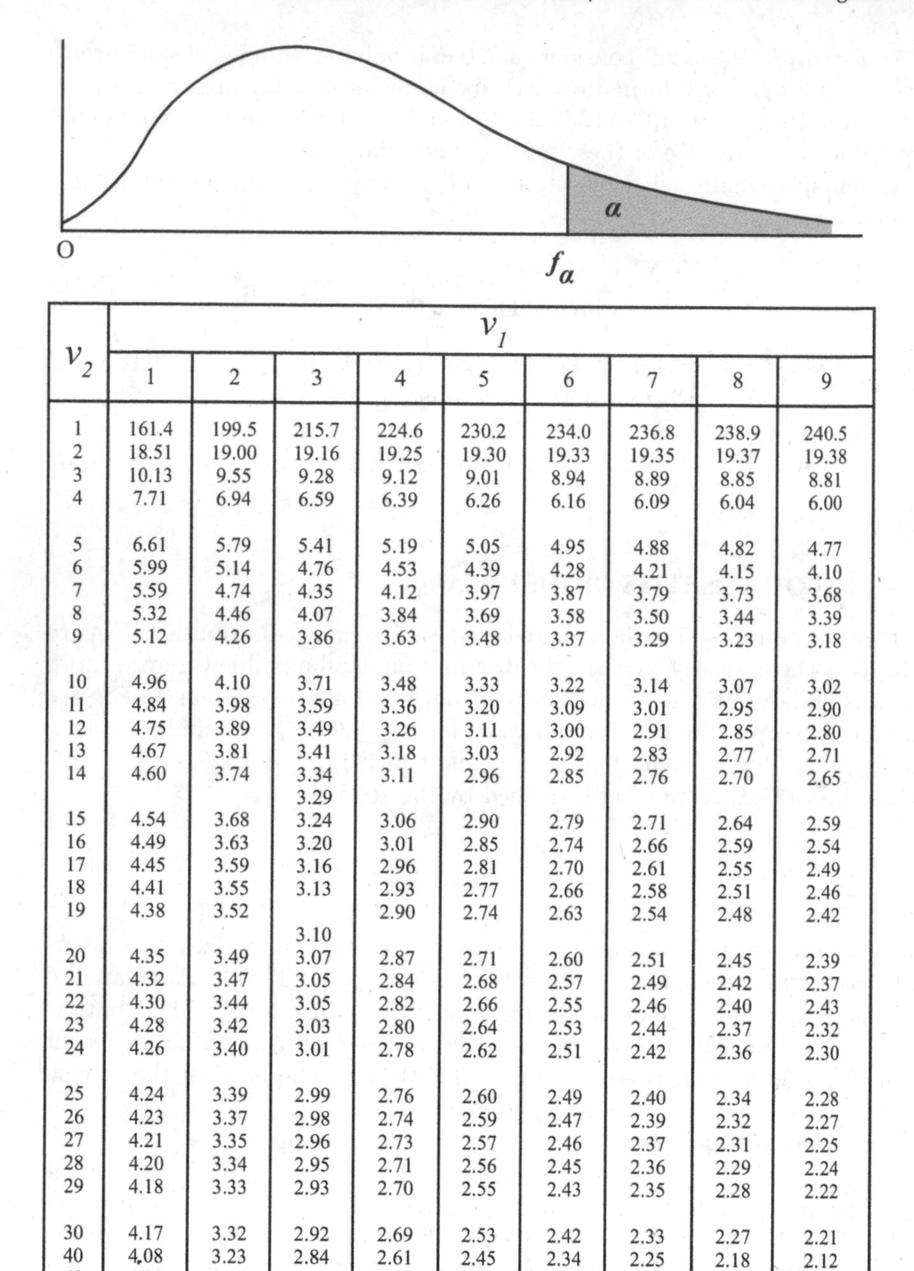


| $\nu_2$ | $\nu_1$ | | | | | | | | |
|---|---|---|---|---|---|---|---|---|---|
| | 1 | 2 | 3 | 4 | 5 | 6 | 7 | 8 | 9 |
| 1 | 161.4 | 199.5 | 215.7 | 224.6 | 230.2 | 234.0 | 236.8 | 238.9 | 240.5 |
| 2 | 18.51 | 19.00 | 19.16 | 19.25 | 19.30 | 19.33 | 19.35 | 19.37 | 19.38 |
| 3 | 10.13 | 9.55 | 9.28 | 9.12 | 9.01 | 8.94 | 8.89 | 8.85 | 8.81 |
| 4 | 7.71 | 6.94 | 6.59 | 6.39 | 6.26 | 6.16 | 6.09 | 6.04 | 6.00 |
| 5 | 6.61 | 5.79 | 5.41 | 5.19 | 5.05 | 4.95 | 4.88 | 4.82 | 4.77 |
| 6 | 5.99 | 5.14 | 4.76 | 4.53 | 4.39 | 4.28 | 4.21 | 4.15 | 4.10 |
| 7 | 5.59 | 4.74 | 4.35 | 4.12 | 3.97 | 3.87 | 3.79 | 3.73 | 3.68 |
| 8 | 5.32 | 4.46 | 4.07 | 3.84 | 3.69 | 3.58 | 3.50 | 3.44 | 3.39 |
| 9 | 5.12 | 4.26 | 3.86 | 3.63 | 3.48 | 3.37 | 3.29 | 3.23 | 3.18 |
| 10 | 4.96 | 4.10 | 3.71 | 3.48 | 3.33 | 3.22 | 3.14 | 3.07 | 3.02 |
| 11 | 4.84 | 3.98 | 3.59 | 3.36 | 3.20 | 3.09 | 3.01 | 2.95 | 2.90 |
| 12 | 4.75 | 3.89 | 3.49 | 3.26 | 3.11 | 3.00 | 2.91 | 2.85 | 2.80 |
| 13 | 4.67 | 3.81 | 3.41 | 3.18 | 3.03 | 2.92 | 2.83 | 2.77 | 2.71 |
| 14 | 4.60 | 3.74 | 3.34 | 3.11 | 2.96 | 2.85 | 2.76 | 2.70 | 2.65 |
| | | | 3.29 | | | | | | |
| 15 | 4.54 | 3.68 | 3.24 | 3.06 | 2.90 | 2.79 | 2.71 | 2.64 | 2.59 |
| 16 | 4.49 | 3.63 | 3.20 | 3.01 | 2.85 | 2.74 | 2.66 | 2.59 | 2.54 |
| 17 | 4.45 | 3.59 | 3.16 | 2.96 | 2.81 | 2.70 | 2.61 | 2.55 | 2.49 |
| 18 | 4.41 | 3.55 | 3.13 | 2.93 | 2.77 | 2.66 | 2.58 | 2.51 | 2.46 |
| 19 | 4.38 | 3.52 | | 2.90 | 2.74 | 2.63 | 2.54 | 2.48 | 2.42 |
| | | | 3.10 | | | | | | |
| 20 | 4.35 | 3.49 | 3.07 | 2.87 | 2.71 | 2.60 | 2.51 | 2.45 | 2.39 |
| 21 | 4.32 | 3.47 | 3.05 | 2.84 | 2.68 | 2.57 | 2.49 | 2.42 | 2.37 |
| 22 | 4.30 | 3.44 | 3.05 | 2.82 | 2.66 | 2.55 | 2.46 | 2.40 | 2.43 |
| 23 | 4.28 | 3.42 | 3.03 | 2.80 | 2.64 | 2.53 | 2.44 | 2.37 | 2.32 |
| 24 | 4.26 | 3.40 | 3.01 | 2.78 | 2.62 | 2.51 | 2.42 | 2.36 | 2.30 |
| 25 | 4.24 | 3.39 | 2.99 | 2.76 | 2.60 | 2.49 | 2.40 | 2.34 | 2.28 |
| 26 | 4.23 | 3.37 | 2.98 | 2.74 | 2.59 | 2.47 | 2.39 | 2.32 | 2.27 |
| 27 | 4.21 | 3.35 | 2.96 | 2.73 | 2.57 | 2.46 | 2.37 | 2.31 | 2.25 |
| 28 | 4.20 | 3.34 | 2.95 | 2.71 | 2.56 | 2.45 | 2.36 | 2.29 | 2.24 |
| 29 | 4.18 | 3.33 | 2.93 | 2.70 | 2.55 | 2.43 | 2.35 | 2.28 | 2.22 |
| 30 | 4.17 | 3.32 | 2.92 | 2.69 | 2.53 | 2.42 | 2.33 | 2.27 | 2.21 |
| 40 | 4.08 | 3.23 | 2.84 | 2.61 | 2.45 | 2.34 | 2.25 | 2.18 | 2.12 |
| 60 | 4.00 | 3.15 | 2.76 | 2.53 | 2.37 | 2.25 | 2.17 | 2.10 | 2.04 |
| 120 | 3.92 | 3.07 | 2.68 | 2.45 | 2.29 | 2.17 | 2.09 | 2.02 | 1.96 |
| ∞ | 3.84 | 3.00 | 2.60 | 2.37 | 2.21 | 2.10 | 2.01 | 1.94 | 1.88 |

**Figure 4.14** Critical values of the $f$-distribution tabulated at the significance level $\alpha = 0.05$ or 5% according to degrees of freedom $\nu_1 = 1, 2, \ldots, 9$ and $\nu_2$. **Figure 4.15** continues this tabulation for $\nu_1 = 10, 12, \ldots, \infty$ [e.g., **[99]**, MATLAB, Mathematica].

| $\nu_2$ | $\nu_1$ | | | | | | | | | |
|---|---|---|---|---|---|---|---|---|---|---|
| | 10 | 12 | 15 | 20 | 24 | 30 | 40 | 60 | 120 | $\infty$ |
| 1 | 241.9 | 243.9 | 245.9 | 248.0 | 249.1 | 250.1 | 251.1 | 252.2 | 253.3 | 254.3 |
| 2 | 19.40 | 19.41 | 19.43 | 19.45 | 19.45 | 19.46 | 19.47 | 19.48 | 19.49 | 19.50 |
| 3 | 8.79 | 8.74 | 8.70 | 8.66 | 8.64 | 8.62 | 8.59 | 8.57 | 8.55 | 8.53 |
| 4 | 5.96 | 5.91 | 5.86 | 5.80 | 5.77 | 5.75 | 5.72 | 5.69 | 5.66 | 5.63 |
| 5 | 4.74 | 4.68 | 4.62 | 4.56 | 4.53 | 4.50 | 4.46 | 4.43 | 4.40 | 4.36 |
| 6 | 4.06 | 4.00 | 3.94 | 3.87 | 3.84 | 3.81 | 3.77 | 3.74 | 3.70 | 3.67 |
| 7 | 3.64 | 3.57 | 3.52 | 3.44 | 3.41 | 3.38 | 3.34 | 3.30 | 3.27 | 3.23 |
| 8 | 3.35 | 3.28 | 3.22 | 3.15 | 3.12 | 3.08 | 3.04 | 3.01 | 2.97 | 2.93 |
| 9 | 3.14 | 3.07 | 3.01 | 2.94 | 2.90 | 2.86 | 2.83 | 2.79 | 2.75 | 2.71 |
| 10 | 2.98 | 2.91 | 2.85 | 2.77 | 2.74 | 2.70 | 2.66 | 2.62 | 2.58 | 2.54 |
| 11 | 2.85 | 2.79 | 2.72 | 2.65 | 2.61 | 2.57 | 2.53 | 2.49 | 2.45 | 2.40 |
| 12 | 2.75 | 2.69 | 2.62 | 2.54 | 2.51 | 2.47 | 2.43 | 2.38 | 2.34 | 2.30 |
| 13 | 2.67 | 2.60 | 2.53 | 2.46 | 2.42 | 2.38 | 2.34 | 2.30 | 2.25 | 2.21 |
| 14 | 2.60 | 2.53 | 2.46 | 2.39 | 2.35 | 2.31 | 2.27 | 2.22 | 2.18 | 2.13 |
| 15 | 2.54 | 2.48 | 2.40 | 2.33 | 2.29 | 2.25 | 2.20 | 2.16 | 2.11 | 2.07 |
| 16 | 2.49 | 2.42 | 2.35 | 2.28 | 2.24 | 2.19 | 2.15 | 2.11 | 2.06 | 2.01 |
| 17 | 2.45 | 2.38 | 2.31 | 2.23 | 2.19 | 2.15 | 2.10 | 2.06 | 2.01 | 1.96 |
| 18 | 2.41 | 2.34 | 2.27 | 2.19 | 2.15 | 2.11 | 2.06 | 2.02 | 1.97 | 1.92 |
| 19 | 2.38 | 2.31 | 2.23 | 2.16 | 2.11 | 2.07 | 2.03 | 1.98 | 1.93 | 1.88 |
| 20 | 2.35 | 2.28 | 2.20 | 2.12 | 2.08 | 2.04 | 1.99 | 1.95 | 1.90 | 1.84 |
| 21 | 2.32 | 2.25 | 2.18 | 2.10 | 2.05 | 2.01 | 1.96 | 1.92 | 1.87 | 1.81 |
| 22 | 2.30 | 2.23 | 2.15 | 2.07 | 2.03 | 1.98 | 1.94 | 1.89 | 1.84 | 1.78 |
| 23 | 2.37 | 2.20 | 2.13 | 2.05 | 2.01 | 1.96 | 1.91 | 1.86 | 1.81 | 1.76 |
| 24 | 2.25 | 2.18 | 2.11 | 2.03 | 1.98 | 1.94 | 1.89 | 1.84 | 1.79 | 1.73 |
| 25 | 2.24 | 2.16 | 2.09 | 2.01 | 1.96 | 1.92 | 1.87 | 1.82 | 1.77 | 1.71 |
| 26 | 2.22 | 2.15 | 2.07 | 1.99 | 1.95 | 1.90 | 1.85 | 1.80 | 1.75 | 1.69 |
| 27 | 2.20 | 2.13 | 2.06 | 1.97 | 1.93 | 1.88 | 1.84 | 1.79 | 1.73 | 1.67 |
| 28 | 2.19 | 2.12 | 2.04 | 1.96 | 1.91 | 1.87 | 1.82 | 1.77 | 1.71 | 1.65 |
| 29 | 2.18 | 2.10 | 2.03 | 1.94 | 1.90 | 1.85 | 1.81 | 1.75 | 1.70 | 1.64 |
| 30 | 2.16 | 2.09 | 2.01 | 1.93 | 1.89 | 1.84 | 1.79 | 1.74 | 1.68 | 1.62 |
| 40 | 2.08 | 2.00 | 1.91 | 1.84 | 1.79 | 1.74 | 1.69 | 1.64 | 1.58 | 1.51 |
| 60 | 1.99 | 1.92 | 1.84 | 1.75 | 1.70 | 1.65 | 1.59 | 1.53 | 1.47 | 1.39 |
| 120 | 1.91 | 1.83 | 1.75 | 1.66 | 1.61 | 1.55 | 1.50 | 1.43 | 1.35 | 1.25 |
| $\infty$ | 1.83 | 1.75 | 1.67 | 1.57 | 1.52 | 1.46 | 1.39 | 1.32 | 1.22 | 1.00 |

**Figure 4.15** Critical values of the $f$ distribution tabulated at the significance level $\alpha = 0.05$ or 5% according to degrees of freedom $\nu_1 = 10, 12, 15, \ldots, 120, \infty$ and $\nu_2$ continued from **Figure 4.14** [e.g., **[99]**, MATLAB, Mathematica].

the $f$ distribution appropriate to ***Example 04.5.4a***. Here, the shaded area represents 42% of the total area [= 100%] under the distribution or the probability $p$ of obtaining estimates $\geq F = 1.14$ at the $\alpha = 5\%$ significance level. Thus, $p > p' = 0.05$ so that the $F$-test provides no evidence for rejecting the null hypothesis. In the bottom panel, by contrast, the probability of obtaining

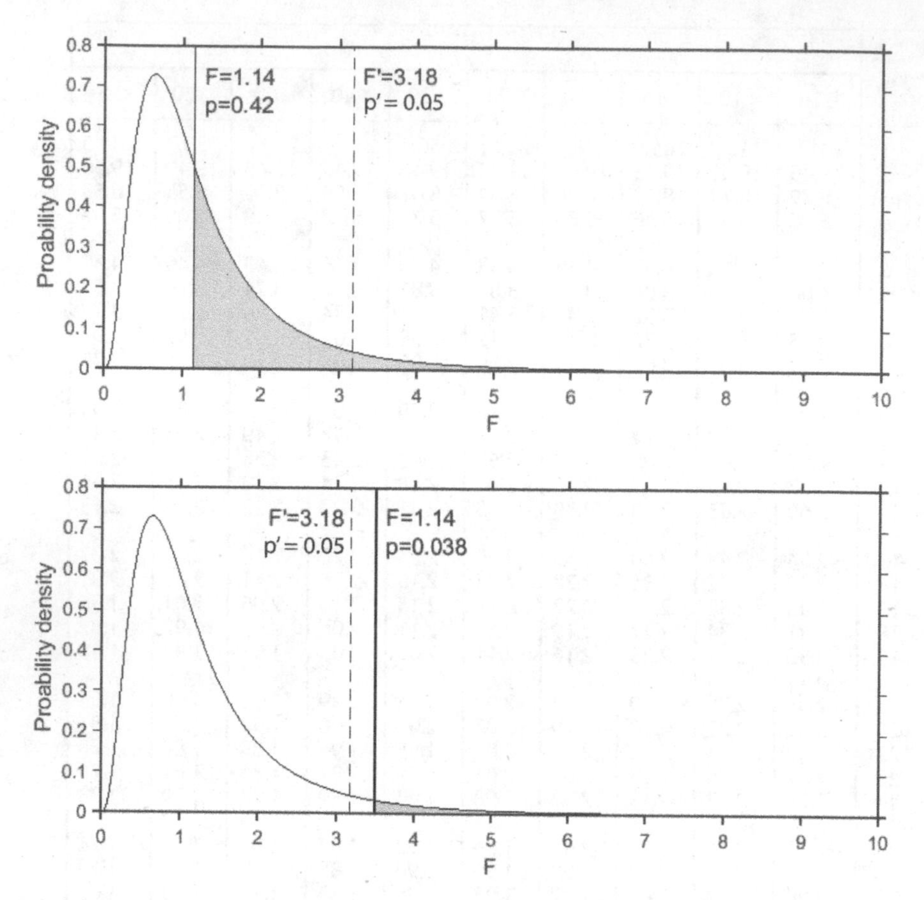


**Figure 4.16** Comparing respective estimated and critical $F$- and $F'$-values of the $f$ distribution with degrees of freedom $\nu_1 = 9$ and $\nu_2 = 9$. The grey-shading gives the probability $p$ for obtaining values $\geq F$. Thus, the null hypothesis is rejected where $p \leq p'$ [bottom panel] and not rejected otherwise [top panel].

an estimate of $F = 3.5$ or greater is only $p = 0.038$ so that $p < p'$ and the null hypothesis is rejected at the 5% significance level.

### 4.5.5 ANALYSIS OF VARIANCE

The previous section compared the variances of only the two sample datasets $x1$ and $x2$. ANOVA [**AN**alysis **O**f **VA**riance] extends this analysis to the $J$ samples $x1, x2, \ldots, xJ$ by arranging the $I$ measurements per sample into the array $\mathbf{B}_{(\mathbf{I}\times\mathbf{J})}$ with elements $b_{ij}$ $\forall$ $i = 1, 2, \ldots, I$ and $j = 1, 2, \ldots, J$. In this format, the sample datasets $xj$ are redefined as the column vectors $\mathbf{bc_j}$ - i.e., $x1 \equiv \mathbf{bc_1}, x2 \equiv \mathbf{bc_2}, \ldots, xJ \equiv \mathbf{bc_J}$.

In general, ANOVA classifies into meaningful components the variability of observations $b_{ij} \in \mathbf{B}_{(\mathbf{I} \times \mathbf{J})}$ obtained from $J$-sets of presumably related samples represented by the column vectors $\mathbf{bc_j} = (b_{i1}\ b_{i2}\ \cdots\ b_{iJ})^t$, with the $I$ measurements or *replicates* for each sample in the row vectors $\mathbf{br_i} = (b_{1j}\ b_{2j}\ \cdots\ b_{Ij})$. In applying ANOVA, it is assumed that each set of replicates is randomly collected from normal parent populations with the same variance. This approach uses several measures of variance including the *total variance* [i.e., the variation of all measurements about their grand mean], *among-sample variance* [i.e., the variation of the sample means about the grand mean], and the *within-sample variance* [i.e., the variation of measurements from each sample about the sample's mean].

### 4.5.5.1 *One-way ANOVA*

*One-way* ANOVA tests the sample means $< \mathbf{bc_j} >$ for the hypothesis $H_0$ : $< \mathbf{bc_1} > = < \mathbf{bc_2} > = \cdots = < \mathbf{bc_J} >$ against the alternate $H_1$ : *at least one sample mean is different.* Ultimately, the $F$-test is used to see if the among-sample-variance matches the within-sample-variance poorly or well enough at a given significance level to conclude that there is or is not reason to reject the notion that the samples are all representative of the same population, and thus probably have equal means even though the sample means are not equal. **Table 4.3** summarizes the one-way ANOVA procedure.

For ANOVA applications, the data are typically arranged or formatted in an array of $\mathbf{bc_j}$ column vectors for the $J$ samples, and $\mathbf{br_i}$ row vectors of the $I$ replicates or measurements per sample. Thus, the observations may be expressed as the array or matrix $\mathbf{B}$ with elements $b_{ij}$ for all rows $i = 1, 2, \ldots, I$ and columns $j = 1, 2, \ldots, J$, with the total number of elements or observations given by $n = I \times J$. ***Example 04.5.5.1a*** illustrates in the

**Table 4.3**
**One-way analysis of variance (ANOVA) in tabular form including the CSS [corrected sums of squares] and $\nu$ [degrees of freedom] for all [$n = I \times J$]-observations involving $J$-samples with $I$-replicates or measurements per sample.**

| VARIATION SOURCE | CSS | $\nu$ | MEAN SQUARES | F-TEST |
|---|---|---|---|---|
| Among Samples | $SS_A$ | $J - 1$ | $MS_A = \frac{SS_A}{\nu}$ | $F_A = \frac{MS_A}{MS_W}$ |
| Within Samples | $SS_W$ | $n - J$ | $MS_W = \frac{SS_W}{\nu}$ | |
| Total Variation | $SS_T$ | $n - 1$ | $MS_T = \frac{SS_T}{\nu}$ | |

**Table 4.4**
**Randomly measured %-concentrations $b_{ij}$ of carbonate cement for [$J = 5$]-sandstone samples with [$I = 6$]-measurements or replicates per sample. The means and standard deviations of the replicate $br_i$-row and sample $bc_j$-column vectors are also listed $\forall\ i = 1, 2, \ldots, I$ and $j = 1, 2, \ldots, J$. Adapted from [20].**

| | $bc_1$ | $bc_2$ | $bc_3$ | $bc_4$ | $bc_5$ | $< br_i >$ | $s_{br_i}$ |
|---|---|---|---|---|---|---|---|
| $br_1$ | 19.2 | 18.7 | 12.5 | 20.3 | 19.9 | **18.1** | **3.2** |
| $br_2$ | 18.7 | 14.3 | 14.3 | 22.5 | 24.3 | **18.8** | **4.6** |
| $br_3$ | 21.3 | 20.2 | 8.7 | 17.6 | 17.6 | **17.1** | **5.0** |
| $br_4$ | 16.5 | 17.6 | 11.4 | 18.4 | 20.2 | **16.8** | **3.3** |
| $br_5$ | 17.3 | 19.3 | 9.5 | 15.9 | 18.4 | **16.1** | **3.9** |
| $br_6$ | 22.4 | 16.1 | 16.5 | 19.0 | 19.1 | **18.6** | **2.5** |
| $< bc_j >$ | **19.2** | **17.7** | **12.2** | **19.0** | **19.9** | ... | ... |
| $s_{bc_j}$ | **2.3** | **2.2** | **2.9** | **2.3** | **2.4** | ... | ... |

non-bold numerical elements of **Table 4.4** the array format for 30 randomly measured %-concentrations of carbonate cement of five sandstone samples. Accordingly, $b_{ij}$ is the $i$th replicate of the $j$th sample, where for example, $b_{24} = 22.5$ and $b_{42} = 17.6$.

Now, the total variance of all replicates for all samples [i.e., all observations] is based on the degrees of freedom average of the total corrected sum of squares (CSS) given by

$$SS_T = [\sum_{j=1}^{J}\sum_{i=1}^{I} b_{ij}^2] - [\frac{(\sum_{j=1}^{J}\sum_{i=1}^{I} b_{ij})^2}{n}] = \sum_{j=1}^{J}\sum_{i=1}^{I}[b_{ij} - < \mathbf{B} >]^2, \quad (4.59)$$

where the first bracket is the total sum of squares of the data, the second bracket gives the correction, and $< \mathbf{B} >$ is the grand mean of the matrix $\mathbf{B}$. Dividing **Equation 4.59** by the relevant degrees of freedom [$\nu = n - 1$] gives the array equivalent of the total variance in **Equation 4.56**, which for the data in **Table 4.4** is

$$s_{\mathbf{B}}^2 = \frac{SS_T}{n-1} = \frac{383.79}{29} = 13.23 = MS_T, \quad (4.60)$$

where $MS_T$ is the *total mean squares* that in the ANOVA **Table 4.3** also refers to the *total variation* or *variance.*

In one-way ANOVA, the total variance is assumed to be divided between the variance across the samples [$\text{MS}_A$] and the variance within the replications [$\text{MS}_W$] so that the $F$-test evaluates the likelihood that these two variances reflect a common parent population. Accordingly, the variance among samples

[MS$_A$] is based on the corrected sum of squares

$$SS_A = [\sum_{j=1}^{J} \frac{(\sum_{i=1}^{I} b_{ij})^2}{n}] - [\frac{(\sum_{j=1}^{J}\sum_{i=1}^{I} b_{ij})^2}{n}] = \sum_{j=1}^{J}[< \mathbf{bc_j} > - < \mathbf{B} >]^2, \tag{4.61}$$

whereas the variance within samples [MS$_W$] involves the CSS

$$\begin{aligned} SS_W = & \quad [\textstyle\sum_{j=1}^{J}\sum_{i=1}^{I} b_{ij}^2] - [\sum_{j=1}^{J} \frac{(\sum_{i=1}^{I} b_{ij})^2}{n}] \\ = & \quad \textstyle\sum_{j=1}^{J}[\sum_{i=1}^{I} b_{ij} - < \mathbf{bc_j} >]^2 = SS_T - SS_A. \end{aligned} \tag{4.62}$$

The degrees of freedom [$\nu$] relevant to calculating the total variance for all $n$ measurements are [$n-1$], whereas for calculating the variance among the $J$-samples, they are [$J-1$] so that by subtraction [e.g., **Equation 4.62**], the leftover degrees of freedom associated with the variance within samples are [$n-1-(J-1) = n-J$]. Thus, for the data in **Table 4.4**, the variance among samples is

$$MS_A = \frac{SS_A}{J-1} = \frac{237.42}{4} = 59.35, \tag{4.63}$$

with the variance within samples of

$$MS_W = \frac{SS_W}{n-J} = \frac{146.37}{25} = 5.85. \tag{4.64}$$

Performing the $F$-test at the 5% significance level reveals

$$F_A = \frac{MS_A}{MS_W} = 10.14 > F' = f_{0.05,4,25} = 2.76, \tag{4.65}$$

and thus the hypothesis $H_0 : MS_A = MS_W$ is rejected so that other sources [e.g., sandstone permeability anisotropies] for the variations in the data of **Table 4.4** may be considered.

The ANOVA rejection indicates only that at least one of the $J$ group means differs from the other means, but not which group mean(s) is(are) different. To identify the offending mean(s), the [$\frac{J[J-1]}{2} = \frac{J!}{2!(J-2)!}$]-pairs of means may be individually scrutinized for a difference greater than the *least significant difference* [LSD] established via the $t$-test. The LSD reflects the fact that when the null hypothesis is true for the difference of, say, the means $< \mathbf{bc_1} >$ and $< \mathbf{bc_2} >$, then the modification of **Equation 4.50** in **Equation 4.51** becomes

$$Z_{t(\mathbf{bc_1},\mathbf{bc_2})} = \frac{< \mathbf{bc_1} > - < \mathbf{bc_2} >}{\sqrt{MS_W(\frac{1}{I_{\mathbf{bc_1}}} + \frac{1}{I_{\mathbf{bc_2}}})}}, \tag{4.66}$$

which simplifies to

$$Z_{t(\mathbf{bc_1},\mathbf{bc_2})} = \frac{< \mathbf{bc_1} > - < \mathbf{bc_2} >}{\sqrt{MS_W(\frac{2}{I})}} \tag{4.67}$$

if the number of replicates per sample is the same so that $I_{\mathbf{bc_1}} = I_{\mathbf{bc_2}} = I$.

Now, the null hypothesis $H_0 : <\mathbf{bc_1}> = <\mathbf{bc_2}>$ is rejected if $|Z_{t(\mathbf{bc_1},\mathbf{bc_2})}| > |Z'_{t(\mathbf{bc_1},\mathbf{bc_2})}| = |t_{\alpha(n-J)}|$ for the desired significance level $\alpha$ and degrees of freedom $\nu = [n - J]$. Thus, rewriting the test for samples with $I$-replicates, for example, shows that the difference in the means is significant if

$$|<\mathbf{bc_1}> - <\mathbf{bc_2}>| > LSD = t_{\alpha(\nu)}\sqrt{MS_W(\frac{2}{I})}. \tag{4.68}$$

As ***Example 04.5.5.1b***, the LSD for the data in **Table 4.4** at the significance level $\alpha = 0.05$ and degrees of freedom $\nu = 30 - 5 = 25$ is $LSD = 2.39 = (1.71\sqrt{5.85 \times 2/6})$. Scrutiny of the sample means $<\mathbf{bc_j}>$ in **Table 4.4** reveals that the pairwise differences involving $<\mathbf{bc_3}>$ are all greater than the LSD, whereas the rest of the pairwise differences are less than the LSD. Thus, the LSD analysis has highlighted the statistically anomalous properties of sample $\mathbf{bc_3}$ for further study.

#### 4.5.5.2 *Two-way ANOVA*

*Two-way* ANOVA tests the statistical significance for one or both of the independent controlled factors $\boldsymbol{fI}$ and $\boldsymbol{fJ}$ to impact the observations. In this case, the observations are typically arranged as the matrix $\mathbf{B_{(I\times J)}}$ where the controlled factor $\boldsymbol{fI}$ is represented by the $\mathbf{br_i}$-row vectors $\forall\ i = 1, 2, \ldots, I$ that hold the $J$ measurements or *treatments* of the controlled factor $\boldsymbol{fJ}$ represented by the $\mathbf{bc_j}$-column vectors $\forall\ j = 1, 2, \ldots, J$.

Two hypotheses are tested: $H_0$(*sample means*): $<\mathbf{br_1}> = <\mathbf{br_2}> = \cdots = <\mathbf{br_I}>$ against the alternate $H_1$(*sample means*): *at least one sample mean is different*; and $H_0$(*controlled factor means*): $<\mathbf{bc_1}> = <\mathbf{bc_2}> = \cdots = <\mathbf{bc_J}>$ against the alternate $H_1$(*controlled factor means*): *at least one controlled factor mean is different* so that the statistically significant influence of a controlled factor cannot be excluded. As ***Example 04.5.5.2a***, the non-bold numerical entries in **Table 4.5** illustrate the matrix arrangement for the

**Table 4.5**

**Vertical [$bc_1$] and horizontal [$bc_2$] permeabilities in millidarcies measured on $I = 5$ randomly selected sandstone core samples. Also listed to the nearest integer are the respective $fI$ [sample] and $fJ$ [treatment] means $< br_i >$ and $< bc_j >$ along with the array's grand mean $< B >$. Adapted from [20].**

| sample | $bc_1$ | $bc_2$ | $< br_i >$ |
|---|---|---|---|
| $\mathbf{br_1}$ | 1037 | 1124 | **1081** |
| $\mathbf{br_2}$ | 963 | 960 | **962** |
| $\mathbf{br_3}$ | 842 | 921 | **882** |
| $\mathbf{br_4}$ | 1121 | 1202 | **1162** |
| $\mathbf{br_5}$ | 1043 | 1028 | **1036** |
| $< bc_j >$ | **1001** | **1047** | $< B > \approx$ **1024** |

**Table 4.6**
**Two-way analysis of variance (ANOVA) without replicates in tabular form including the CSS [corrected sums of squares] and $\nu$ [degrees of freedom] for all [$n = I \times J$]-observations involving $I$-samples with $J$-treatments or measurements per sample.**

| VARIATION SOURCE | CSS | $\nu$ | MEAN SQUARES | F-TEST |
|---|---|---|---|---|
| Among ***fI***-elements | $SS_I$ | $I - 1$ | $MS_I = \frac{SS_1}{\nu}$ | $F_I = \frac{MS_I}{MS_W}$ |
| Among ***fJ***-elements | $SS_J$ | $J - 1$ | $MS_J = \frac{SS_2}{\nu}$ | $F_J = \frac{MS_J}{MS_W}$ |
| Within Factors | $SS_W$ | $(I-1) \times (J-1)$ $= n - I - J + 1$ | $MS_W = \frac{SS_W}{\nu}$ | |
| Total Variation | $SS_T$ | $n - 1$ | $MS_T = \frac{SS_T}{\nu}$ | |

$n = I \times J = 10$ measurements of the vertical and horizontal permeabilities for $I = 5$ randomly chosen sandstone core samples.

For treatments involving multiple sets of ***fJ*** measurements or replicates per ***fI***-element, the two-way ANOVA *with replicates* is applicable. However, the treatments in **Table 4.5** involve only a single set of permeability measurements per sample where the two-way ANOVA *without replicates* is appropriate.

In general, **two-way ANOVA without replicates** has the *total variance* that involves the *among **fI**-elements variance* [i.e., the variance of the $I$ ***fI*** elements], *among **fJ**-elements variance* [i.e., the variance of the $J$ ***fJ*** elements], and the *residual error or within factors variance.* **Table 4.6** summarizes the procedure for two-way ANOVA without replicates.

Analogous to **Equation 4.59**, the total variance here of all observations is based on the degrees of freedom average of the total corrected sum of squares (CSS) given by

$$SS_T = [\sum_{j=1}^{J}\sum_{i=1}^{I} b_{ij}^2] - [\frac{(\sum_{j=1}^{J}\sum_{i=1}^{I} b_{ij})^2}{n}] = \sum_{j=1}^{J}\sum_{i=1}^{I}(b_{ij} - <\mathbf{B}>)^2. \quad (4.69)$$

where $<\mathbf{B}>$ is the grand mean of all the measurements. In addition, the variance among data in the ***fI*** elements or levels is

$$SS_I = I\sum_{j=1}^{J}(<\mathbf{bc_j}> - <\mathbf{B}>)^2, \quad (4.70)$$

whereas the variance among data in the ***fJ*** elements or treatments is

$$SS_J = J\sum_{i=1}^{I}(<\mathbf{br_i}> - <\mathbf{B}>)^2. \tag{4.71}$$

The variance due to residuals, which is analogous to the *within sample* variance in one-way ANOVA [**Equation 4.61**], is

$$SS_W = \sum_{j=1}^{J}\sum_{i=1}^{I}(b_{ij} - <\mathbf{br_i}> - <\mathbf{bc_j}> + <\mathbf{B}>)^2. \tag{4.72}$$

The $F_I$-test in **Table 4.6** is used to see if the variances for data collected at different ***fI*** levels match the *within factors* or residuals variance *poorly* or *well* enough at a given $\alpha$ significance level to conclude that there *is* or *is not* reason to reject the notion that the data are all representative of the same population regardless of the ***fI*** level, and thus should have equal means even though the data means are not equal. In addition, the $F_J$-test checks to see if the variances for data or treatments collected at different ***fJ*** levels match the residuals variance *poorly* or *well* enough at a given $\alpha$ significance level to conclude that there *is* or *is not* reason to reject the notion that the treatments are all representative of the same population regardless of the ***fJ*** level, and thus should have equal means even though the treatment means are not equal.

**Table 4.7** gives the two-way ANOVA for the vertical and horizontal permeability measurements on the five different sandstone core samples listed in **Table 4.5**. At the $\alpha = 5\%$ signifance level, $F_I = 18.37 > F'_{(0.05,4,4)} = 6.39$ so that the null hypothesis is rejected that the mean permeabilities of the samples are equal. Here, the critical $F'_{(\alpha,\mu_1,\mu_2)}$ is taken from the $f$-distribution [**Figure 4.14**] for the respective *among samples* and *within factors* degrees of freedom $\mu_1 = 4$ and $\mu_2 = 4$. Thus, the samples probably are from different populations with unequal permeability means.

**Table 4.7**

**Two-way analysis of variance (ANOVA) without replicates of Table 4.5 that lists the $J = 2$ directional permeability measurements for each of the $I = 5$ sandstone samples from different cores.**

| VARIATION SOURCE | CSS | $\nu$ | MEAN SQUARES | F-TEST |
|---|---|---|---|---|
| Among Samples | $92,866$ | $5-1=4$ | $23,222$ | $F_I = 18.37$ |
| Among Directions | $5,244$ | $2-1=1$ | $5,244$ | $F_J = 4.15$ |
| Within Factors | $5,058$ | $(5-1)\times(2-1) =$ $10-5-2+1=4$ | $1,265$ | |
| Total Variation | $103,189$ | $10-1=9$ | | |

In addition, the experimental $F_J = 4.15 < F'_{(0.05,1,4)} = 7.71$, and thus the null hypothesis cannot be rejected that the mean vertical and horizontal permeabilities are equal. Here the *among directions* degrees of freedom $\mu_3 = 1$ was used to select the critical $F'_{(\alpha,\mu_3,\mu_2)}$-value in the $f$-distribution [**Figure 4.14**]. A possible interpretation of the $F_I$ and $F_J$ results is that the permeability of the sandstone probably varies spatially across the wells, but otherwise is relatively isotropic vertically and horizontally.

Interpretations of an ANOVA tabulation [e.g., **Table 4.7**] are often made more explicit by including a further column that lists the probabilities $p_I$ and $p_J$ for obtaining higher values of $F_I$ and $F_J$, respectively, as described in **Chapter 4.5.4** above. Accordingly, the respective fractional probabilities of obtaining higher $F_I$- and $F_J$-values in **Table 4.7** are $p_I = 0.0078$ and $p_J = 0.1114$.

Where any of these probabilities are smaller than the significance level of the $F_I$- and $F_J$-estimates [i.e., $\alpha = 0.05$ in **Table 4.7**], the corresponding null hypothesis is rejected. Thus, as previously concluded, the null hypothesis is rejected for the equality of the *among sample* permeability means, but not for the *among directions* equality of the permeability means.

In **two-way ANOVA with replicates**, the measurements are commonly arranged with each ***fI*** element subdivided into $K$ replicate or repeated sets of ***fJ*** elements or treatments. As ***Example 04.5.5.2b***, the non-bold numerical entries in **Table 4.8** illustrate the matrix arrangement for the $n = I \times J \times K = 20$ measurements of $K = 2$ sets of vertical and horizontal permeabilities for each of $I = 5$ randomly chosen sandstone core samples.

**Table 4.8**

**Vertical [$\mathrm{bc_1}$] and horizontal [$\mathrm{bc_2}$] permeabilities in millidarcies measured in sets of $K = 2$ replicates on each of $I = 5$ randomly selected sandstone core samples. Also listed to the nearest integer are the respective $fI$ [sample] and $fJ$ [treatment] means $< \mathrm{br_i(K)} >$ and $< \mathrm{bc_j} >$ along with the array's grand mean $< \mathrm{B} >$. Adapted from [20].**

| sample | $\mathrm{bc_1}$ | $\mathrm{bc_2}$ | $< \mathrm{br_i(k)} >$ | ... | $< \mathrm{br_i(K)} >$ |
|---|---|---|---|---|---|
| $\mathrm{br_1(1)}$ | 1037 | 1124 | **1081** | ... | ... |
| $\mathrm{br_1(2)}$ | 1205 | 1290 | **1248** | → | **1165** |
| $\mathrm{br_2(1)}$ | 963 | 960 | **962** | ... | ... |
| $\mathrm{br_2(2)}$ | 910 | 960 | **935** | → | **949** |
| $\mathrm{br_3(1)}$ | 842 | 921 | **882** | ... | ... |
| $\mathrm{br_3(2)}$ | 850 | 990 | **920** | → | **901** |
| $\mathrm{br_4(1)}$ | 1121 | 1167 | **1144** | ... | ... |
| $\mathrm{br_4(2)}$ | 1009 | 1190 | **1100** | → | **1122** |
| $\mathrm{br_5(1)}$ | 1043 | 1028 | **1036** | | |
| $\mathrm{br_5(2)}$ | 907 | 1031 | **969** | → | **1003** |
| $< \mathrm{bc_j} >$ | **989** | **1066** | **1028 ≈** | $< \mathrm{B} >$ | **≈ 1028** |

**Table 4.9**
**Two-way analysis of variance (ANOVA) with replicates in tabular form including the CSS [corrected sums of squares] and $\nu$ [degrees of freedom] for all $[n = I \times J \times K]$-observations involving $I$-samples with $K$ sets of $J$-treatments or measurements per sample.**

| VARIATION SOURCE | CSS | $\nu$ | MEAN SQUARES | F-TEST |
|---|---|---|---|---|
| Among ***fI***-elements | $SS_{IR}$ | $I-1$ | $MS_{IR} = \frac{SS_{IR}}{\nu}$ | $F_{IR} = \frac{MS_{IR}}{MS_{WR}}$ |
| Among ***fJ***-elements | $SS_{JR}$ | $J-1$ | $MS_{JR} = \frac{SS_{JR}}{\nu}$ | $F_{JR} = \frac{MS_{JR}}{MS_{WR}}$ |
| Interaction | $SS_{Int}$ | $(I-1)\times(J-1)$ | $MS_{Int} = \frac{SS_{Int}}{/\nu}$ | $F_{Int} = \frac{MS_{Int}}{MS_{WR}}$ |
| Within Factors | $SS_{WR}$ | $n(I \times J)$ | $MS_{WR} = \frac{SS_{WR}}{\nu}$ | |
| Total Variation | $SS_T$ | $n-1$ | $MS_T = \frac{SS_T}{\nu}$ | |

For this case, the variance within replicate samples can be directly assessed. Contributions to the *total variance* include the *among **fI**-elements variance* [e.g., variance of the $(I \times K)$ ***fI***-samples], *among **fJ**-elements variance* [e.g., variance of the $(J \times I)$ ***fJ***-treatments], *interactions between factors variance*, and the *within replicates variance.*

**Table 4.9** summarizes the procedure for two-way ANOVA with replicates. Here, the *among **fI**-elements with replicates variance* is based on the sum of the squared differences between the average value at each level of ***fI*** [including all replicates and levels of ***fJ***] given by

$$SS_{IR} = \sum_{i=1}^{I}(<\mathbf{br_i(K)}> - <\mathbf{B}>)^2, \quad (4.73)$$

whereas the *among **fJ**-elements with replicates variance* is

$$SS_{JR} = \sum_{j=1}^{J}(<\mathbf{bc_j}> - <\mathbf{B}>)^2. \quad (4.74)$$

The *residual* or *within sample with replicates variance* is

$$SS_{WR} = \sum_{k=1}^{K}\sum_{j=1}^{J}\sum_{i=1}^{I}(b_{ijk} - <\mathbf{br_i(k)}>)^2. \quad (4.75)$$

With two controlled factors and replicate measurements, the total variance includes the additional contribution of the *interaction variance* given by

$$SS_{Int} = \sum_{k=1}^{K}\sum_{i=1}^{I}\sum_{j=1}^{J}(<\mathbf{br_i(k)}> - <\mathbf{br_i(K)}> - <\mathbf{bc_j}> - <\mathbf{B}>)^2. \quad (4.76)$$

The $F_{IR}$-test is used to see if the among data variance collected at different $\boldsymbol{fI}$-levels matches the residual variance poorly or well enough at a given significance level to conclude that there is or is not reason to reject the notion that the samples are all representative of the same population regardless of the $\boldsymbol{fI}$-level, and thus should have equal means even though the sample means are not equal. The $F_{JR}$-test checks to see if the among data variance collected at different $\boldsymbol{fJ}$-levels matches the residual variance poorly or well enough at a given significance level to conclude that there is or is not reason to reject the notion that the samples are all representative of the same population regardless of the $\boldsymbol{fJ}$-level, and thus should have equal means even though the sample means are not equal. The $F_{Int}$-test assesses whether the interaction between factors $\boldsymbol{fI}$ and $\boldsymbol{fJ}$ matches the residual variance poorly or well enough at a given significance level to conclude that there is or is not reason to reject the notion that the interaction between factors $\boldsymbol{fI}$ and $\boldsymbol{fJ}$ accounts for a significant fraction of the total variance.

**Table 4.10** gives the two-way ANOVA for the $K = 2$ replicate sets of vertical and horizontal permeability measurements [i.e., $J = 2$] on each of $I = 5$ different sandstone core samples that are listed in **Table 4.8**. At the $\alpha = 5\%$ signifance level, $F_{IR} = 10.63 > F'_{(0.05,4,10)} = 3.48$, and thus the null hypothesis is rejected that the mean permeabilities of the samples are equal. Here, the critical $F'_{(\alpha,\mu_1,\mu_2)}$ is taken from the $f$-distribution [**Figure 4.14**] for the respective *among samples* and *within factors* degrees of freedom $\mu_1 = 4$ and $\mu_2 = 10$. Thus, the samples probably are from different populations with unequal permeablility means. Furthermore, the experimental

**Table 4.10**
**Two-way analysis of variance (ANOVA) with replicates of Table 4.8 that lists $K = 2$ replicate sets of $J = 2$ directional permeability measurements for each of the $I = 5$ sandstone samples from different cores.**

| VARIATION SOURCE | CSS | $\nu$ | MEAN SQUARES | F-TEST |
|---|---|---|---|---|
| Among Samples | 201,996 | $5-1=4$ | 50,499 | $F_{IR} = 10.63$ |
| Among Directions | 29,953 | $2-1=1$ | 29,953 | $F_{JR} = 6.31$ |
| Interaction | 5,837 | $(5-1)\times(2-1)=4$ | 1,459 | $F_{Int} = 0.31$ |
| Within Factors | 47,406 | $20-(5\times 2)=10$ | 4,750 | |
| Total Variation | | $20-1=19$ | | |

$F_{JR} = 6.31 > F'_{(0.05,1,10)} = 4.96$, and thus rejects the null hypothesis that the mean vertical and horizontal permeabilities are equal. Here the *among directions* degrees of freedom $\mu_3 = 1$ was used to select the critical $F'_{(\alpha,\mu_3,\mu_2)}$-value in the $f$-distribution [**Figure 4.14**]. These $F_{IR}$ and $F_{JR}$ results with replicates suggest that the permeabilities of the sandstone probably vary spatially across the wells, and also are anisotropic. Finally, $F_{Int} = 0.31 < F'_{(0.05,4,10)} = 3.48$ so that the interaction between the sample distribution and permeability directions is probably not significant.

To make these interpretations more explicit in the ANOVA tabulation, the respective probabilities $p_{IR}, p_{JR}$ and $p_{Int}$ for obtaining higher values of $F_{IR}$, $F_{JR}$, and $F_{Int}$ could be added to **Table 4.10**. Scrutinizing these probabilities for values smaller than the significance level of the $F_{IR}$-, $F_{JR}$-, and $F_{Int}$-estimates [i.e., $\alpha = 0.05$ in **Table 4.10**] identifies the rejectable null hypotheses.

For the $F_{IR}$-, $F_{JR}$- and $F_{Int}$-values in **Table 4.10** in particular, the fractional probabilities of obtaining higher values are $p_{IR} = 0.0013$, $p_{JR} = 0.0308$, and $p_{Int} = 0.8667$, respectively. Thus, as previously concluded, the null hypotheses are rejected for the equalities of the *among sample* permeability means and the *among directions* permeability means, but not for the interaction between the sample distribution and permeability directions.

The above discussion effectively introduces the basic one- and two-way ANOVA concepts. However, further elaborations of the ANOVA table can accommodate more sources of variability in environmental data [e.g., [**20, 35, 42, 56, 99**]].

### 4.5.6 DISTRIBUTION TESTING

For applications of the $Z_t$-test or the $F$-test, data are required that follow a normal distribution. The $X^2$-test that follows the $\chi^2$ function provides a *Goodness of Fit* test for comparing the probability properties of a data distribution against any other distribution. For this application, the $X^2$-test is defined by

$$X^2 = \sum_{i=1}^{K} \frac{[O_i - E_i]^2}{E_i}, \tag{4.77}$$

where $O_i$ and $E_i$ are the respective observed and expected frequencies for the $i$-th class and there are $K$ classes or subdivisions of the distribution.

In general, theory and numerical experiments show that effective applications require at least 5 occurences or expected frequencies within each class or interval of the distribution. Otherwise, the size of the class interval should be enlarged so that $E_i \geq 5$ for all $i$-classes [e.g., [**99**]].

To perform the $X^2$-test, the number of degrees of freedom $\nu = K - m$ must be determined, where $K << n$ is the number of classes or subdivisions of the distribution, and $m$ is the number of parameters that must be calculated from the data. For example, $m = 3$ if the normal distribution is being tested

**Table 4.11**
**Observed $[O_i]$ and expected $[E_i]$ frequencies for 120 tosses of a single die, where each $K$-th face or outcome is expected to occur with equal probability [= $(1/6) \times 100\%$].**

| K | 1 | 2 | 3 | 4 | 5 | 6 |
|---|---|---|---|---|---|---|
| $O_i$ | 20 | 22 | 17 | 18 | 19 | 24 |
| $E_i$ | 20 | 20 | 20 | 20 | 20 | 20 |

for which the parameters $\bar{x}, s_x^2$, and the total frequency for computing the expected $E_i$ are needed.

The calculated $X^2$ is compared against the critical value $X^{2'} = \chi^2_{(\alpha,\nu)}$ for the desired level of significance $\alpha$ and degrees of freedom $\nu$. If $X^2 > X^{2'}$, then the hypothesis $H_0 : O_i = E_i$ is rejected - i.e., the observed frequencies differ significantly for the expected frequencies so that $X^2$ is large and the fit is poor. On the other hand, if $X^2 < X^{2'}$, then there is no reason to reject the null hypothesis - i.e., the observed frequencies are close enough to the expected frequencies that the fit is acceptable.

As ***Example 04.5.6a***, consider the 120 tosses of a single die shown by the outcomes $O_i$ in **Table 4.11**. If the die is honest, then the outcomes should follow the uniform distribution where each of the faces has an equal probability of being observed - i.e., on average, each of the faces should have come up $[E_i = (1/6) \times 120 = 20]$ times. In other words, are the descrepancies between $O_i$ and $E_i$ the result of random sampling flucations and the die is <u>*honest*</u>, or is it <u>*dishonest*</u>, but biased because the distribution $O_i$ is not uniform?

Applying **Equation 4.77** to test the hypothesis $H_0 : O_i = E_i$ yields

$$X^2 = \frac{[20-20]^2}{20} + \frac{[22-20]^2}{20} + \cdots + \frac{[24-20]^2}{20} = 1.7. \tag{4.78}$$

In this case, the applicable degrees of freedom are $\nu = 6 - 1 = 5$ because the only parameter provided by the data is the total frequency [e.g., $n = 120$] needed for estimating $E_i$. Using the significance level $\alpha = 5\%$ yields the critical value $X^{2'} = \chi^2_{(0.05,5)} = 11.070$ [e.g., **Figure 4.13**]. Accordingly, $X^{2'} > X^2$ so that there is no reason to believe that observations are not distributed as expected - i.e., the $X^2$-test fails to reject the notion that the distribution of outcomes is not uniform, and hence it may be concluded that the die is honest at the 95% confidence level.

As another example [i.e., ***Example 04.5.6b***], consider testing the distibution of 40 observations of a given quantity, which should follow the normal distribution if the observational errors are random. Dividing the normal distribution into eight $K$-classes centered on the data mean $\bar{x} \approx \mu_x$ at interval widths of one standard deviation $s_x \approx \sigma_x$ as shown in **Figure 4.8** allows the use of the results in **Table 4.1** to easily compute the expected frequencies or probabilities of the observations [i.e., $E_i$] for each class. For example, the area between $\mu_x$ and $\mu_x + \sigma_x$ is about 34%, so that the number of observations expected to fall into this interval is $E_5 = 0.34 \times 40 \approx 14$, whereas the expectation for the interval between $\mu_x + \sigma_x$ and $\mu_x + 2\sigma_x$ is $E_6 = 0.135 \times 40 \approx 6$, etc.

The expected probabilities $[E_i]$ may be compared against the observed frequencies $[O_i]$ via the $X^2$-test for evidence that rejects or fails to reject the null hypothesis at a desired significance level $\alpha$. In this case, the degrees of freedom $\nu = 8 - 3 = 5$ because the parameters $\bar{x}, s_x^2$ and the total frequency $n = 40$ had to be estimated from the data to implement the test.

### 4.5.7 NON-PARAMETRIC TESTING

The procedures considered thus far for making statistical inferences are valid only if the samples are large or approximately Gaussian. To make inferences about small independent samples selected from non-normal populations, *non-parametric* or *distribution-free* tests can be applied which assume no knowledge of the distribution.

Non-parametric tests rank or otherwise characterize data within a sample and then evaluate the results in terms of how probable or improbable are the outcomes. As ***Example 04.5.7a***, consider the petrological controversy concerning the magnetic properties of the mantle. If the Moho is a magnetic boundary essentially separating a magnetic crust from a non-magnetic mantle, then a magnetic anomaly difference between the oceans and continents should be observed. Specifically, the oceans should show an average negative magnetic anomaly and the continents a mean positive anomaly because the crust of the oceans is significantly thinner than that of the continents. This magnetic difference is not directly observed in the magnetic anomaly maps from NASA's Magsat mission [**44**]. However, it is evident statistically from the individual histograms of continental and oceanic satellite anomalies shown in [**Figure 4.17**].

Note that as datasets become large [i.e., $n \to 25 - 30+$], the $X^2$-test becomes excessively sensitive and rejects any of the distributions in **Figure 4.17** as following a Gaussian function. However, the central limits theorem guarantees that the means and variances of large datasets can be tested by the respective $Z$ and $F$ statistics. Indeed, applying these standard tests to the grand histograms of all continental and oceanic observations [**Figure 4.18**] rejects the hypothesis $H_0 : \mu_c = \mu_o$ in favor of the alternative that the continents are more magnetic than the oceans.

Somewhat weaker non-parametric tests also can be constructed to support this magnetic difference. For example, given the four continental and two oceanic histograms, the chances that 4 of the regional histograms have positive means with the remaining 2 having negative means is equivalent to the number of combinations of 6 objects taken 2 at a time [e.g., **Equation 4.33**].

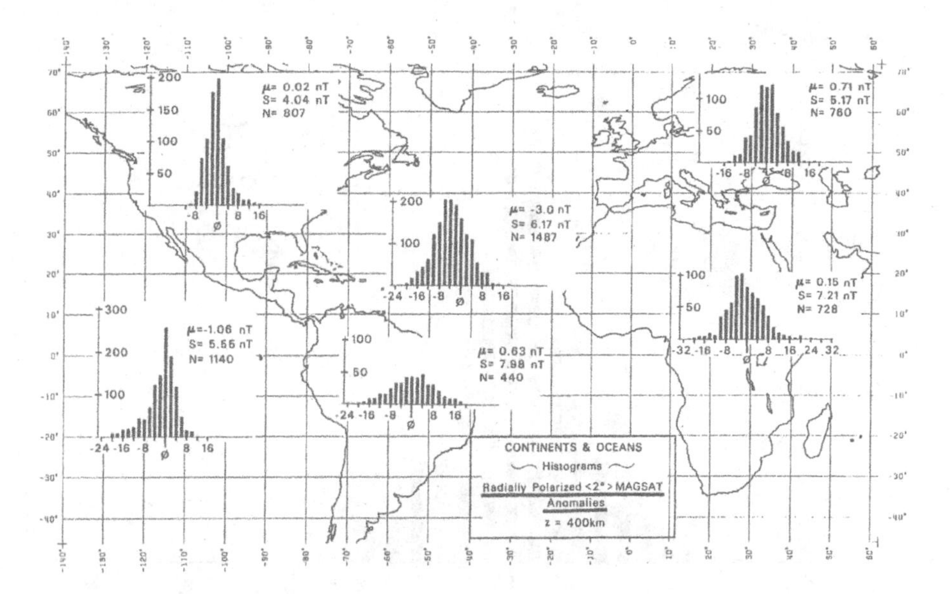


**Figure 4.17** Histograms of $2^o$-averaged magnetic anomalies in nanoTesla [nT] of the Earth's continents and oceans observed by NASA's Magsat satellite mission at altitudes of roughly 400 km above sea level. [Adapted from [**44**].]

Accordingly, the chances are 1 in 15 of observing the pattern of histograms in **Figure 4.17** if the histograms were assigned the regions by random chance alone.

A stonger improbability will result if the sign of the means is treated like the flipping of a coin where the anomaly mean of each region has a 50% chance of being positive or negative. Applying this uniform probability assumption via **Equation 4.26** suggests, accordingly, that there is 1 chance in 64 of obtaining the histogram pattern in **Figure 4.17**. Indeed, the improbability can be further hyped by having the means take on either a zero, positive, or negative value with equal probability.

Widely used non-parametric tests include the *Wilcoxon Two-Sample Test* of the hypothesis $H_0 : \mu_1 = \mu_2$ that the means of two non-normal populations are equal when only small [i.e., $n \leq 20$] independent samples are available. This test assigns a ranking from largest to smallest values of the observations and compares the sums of the ranking against probable or improbable ranking outcomes. The test is always superior to the $t$-test for small, non-normal populations [e.g., [**20, 99**]].

The *Wilcoxon Test for Paired Observations [Sign Test]*, on the other hand, tests $n$ matched pairs of observations selected from two small non-normal populations. It tests the hypothesis $H_0 : \mu_1 = \mu_2$ [or $\mu_1 - \mu_2 = 0$] by examining the signs of the residuals between matched pairs of observations. Where $H_0$ is true, the sum of the plus signs should be approximately equal to the sum

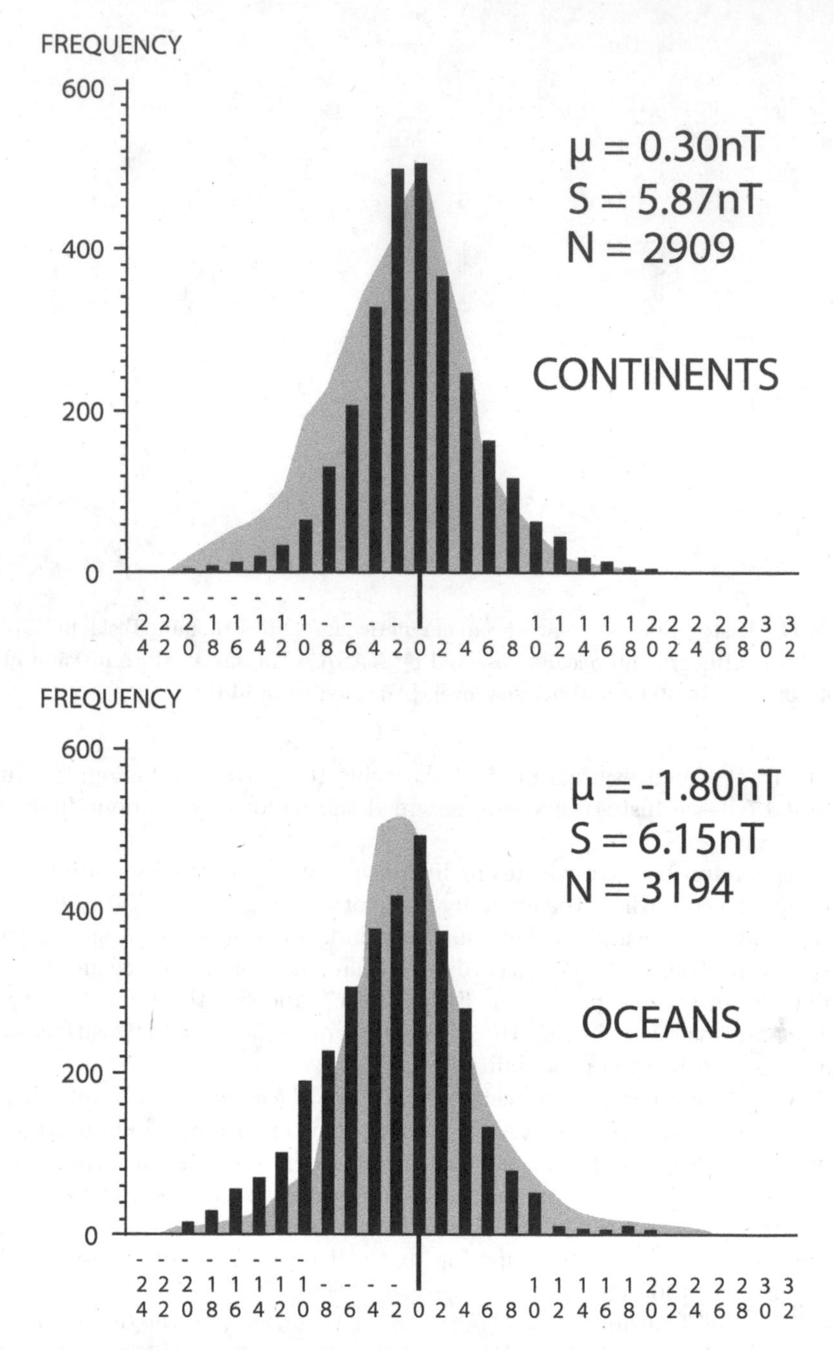


**Figure 4.18** Grand histograms of the Magsat magnetic anomaly histograms of the Earth's continents and oceans from **Figure 4.17**. For each distribution, a shaded representation of the other histogram is given to facilitate comparison. [Adapted from [**44**].]

of the minuses. When one sign occurs more often that it should based on chance alone, then the hypothesis that the population means are equal is rejected [e.g., **[20, 99]**]. The test is appropriate, for example, in checking measurements made by two or more people for reading errors. The reading differences should be random in contrast to the presence of systematic errors [e.g., operator reading bias].

Additional non-parametric methods include the *Kurskal-Wallis* and the *Kolmogorov-Smirnov Tests*. The first non-parmetric test is an alternative to one-way ANOVA, whereas the second provides a goodness of fit alternative to the $X^2$-test [e.g., **[20, 99]**].

## 4.6 KEY CONCEPTS

1. Modern data analysis yields results together with relevant statistical error measures based on the data mean and variance. For small datasets with $n \leq 15$, the unbiased variance is taken with the sum of squared deviations divided by the degrees of freedom $\nu = n - 1$, rather than by $n$ that is appropriate for larger datasets.
2. The uncertainty or error of a function with measured parameters and parameter errors can be estimated using Taylor's theorem. The resulting variance propagation rule, for example, reveals the function's variance as the sum of each parameter's variance times the squared variation of the function with the parameter. The standard error of the mean, on the other hand, is the ratio of the data's standard deviation to the root squared number of data.
3. Approximated by the histogram, the statistical distribution of the data summarizes their integral and differential properties in terms of the mean and variance. Statistical error bars and inferences for the data are obtained assuming that the probability of finding the dependent variable's value within an interval of the independent variable is given by the area under curve over the interval, where the distribution's total area is the total probability or 100%.
4. Widely used distributions in the environmental sciences and engineering include the binomial distribution that governs discrete processes involving a fixed number of independent trials where each has only two possible outcomes. It evaluates the chances of obtaining various combinations of $n$ things taken $x$ at a time.
5. Perhaps the most widely used distribution of all is the mean-centered bell-shaped Gaussian or normal distribution where roughly 68.3%, 95.5%, and 99.7% of the data are respectively contained within $\pm 1$, $\pm 2$, and $\pm 3$ standard deviations of the mean
6. As consequences of the central limits theorem for well-sampled datasets approximately numbering $n \geq 25$, the means follow the normal distribution with standard error-based confidence limits. These

confidence limits also may be compared to infer disparate origins for the datasets.

7. For poorly sampled data roughly numbering $n < 25$ with approximately normal or bell-shaped distributions, the statistical error bars and inference testing are established using the weaker Student t-test.
8. Confidence intervals on the variance of normally distributed data are obtained using the $\chi^2$ distribution. These data variances may also be tested to infer disparate dataset origins by Fisher's f distribution of the variance ratios for datasets drawn from a normal population.
9. Analysis of variance [ANOVA] subdivides the total variance between two or more datasets into within-sample, among-sample, and other meaningful components using the f distribution.
10. The goodness of fit test uses the $\chi^2$ distribution to establish the data's statistical distribution by comparing the observed and expected data densities across 8 or more probability intervals of the expected distribution.
11. Somewhat weaker non-parametric or distribution-free statistical inferences can be established by ranking or otherwise organizing data within a sample into parameters that are independent of the shape or form of the data's probability distribution. For example, the inference of a mean-centered data distribution can be tested by subtracting the mean from the data to see if the numbers of negative and positive residuals are roughly equal.

# 5 Data Sampling

## 5.1 OVERVIEW

*Successful environmental data analysis is based on collecting appropriate and defensible data that accurately portray the problem of interest. The choice of a sampling method in earth and environmental investigations commonly considers sufficiency issues like the sample's capacity to resolve variations of interest and estimate the desired parameters within tolerable error limits. It may also address biasing issues like operator reading uncertainties and other systematic errors in the sample.*

*Common sampling strategies in earth and environmental applications include collecting data on a uniform grid where the smallest data feature is defined by the grid's Nyquist wavelength of two station intervals or three consecutive data points. Data collected over a random distribution of sample coordinates also are common, but they are always submitted for electronic analysis in a grid format. Thus, to ensure that critical data variations are effectively represented in electronic computing, the investigator should explicitly grid the random data sample.*

*Cluster sampling, and adaptive cluster sampling when combined with random sampling, also have important earth and environmental applications. Clustered samples are efficient to measure and map small wavelength features and unbiased means of the population.*

## 5.2 INTRODUCTION

The method used in selecting samples is extremely important for scientifically based decision making and the results of the investigation [e.g., [**22, 92, 95**]]. This issue is particularly significant to the success of environmental studies where data accessibility, survey logistics, and other complications severely constrain the sampling procedures that may be applied. Although there are no easy answers to these problems, some general principles may be offered on how to choose an effective sampling procedure and see if the collected sample is representative of the population.

## 5.3 SAMPLING CONSIDERATIONS

The choice of an effective sampling method is critical for collecting appropriate and defensible data to accurately represent the problem of interest. It commonly requires consideration of a number of mapping issues including data sufficiency, bias, distribution, and other error sources.

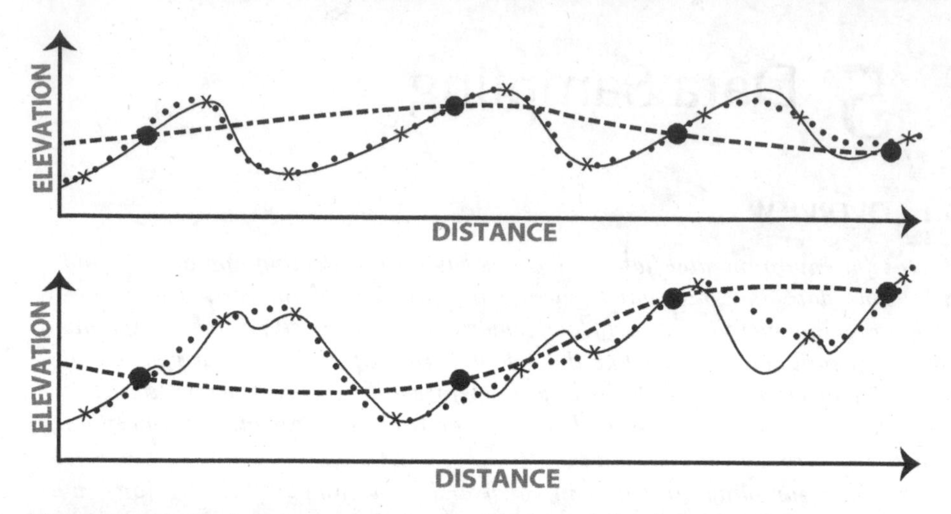


**Figure 5.1** Examples of mapping the topographic profiles [solid lines] of sand dunes with one station or sample per dominant wavelength [solid dots] and with three stations per dominant wavelength [crosses]. The dashed curves through the one-station per dominant wavelength samples yield completely incorrect renderings or aliases of the profiles. The dotted curves through the three-stations per dominant wavelength samples, however, represent the profiles with significantly greater accuracy.

### 5.3.1 SUFFICIENCY

To be effective in analysis, the sample must closely approximate the variations of the population. For example, the sample must account for the significant wavelengths of the population, which involves the choice of an effective station interval for the observations. As ***Example 05.2.1a***, consider the sand dune profile in **Figure 5.1** that has been sampled at one and three stations per dominant wavelength.

In general, the sand dune profile must be sampled at least 3 or more times [i.e., spanning 2 or more station intervals] per shortest wavelength component in the profile to accurately render or approximate the profiles. For a uniformly gridded dataset, this limit is called the *Nyquist wavelength* with its inverse called the *Nyquist frequency*. A smaller number of observations per dominant wavelength yields the *aliased* samples rendered by the smoothed dashed line profile as an approximation of the true solid line profile. The true signal obviously cannot be recovered from the aliased signal, which introduces information about the true signal into the analysis that is completely incorrect except at the points of measurement.

Minimizing potential aliasing effects is difficult in practice because the population's variations commonly are poorly known. One possibility is to collect the sample at an excessively high rate, but this strategy can be impractical where, for example, field conditions limit data accessibility. Another approach, which is particularly effective in data gridding [e.g., see **Chapter 10.4.4.C**], is to collect successive samples at higher and higher rates until the changes in the output signals become insignificant. The largest station interval at which the output changes are insignificant marks the maximum station spacing for obtaining an unalised sample.

A further potential sampling design requirement may be to determine the number of data that is sufficient to obtain results within prescribed error limits. This issue is considered in **Chapter 4.5**, where the data numbers are derived from the relevant confidence intervals that establish with tolerable error bars the sample means, variances and other parameters involving these elementary statistical attributes.

### 5.3.2 BIASING

Choosing samples on a subjective basis can easily lead to biased results in analysis. A possible remedy is to employ a random sampling technique such as a random number generator or table to choose the locations at which samples are taken. Bias can be tested for by collecting several samples and analyzing each to see that the related sample means, variances, and other elementary statistical parameters give consistent results.

Collecting an insufficient number of samples leads to other sources of bias that may take the form of aliasing, or in cases where random fluctuations are large, incorrect estimates of the mean and other critical statistical parameters. This type of bias can be evaluated by collecting samples of different sizes and comparing confidence intervals and hypothesis tests for consistency.

### 5.3.3 DISTRIBUTION

Observation sites must be established to ensure the collection of an adequate sample of the population. Circumstances sometimes force the mapping of a population at irregular or inapproriate coordinates of time, space, or other independent variables. A test to see if the distribution of observations is introducing bias is to collect several samples with different distributions and check the results of the analyses for consistency.

### 5.3.4 ERRORS

As noted in **Chapter 3.2**, sampling errors include random and systematic fluctuations that can degrade a sample's capacity to represent the population. This problem may be treated by estimating the errors that are present and calculating confidence intervals, which in turn may be propagated through the final results of the analysis. Collection of more data [e.g., via **Equation 4.46**] can reduce the amount of error by allowing for better estimates of the mean and other critical statistical parameters.

## 5.4 SAMPLING METHODS

Because of sampling's critical role in the success of earth and environmental science and engineering investigations, a large body of literature exists on sampling strategies and their related performance statistics [e.g., see **[92, 95]** and references therein]. Common patterns of sampling in earth and environmental investigations [**Figure 5.2**] obtain systematic, random, and clustered collections of observations, each with relative advantages and limitations.

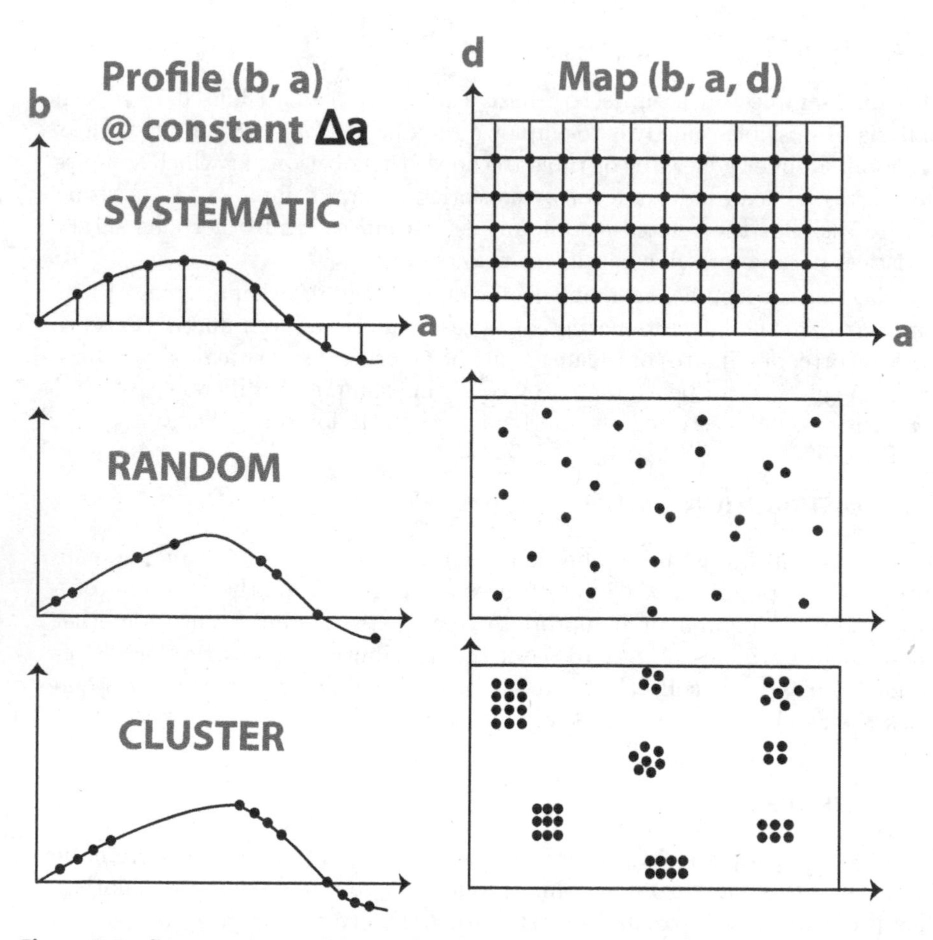


**Figure 5.2** Systematic, random, and cluster sampling patterns in one [profile] and two [map] dimensions. In systematic map [i.e., $2D$] and higher dimensional sampling of the $b$-data, the sampling intervals $\Delta a$, $\Delta d$, etc. must be uniform, but not necessarily equal.

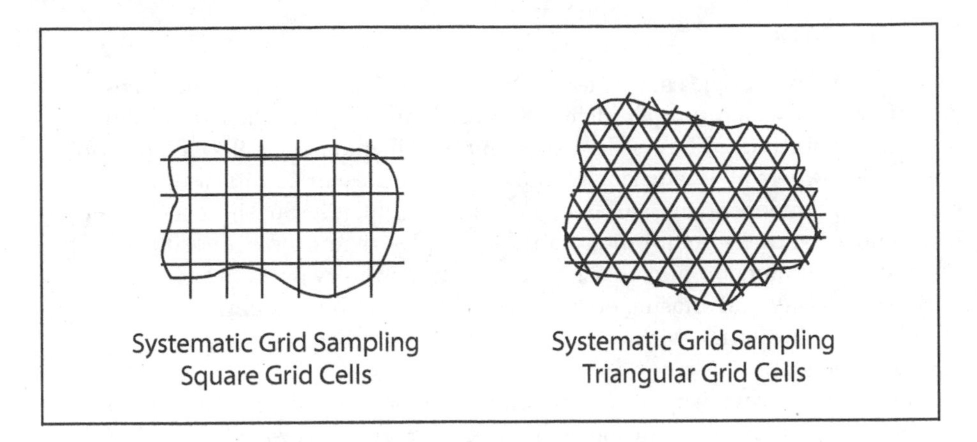


**Figure 5.3** Examples of grid sampling. [Adapted from [**95**].]

### 5.4.1 SYSTEMATIC

These samples [e.g., **Figures 5.2** and **5.3**] are taken at uniform spacing of the independent variable [e.g., space, time, etc.] along a profile or on a grid. Systematic sampling permits making definitive statements concerning the wavelength/frequency properties of the collected data. The procedure is relatively straightforward to implement and the collected data have minimal computer storage requirements because data coordinates can be generated in-core as the analysis progresses. It also helps to minimize biasing effects and sufficiency problems. The most serious problem with systematic sampling is aliasing where the sample interval is too large.

### 5.4.2 RANDOM

This strategy collects samples at random coordinates of the independent variable [**Figure 5.2**]. Its advantages include simplicity, objectivity, and relatively uniform coverage of independent variable space. Random sampling is most suited to point counting estimates [e.g., areal means, population densities, etc.]. It may be the only feasible approach in many problems where data access and mapping resources are limited. The most serious problem with this strategy is aliasing where the interval variations between samples become too great.

In electronic analysis, randomly distributed data are always processed in pixelated or gridded formats. Furthermore, machine computations generate the affiliated grid internally if the investigator does not explicitly submit a grid. Thus, it is good practice to grid the data explicitly to ensure that the fundamental data variations are appropriately represented for electronic computing. As noted in **Chapter 15.2**, numerous gridding schemes are available to effectively represent data for electronic analysis.

### 5.4.3 CLUSTER

In this strategy, samples are collected in clusters [**Figure 5.2**], sometimes by design, but also as a result of difficulties in accessing the data, lack of resources, or other limitations. The method's advantages include the relative speed of application where it is efficient to make several measurements over a small range of the independent variable. Potential problems, on the other hand, include biasing due to non-representative samples or improper distribution of sample measurements, and aliasing due to large intervals between clusters. Potential biasing and aliasing errors may be detected by comparing the short-wavelength information with each cluster with the information represented by all the clusters.

Combining cluster with random sampling yields an *adaptive cluster sampling* strategy that is useful for estimating and mapping subtle details of a population [e.g., [**95**]]. An example of adaptive cluster sampling of contamination plumes is shown in **Figure 5.4**. Here, a random sample of 10 observations [bold squares] has identified 3 contaminated samples [left map]. Additional sampling of the units adjacent to the contaminated samples [crosses] yields clusters of samples that map out the spatial details of the plumes. The clustered samples also provide unbiased estimates of the plumes' contamination means.

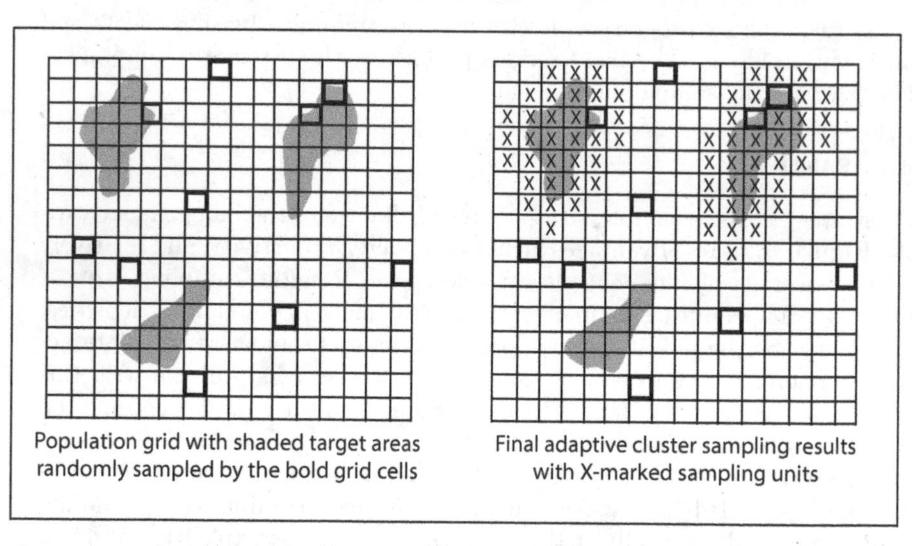


**Figure 5.4** Adaptive cluster sampling of contamination plumes [grey shaded] involving an initial random search [left map] that is followed up by producing cluster samples around the randomly selected observations that exceed some threshold contamination value [right map]. [Adapted from [**95**].]

### 5.4.4 SELECTING A METHOD

The choice of an appropriate sampling method for an application is fundamentally based on the anticipated attributes of the data variations and related statistical parameters. A test sample or simulations of the data variations can effectively constrain this choice.

Upon adopting a sampling method taking aliasing, biasing, and other error sources into account, the number and coordinates of necessary samples must be determined. The adopted procedure also must be tested using appropriate statistical measures or by collecting data via alternate sampling procedures and comparing the analytical results of the datasets for consistency.

## 5.5 KEY CONCEPTS

1. Effective data sampling involves collecting data with variations that accurately represent the problem of interest.
2. The sampling strategy considers data sufficiency issues like minimizing aliasing, where erroneous data variations [i.e., aliased signals] are introduced into the analysis because the sampling interval is too large to capture significant features of the population - i.e., the Nyquist sampling wavelength of two station intervals or three data points is longer than the Nyquist wavelength of the population.
3. Another sufficiency issue concerns the number of samples required to estimate parameters of the population within tolerable error limits. This problem can be addressed using the confidence intervals based on probability density distributions of the data.
4. A further sampling issue is the presence of biased data resulting from operator reading errors, and other systmatic uncertainties encountered in the mapping process. Biasing is often difficult to assess, except perhaps by collecting several sample sets and comparing them for consistency.
5. Systematic sampling is a common data collection method in earth and environmental applications where data are taken at uniformly spaced nodes of one or more dimensional grids. Grid sampling is relatively straightforward to implement, and resolves data variations down to the minimum Nyquist wavelength of the sampling grid. However, where the sampling Nyquist wavelength is larger than the population's Nyquist wavelength, the gridded sample will contain erroneous aliased signals.
6. Random sampling also is common in earth and environmental applications where data are taken at randomly spaced independent variable coordinates. Despite its relative advantages of simplicity and objectivity, the randomly spaced samples need to be gridded for electronic analysis. Thus, it may suffer from aliasing where the interval variations between samples become too large.

7. Cluster sampling and adaptive cluster sampling also have major earth and environmental applications. Sample clusters are efficient to measure and evaluate fine structures and unbiased means of the population.
8. Choosing a sampling method is fundamentally based on the anticipated data variations, which might be tested by computer simulations and/or pilot field studies. The choice takes into account aliasing, biasing, and other errors, as well at the number and coordinates of samples that effectively map the data variations targeted by the investigation.

# 6 Algebraic Linear Regression

## 6.1 OVERVIEW

*Least squares linear regression involves fitting data to the straight-line equation with data-derived intercept and slope parameters so that the sum of the squared deviations of the data from their straight-line predictions is minimum.*

*The least squares intercept and slope estimates are determined from the residual function defined by the sum of the squared differences between the data and their straight-line predictions. Differentiating the residual function with respect to the intercept and slope parameters yields normal equations that, when set to zero, can be solved for the least squares parameter estimates.*

*Measures of the straight-line equation's fit of the data commonly invoke statistically based goodness-of-fit parameters like minimum variances and confidence intervals on the intercept, slope, and predictions, as well as a strong correlation coefficient and high-significance ANOVA testing on the match between predictions and data.*

*Intercept and slope variances may be evaluated for uniform and normal uncertainties in the data's dependent variable magnitudes. For equal observational uncertainties, the parameter variances can be estimated using the uniform data variance that is either known or evaluated from the residual function divided by the degrees of freedom $[n-2]$. For unequal observational uncertainties, the maximum likelihood assumption is invoked where errors in each observation follow the normal distribution, and thus each term of the residual function is multiplied by the inverse variance of the corresponding data point's error distribution.*

*Statistically probable error bars or confidence intervals on the intercept and slope parameters can be estimated from the weighted parameter standard deviations multiplied by the t-distribution value $t_{(\alpha/2;n-2)}$ at the desired $\alpha$-significance level for $[n-2]$ degrees of freedom. Confidence intervals on the predictions are similarly estimated from a weighted standard deviation of the differences between the predictions and data multiplied by the desired t-value.*

*The correlation coefficient is the square root of the coherency coefficient that gives the percent of the straight-line equation predictions which fit the observations. It trends positive to a maximun of +1 as more of the predictions directly follow the observations, and negative to a minimum of -1 as more of the predictions inversely follow the observations. Its statistical significance, which is a direct function of the number of data, can be tested using the normal distribution.*

*The ANOVA table also can test the statistical significance of linear regression using the ratio of the mean squared regression residuals of the predictions about the data mean to the mean squared deviations of the predictions about the data. Where the ratio is larger than the critical f-distribution value* $f_{(\alpha;1;n-2)}$ *at the desired* $\alpha$*-significance level and degrees of freedom* 1 *and* $[n-2]$*, the hypothesis that the data are not fit by the estimated straight line is rejected.*

*Classical linear regression minimizes the sum of squared dependent variable magnitude differences, but alternate error minimizing criteria also have earth and environmental applications. These include the weighted least squares line that seeks to minimize variable squared deviations of the dependent data using the inverses of their variances. Where variable uncertainties occur in both dependent and independent variables, reversing the regression variables maps out solution space, the least squares cubic line minimizes data deviations perpendicular to the line, and the reduced major axis line minimizes the area of the right triangle with apex at the data point and sides corresponding to dependent and independent variable deviations of the data point from the line.*

## 6.2 CONCEPT OF LEAST SQUARES

Data are measured for the patterns they describe in terms of the independent variables. The ultimate objective of data mapping is to convert the data patterns into analytical functions that can account for the variations in the data. In **Figure 6.1**, for example, the observations $b_i \in \mathbf{b}$ are measured at independent variables $a_i \in \mathbf{a}$ to produce a systematic pattern that suggests the existence of a function $f[a_i] \ni b_i = f[a_i]$, or more generally $\mathbf{b} = \mathrm{f}[\mathbf{a}]$.

Given that observations always contain errors $\varepsilon_i$, the *least squares principle* is usually implemented to effectively estimate $f[a_i]$. This principle involves 1) defining Laplace's *residual* or *objective function* as the sum of squares of

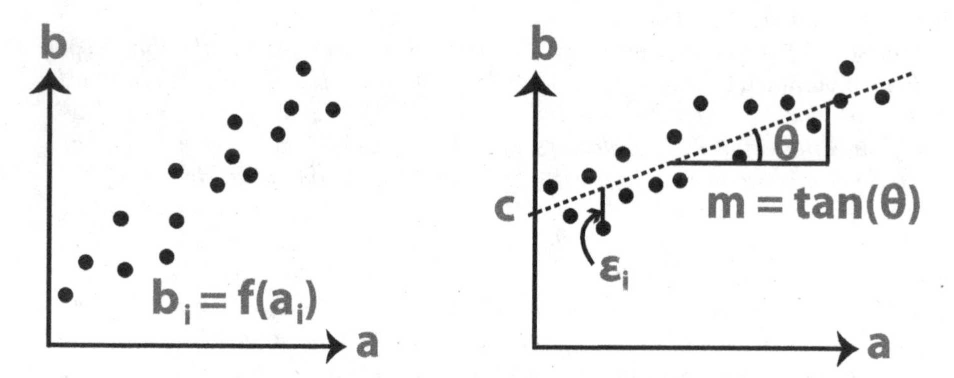


**Figure 6.1** Linear distributions of dependent-variable observations $b_i$ against independent-variable coordinates $a_i$. The observations were obtained with the objective of relating them to an analytical function $f(a_i)$ of the straight line with intercept $c$ and slope $m$ so that the sum of the squared deviations $\varepsilon_i^2$ between the observations and their straight-line predictions is minimum.

differences between the estimated and measured observations - i.e.,

$$R = \sum_{i}^{n} [b_i - (\hat{b}_i = f[a_i])]^2, \tag{6.1}$$

**2)** differentiating the residual function $R$ with respect to the parameters defining $f[a_i]$ and setting the resulting *normal equations* to zero, and **3)** solving the normal equations for the parameters of $f[a_i]$ that yield least squares approximations $\hat{b}_i$ of the observations $b_i$.

Note that no statistical concepts are required in establishing the least squares estimate of the approximating function $f[a_i] = \hat{b}_i$. However, they can be incorporated as necessary via error propagation [e.g., **Chapter** 3.6].

## 6.3 STRAIGHT LINE MODELING

Linear regression commonly refers to the least squares fitting of a straight line through several or more observations, which is perhaps the most widely used application of the least squares principle. The objective is to represent the data by a straight line equation

$$f[a_i] = c + m \times a_i = \hat{b}_i, \tag{6.2}$$

where $c$ is the intercept and $m$ is the slope of the approximating value or estimate $\hat{b}_i$. Note that $\hat{b}_i$ is the modeled value so that the actual functional relationship between $a_i$ and $b_i$ is

$$b_i = c + ma_i + \varepsilon_i = \hat{b}_i + \varepsilon_i, \tag{6.3}$$

where $\varepsilon_i$ represents the error terms implicitly defined or measured in the $[b_i]$-direction.

For two observations $[n = 2]$, the intercept $c$ and slope $m$ are determined exactly by $[b_1, a_1]$ and $[b_2, a_2]$, which accordingly give

$$m = \frac{b_2 - b_1}{a_2 - a_1} = \frac{\Delta_v}{\Delta_h} \text{ and } c = b_1 - ma_1, \tag{6.4}$$

where $m$ is also the ratio of the rise $[\Delta_v]$ to run $[\Delta_h]$ of the data.

In the general case where $n \geq 3$, a least squares fit of $\hat{b}_i = c + ma_i$ with defining unknown parameters $c$ and $m$ can be obtained using the deviation or residual function defined by

$$R \equiv \sum_{i=1}^{n} [b_i - \hat{b}_i]^2 = \sum_{i=1}^{n} [b_i - c - ma_i]^2 = \sum_{i=1}^{n} \varepsilon_i^2. \tag{6.5}$$

This definition implies that the errors $\varepsilon_i$ are defined only in the $[b_i]$-direction - i.e., there is no uncertainty in the $[a_i]$-direction and the $a_i$ are known exactly.

The more general case of uncertainties in both the $[a_i]$- and $[b_i]$-directions is considered further in **Chapter 6.5**.

Next, the residual function $R$ is differentiated with respect to the defining unknowns $c$ and $m$, and the resulting normal equations are set to zero - i.e.,

$$\begin{aligned} \frac{\partial R}{\partial c} &= -2\sum[b_i - c - ma_i] &\equiv 0, \text{ and} \\ \frac{\partial R}{\partial m} &= -2\sum a_i[b_i - c - ma_i] &\equiv 0. \end{aligned} \tag{6.6}$$

The above equations may be rearranged into the following pair of simultaneous equations

$$\begin{aligned} \sum_{i=1}^{n} b_i &= \textstyle\sum_{i=1}^{n} c + \sum_{i=1}^{n} ma_i &= nc + m\sum_{i=1}^{n} a_i \text{ since } \sum_{i=1}^{n} 1 = n, \text{ and} \\ \sum_{i=1}^{n} a_i b_i &= \textstyle\sum_{i=1}^{n} ca_i + \sum_{i=1}^{n} ma_i^2 &= c\sum_{i=1}^{n} a_i + m\sum_{i=1}^{n} a_i^2, \end{aligned} \tag{6.7}$$

with the slope

$$m = \frac{n\sum_{i=1}^{n} a_i b_i - \sum_{i=1}^{n} a_i \sum_{i=1}^{n} b_i}{n\sum_{i=1}^{n} a_i^2 - (\sum_{i=1}^{n} a_i)^2} = \frac{\sum_{i=1}^{n}(a_i - \bar{a})(b_i - \bar{b})}{\sum_{i=1}^{n}(a_i - \bar{a})^2}, \tag{6.8}$$

and the intercept

$$c = \frac{\sum_{i=1}^{n} a_i^2 \sum_{i=1}^{n} b_i - \sum_{i=1}^{n} a_i \sum_{i=1}^{n} a_i b_i}{n\sum_{i=1}^{n} a_i^2 - (\sum_{i=1}^{n} a_i)^2} = \bar{b} - m\bar{a}, \tag{6.9}$$

so that the predictions are

$$\hat{b}_i = \bar{b} + m[a_i - \bar{a}] = c + ma_i. \tag{6.10}$$

As ***Example 06.3a***, the regression line coefficients for the experimental data in **Table 6.1** are

$$m = 2.9 \text{ and } c = 0.3 = \bar{b} - m\bar{a} \tag{6.11}$$

so that the least squares regression line for the sample data is given by

$$\hat{b}_i = 0.3 + 2.9a_i. \tag{6.12}$$

**Table 6.1**

**Experimental observations for *Example 06.3a*.**

| i | 1 | 2 | 3 | 4 | 5 | 6 | 7 | 8 | 9 |
|---|---|---|---|---|---|---|---|---|---|
| $a_i$ | 1.5 | 1.8 | 2.4 | 3.0 | 3.5 | 3.9 | 4.4 | 4.8 | 5.0 |
| $b_i$ | 4.8 | 5.7 | 7.0 | 8.3 | 10.9 | 12.4 | 13.1 | 13.6 | 15.3 |

By substituting any two values of $a_i$ into the above equation, the sample's regression line can be drawn through the resulting two $\hat{b}_i$-estimates - e.g., $\hat{b}_1 = 4.7$ at $a_1 = 1.5$ and $\hat{b}_2 = 14.9$ at $a_2 = 5.0$ so that connecting these two predictions with a straight line renders an effective graphical representation of the least squares model.

## 6.4 MEASURES OF THE FIT

Errors can be attached to the regression's intercept $c$-, slope $m$-, and $\hat{b}_i$-estimates that range from statistical-free uncertainties via Taylor's theorem [e.g., **Equation 3.15**] to statistical characterizations, which since the early 20th century have become pretty much standard practice in environmental science and engineering. Statistical measures of the fit include the intercept $c$ and slope $m$ variances, the confidence intervals on $c$, $m$, and the predictions $\hat{b}_i$, and the correlation coefficient and significance test for $\hat{b}_i$ and the observations $b_i$.

### 6.4.1 VARIANCES OF THE INTERCEPT AND SLOPE

Variance computations as measures of the uncertainties in the intercept $c$ and slope $m$ estimates are predicated on how the uncertainties in the observations are distributed. For example, the above estimates of intercept [**Equation 6.9**] and slope [**Equation 6.8**] assumed the uncertainty in each observation is from a common random distribution with equal observational variances - i.e., $s^2_{b_i}$ is a constant value $[\equiv s^2_{\mathbf{B}}]$ $\forall$ $b_i \in \mathbf{b}$.

However, if observational uncertainty varies between the observations - i,e., $s^2_{b_i}$ is not constant - then an improved *maximum likelihood* estimate of the objective function in **Equation 6.5** is warranted, from which maximum likelihood estimates of the intercept $c_{ml}$ and slope $m_{ml}$ may be obtained. The two subsections below provide further details on estimating the intercept and slope variances from observations containing constant and variable errors, respectively.

#### 6.4.1.1 *Uniform Error Observations*

Where observational uncertainties are all equal, the constant variance is either known or can be estimated from

$$\begin{aligned} s^2_{\mathbf{b}} \approx \frac{R}{[n-2]} &= (\tfrac{1}{[n-2]}) \textstyle\sum_{i=1}^n [b_i - c - ma_i]^2 \qquad (6.13) \\ &= (\tfrac{1}{[n-2]})[\textstyle\sum_{i=1}^n b_i^2 + nc^2 + m^2 \sum_{i=1}^n a_i^2 - 2cm \sum_{i=1}^n b_i \\ &\quad -2m \textstyle\sum_{i=1}^n a_i b_i + 2cm \sum_{i=1}^n a_i], \end{aligned}$$

assuming that the predictions $\hat{b}_i$ are the mean values about which the observations $b_i$ are normally distributed. Here, $s^2_{\mathbf{b}}$ measures the squared deviations of the data from the straight line instead of from the mean $\bar{\mathbf{b}}$ as is the case for the conventional sample variance [e.g., **Equation 4.7**], and the degrees of freedom $[n - 2]$ come from having to calculate both $c$ and $m$ to get $s^2_{\mathbf{b}}$. The conventional sample variance is similar to **Equation 6.13**, but with $m = 0$, $c = \bar{\mathbf{b}}$, and $[n - 2]$ degrees of freedom.

To estimate the variance of the intercept in **Equation 6.10** by the propagation rule in **Equation 4.13** requires the derivative

$$\frac{\partial c}{\partial b_i} = (\frac{1}{\Delta})[\sum_{i=1}^{n} a_i^2 - a_i \sum_{i=1}^{n} a_i] \text{ with } \Delta = n\sum_{i=1}^{n} a_i^2 - [\sum_{i=1}^{n} a_i]^2. \quad (6.14)$$

Thus, the intercept's first-order variance is

$$\begin{aligned} s^2_{c(\mathbf{b})} &\approx \textstyle\sum_{i=1}^{n} (\frac{s_{\mathbf{b}}}{\Delta})^2 [(\sum_{i=1}^{n} a_i^2)^2 - 2a_i \sum_{i=1}^{n} a_i^2 \sum_{i=1}^{n} a_i + a_i^2 (\sum_{i=1}^{n} a_i)^2] \\ &\approx (\tfrac{s_{\mathbf{b}}}{\Delta})^2 [n(\textstyle\sum_{i=1}^{n} a_i^2)^2 - 2(\sum_{i=1}^{n} a_i)^2 \sum_{i=1}^{n} a_i^2 + \sum_{i=1}^{n} a_i^2 (\sum_{i=1}^{n} a_i)^2] \\ &\approx (\tfrac{s_{\mathbf{b}}}{\Delta})^2 (\textstyle\sum_{i=1}^{n} a_i^2)[n \sum_{i=1}^{n} a_i^2 - (\sum_{i=1}^{n} a_i)^2] \\ &\approx (\tfrac{s^2_{\mathbf{b}}}{\Delta}) \textstyle\sum_{i=1}^{n} a_i^2, \end{aligned} \quad (6.15)$$

where $\Delta$ is from **Equation 6.14**.

Similarly, the variance propagation rule estimate of the first-order variance for the slope [**Equation 6.8**] requires the derivative

$$\frac{\partial m}{\partial b_i} = (\frac{1}{\Delta})[na_i^2 - \sum_{i=1}^{n} a_i], \quad (6.16)$$

so that

$$\begin{aligned} s^2_{m(\mathbf{b})} &\approx \textstyle\sum_{i=1}^{n} (\frac{s_{\mathbf{b}}}{\Delta})^2 [n^2 a_i^2 - 2na_i \sum_{i=1}^{n} a_i + (\sum_{i=1}^{n} a_i)^2] \\ &\approx (\tfrac{s_{\mathbf{b}}}{\Delta})^2 [n^2 \textstyle\sum_{i=1}^{n} a_i^2 - 2n(\sum_{i=1}^{n} a_i)^2 + n(\sum_{i=1}^{n} a_i)^2] \\ &\approx n(\tfrac{s_{\mathbf{b}}}{\Delta})^2 (\textstyle\sum_{i=1}^{n} a_i^2)[n \sum_{i=1}^{n} a_i^2 - (\sum_{i=1}^{n} a_i)^2] \\ &\approx n(\tfrac{s^2_{\mathbf{b}}}{\Delta}). \end{aligned} \quad (6.17)$$

#### 6.4.1.2 *Variable Error Observations*

Where the errors in the observations $b_i$ are not equal [i.e., $\sigma^2_{b_i} \neq$ constant], the simple residual function $R$ in **Equation 6.5** can be modified into a *maximum likelihood* weighted residual $R_{ml}$ that yields improved intercept $c_{ml}$ and slope $m_{ml}$ estimates. Subjecting these improved estimates to the variance propagation rule in **Equation 4.13** obtains, in turn, the appropriate intercept and slope variances $s^2_{c(ml)}$ and $s^2_{m(ml)}$, respectively.

In particular, suppose that the observations $b_i$ are normally distributed about the actual value $\hat{b}_i$ for each value of the independent variable $a_i$. Then by **Equation 4.39**, the probability for any $a_i$ of obtaining the observed measurement $b_i$ is

$$P_G[a_i] = \frac{1}{\sigma_i\sqrt{2\pi}} \exp[-0.5(\frac{b_i - \hat{b}_i}{\sigma_i})^2], \tag{6.18}$$

where $\sigma_i^2$ is the variance of the normal distribution about the "true" value $\hat{b}_i$, and not the variance of $b_i$ about its mean $\bar{b}$.

Now, by **Equation 4.26**, the probability of making the observed set of $n$ measurements of $b_i$ from which $c$ and $m$ are determined is

$$\begin{aligned} P_G[c,m] = & \Pi_{i=1}^n P_G[i] = P_G[1] \times P_G[2] \times \cdots \times P_G[n] \\ = & \Pi_{i=1}^n [\tfrac{1}{\sigma_i\sqrt{2\pi}}] \exp[-0.5 \textstyle\sum_{i=1}^n (\frac{b_i - \hat{b}_i}{\sigma_i})^2], \end{aligned} \tag{6.19}$$

where $b_i - \hat{b}_i = b_i - c - ma_i$, and the summation is due to the exponential product rule [e.g., $e^x e^y = e^{(x+y)}$]. The *maximum likelihood* condition imposed by **Equation 6.19** suggests that the best estimates for $c$ and $m$ are those values for which $P_G[c,m]$ is *maximized* - i.e., those values for which the summation term in the exponent of $P_G[c,m]$ is *minimized*. This minimization, however, suggests that the weighted residual function given by

$$R_{ml} \equiv \sum_{i=1}^{n} [\frac{1}{\sigma_i^2}] \times [b_i - c - ma_i]^2 \tag{6.20}$$

should have been used for computing the normal equations rather than the simple unweighted residual from **Equation 6.5**.

Accordingly, to ensure that the observations with the smallest variances have the most effect in the regression requires maximum likelihood estimates for the slope given by

$$m_{ml} = \frac{1}{\Delta_{ml}} [\sum_{i=1}^{n} \frac{1}{\sigma_i^2} \sum_{i=1}^{n} \frac{a_i b_i}{\sigma_i^2} - \sum_{i=1}^{n} \frac{a_i}{\sigma_i^2} \sum_{i=1}^{n} \frac{b_i}{\sigma_i^2}], \tag{6.21}$$

and the intercept

$$c_{ml} = \frac{1}{\Delta_{ml}} [\sum_{i=1}^{n} \frac{a_i^2}{\sigma_i^2} \sum_{i=1}^{n} \frac{b_i}{\sigma_i^2} - \sum_{i=1}^{n} \frac{a_i^2}{\sigma_i^2} \sum_{i=1}^{n} \frac{a_i b_i}{\sigma_i^2}], \tag{6.22}$$

where

$$\Delta_{ml} = \sum_{i=1}^{n} \frac{1}{\sigma_i^2} \sum_{i=1}^{n} \frac{a_i^2}{\sigma_i^2} - (\sum_{i=1}^{n} \frac{a_i}{\sigma_i^2})^2. \tag{6.23}$$

To find the first-order variance of the intercept requires its derivative given by

$$\frac{\partial c_{ml}}{\partial b_i} = \frac{1}{\Delta_{ml}} [\frac{1}{\sigma_i^2} \sum_{i=1}^{n} \frac{a_i^2}{\sigma_i^2} - \frac{a_i}{\sigma_i^2} \sum_{i=1}^{n} \frac{a_i}{\sigma_i^2}], \tag{6.24}$$

so that

$$\sigma^2_{c(ml)} \approx (\frac{1}{\Delta_{ml}})\sum_{i=1}^{n}\frac{a_i^2}{\sigma_i^2}. \quad (6.25)$$

Additionally, the derivative of the maximum likelihood slope estimate is

$$\frac{\partial m_{ml}}{\partial b_i} = \frac{1}{\Delta_{ml}}[\frac{a_i}{\sigma_i^2}\sum_{i=1}^{n}\frac{1}{\sigma_i^2} - \frac{1}{\sigma_i^2}\sum_{i=1}^{n}\frac{a_i}{\sigma_i^2}], \quad (6.26)$$

from which

$$\sigma^2_{m(ml)} \approx (\frac{1}{\Delta_{ml}})\sum_{i=1}^{n}\frac{1}{\sigma_i^2}. \quad (6.27)$$

Applications of the above equations require knowledge of the variances $\sigma_i^2$ of the measurements of $b_i$ about the straight line predictions $\hat{b}_i$. These variances cannot be evaluated from the data, but in practice, they are perhaps most reliably obtained from repeat measurements at each $i$th observation site. **[6]** provides further details concerning the development and applications of the above results.

### 6.4.2 CONFIDENCE INTERVALS ON THE INTERCEPT AND SLOPE

In general, linear regressions performed on several sets of $[b_i, a_i]$-data will yield sample intercepts and slopes that are normally distributed about the population's intercept $\delta$ and slope $\gamma$ coefficients. Accordingly, the confidence limits on the intercept and slope are respectively given by

$$\delta \in c \pm t_{(\alpha/2;n-2)}\sqrt{(s_{\mathbf{b}}^2\sum_{i=1}^{n}a_i^2)/(n \times CSSa)}, \quad (6.28)$$

and

$$\gamma \in m \pm t_{(\alpha/2;n-2)}\sqrt{(s_{\mathbf{b}}^2)/(CSSa)}, \quad (6.29)$$

where $s_{\mathbf{b}}^2$ is the unbiased variance of the prediction's deviations from the data [**Equation 6.13**], $t_{(\alpha/2;n-2)}$ is the value of the $t$-distribution [e.g., **Figure 4.12**] for significance level $\alpha$ and degrees of freedom $\nu = n - 2$, with the *corrected sum of squares of the* **a**-*coefficients* given by

$$CSSa = \sum_{i=1}^{n}a_i^2 - \sum_{i=1}^{n}a_i\sum_{i=1}^{n}\frac{a_i}{n}. \quad (6.30)$$

Further widely applied corrected sums include the *corrected sum of squares of the* **b**-*coefficients*

$$CSSb = \sum_{i=1}^{n}b_i^2 - \sum_{i=1}^{n}b_i\sum_{i=1}^{n}\frac{b_i}{n}, \quad (6.31)$$

and the *corrected sum of cross products*

$$CSCP = \sum_{i=1}^{n} a_i b_i - \sum_{i=1}^{n} a_i \sum_{i=1}^{n} \frac{b_i}{n}. \tag{6.32}$$

Using the corrected sums, the coefficients of the regression may be written as

$$m = \frac{CSCP}{CSSa} \text{ and } c = \bar{b} - m\bar{a}, \tag{6.33}$$

with

$$s_{\mathbf{b}}^2 = \frac{1}{(n-2)}[CSSb - \frac{(CSCP)^2}{CSSa}] \tag{6.34}$$

for the default variance of the system [i.e., **Equation 6.13**].

The confidence intervals above are useful measures of the uncertainties in $c$ and $m$ about the population values. Thus, for a given $q\%$ confidence interval on the slope, for example, the hypothesis $H_0 : m = \delta$ is rejected if the critical slope value $\delta$ falls outside the confidence interval; otherwise there is no reason to reject $H_0$.

As ***Example 06.4.2a***, find the 95% confidence interval for the slope estimate $m = 2.9$ from ***Example 06.3a***. For this case, $t_{(0.025;7)} = 2.4$, and thus the 95% confidence interval is $2.5 \leq 2.9 \leq 3.3$.

### 6.4.3 CONFIDENCE INTERVALS ON THE PREDICTIONS

The error bars on the predictions $\hat{b}_i$ reflect perhaps the most widely reported confidence interval in linear regression. This confidence interval is given by

$$\hat{b}_i \pm t_{(\alpha/2;n-2)} \left( s_{\mathbf{b}} \sqrt{\frac{1}{n} + \frac{(a_i - \bar{a})^2}{CSSa}} \right). \tag{6.35}$$

The confidence band on the regression line can be graphed by drawing a smooth curve through the interval limits computed for several $a_i$ values [e.g., **Figure 6.2**].

The confidence band is always wider at the ends of the fitted line than at the center because there are more data points near the mean value $\bar{b}$. Similarly, the error in estimating $\hat{b}_i$ values increases as the distance from the mean increases.

As ***Example 06.4.3a***, estimate the 95% error bar for the prediction $\hat{b}_i$ at $a_i = 2$ from the solution in ***Example 06.3a***. Accordingly, by **Equation 6.12**, $\hat{b}[2] = 6.1$ with the 95% confidence interval $4.7 \leq 6.1 \leq 7.5$.

### 6.4.4 CORRELATION COEFFICIENT

A standard estimate of the goodness of fit of the regression of $b_i$ on $a_i$ is the *correlation coefficient* [$r$]. It is the dimensionless measure of the deviation of

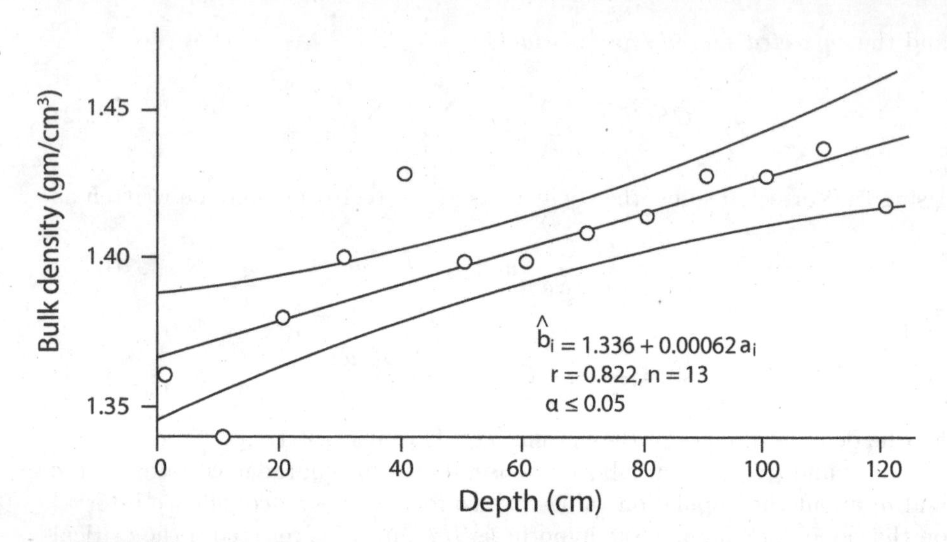


**Figure 6.2** The 95% confidence band for the classical regression predictions $[\hat{b}_i]$ of bulk density measurements $[b_i]$ on depths $[a_i]$ below the ground's surface.

the observations $b_i$ about $\hat{b}_i$ [$= c + ma_i$] derived from the *coherency coefficient* [$r^2$] given by

$$r^2 = \frac{SS_R}{SS_T}, \text{ with } SS_R = \sum_{i=1}^{n} [\hat{b}_i - \bar{b}]^2 \text{ and } SS_T = \sum_{i=1}^{n} [b_i - \bar{b}]^2, \quad (6.36)$$

where $SS_R$ is the sum of squares due to regression [i.e., the variation of the least squares model about the mean of the data], and $SS_T$ is the total sum of squares [i.e., the variation of the data about the mean, which is equivalent to sample variance in **Equation 4.7**, but devoid of the $(1/[n-1])$-term].

The correlation coefficient may be expressed in several ways, including

$$r = \frac{m}{|m|}\sqrt{r^2} = \frac{m}{|m|}\sqrt{\frac{SS_R}{SS_T}} = \frac{\sigma^2_{\mathbf{ab}}}{\sqrt{\sigma^2_{\mathbf{a}}\sigma^2_{\mathbf{b}}}} = m\sqrt{\frac{\sigma^2_{\mathbf{a}}}{\sigma^2_{\mathbf{b}}}}, \quad (6.37)$$

for all $a_i \in \mathbf{a}$ and $b_i \in \mathbf{b}$, where the covariance of $\mathbf{a}$ and $\mathbf{b}$ is

$$\sigma^2_{\mathbf{ab}} = \frac{1}{(n-1)} \sum_{i=1}^{n} [a_i - \bar{a}][b_i - \bar{b}], \quad (6.38)$$

and the variances of $\mathbf{a}$ and $\mathbf{b}$ are respectively

$$\sigma^2_{\mathbf{a}} = \frac{1}{(n-1)} \sum_{i=1}^{n} [a_i - \bar{a}]^2 \quad (6.39)$$

and

$$\sigma^2_{\mathbf{b}} = \frac{1}{(n-1)} \sum_{i=1}^{n} [b_i - \bar{b}]^2. \quad (6.40)$$

The sign of the correlation coefficient is selected according to the sign of the slope coefficient, which can also be expressed as

$$m = \frac{\sigma^2_{\mathbf{ab}}}{\sigma^2_{\mathbf{a}}}. \tag{6.41}$$

Thus, the covariance $\sigma^2_{\mathbf{ab}}$ ultimately controls the sign of $r$ so that $r > 0$ if $m > 0$, and vice versa.

In addition, the magnitude of $r$ ranges between $\pm 1$ where $r = +1$ means perfect *positive* or *direct* correlation with the gradient in the predictions $\hat{b}_i$ directly matching the gradient in the observations $b_i$ on the line $\hat{\mathbf{b}}$ [e.g., **Figure 6.3(A)**]. On the other hand, $r = -1$ indicates perfect *negative* or *inverse* correlation where all predictions inversely match the observations on a line with opposite or orthogonal slope to the line of observations [e.g., **Figure 6.3(B)**].

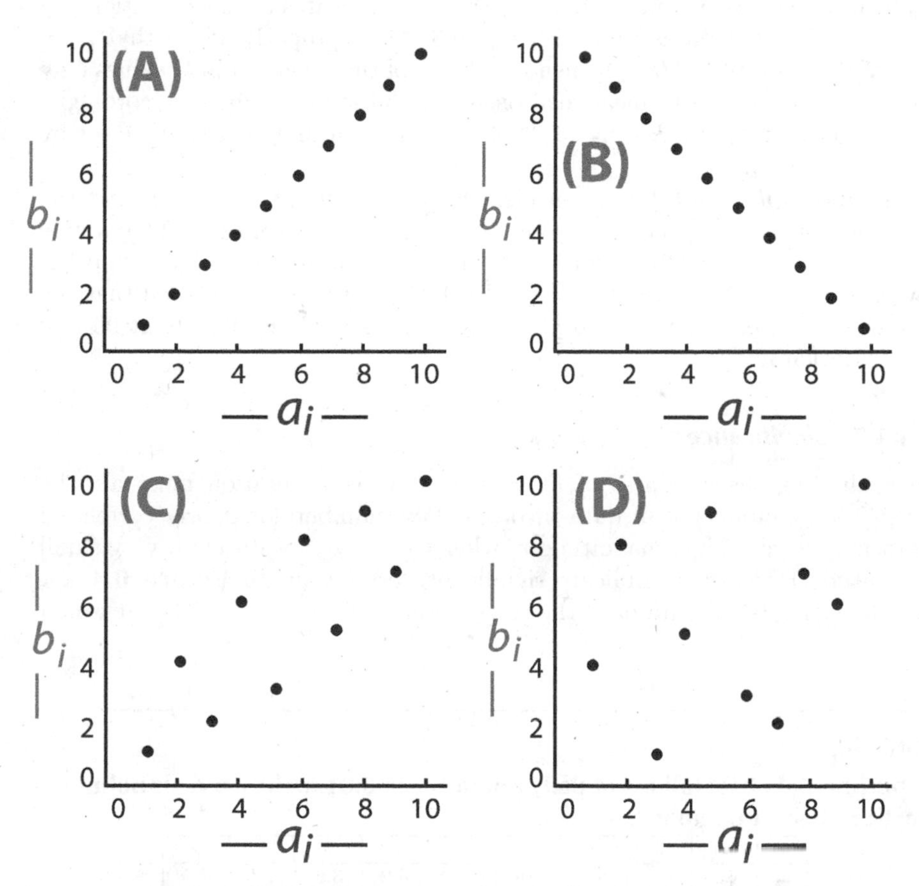


**Figure 6.3** Graphs of $[a_i, b_i]$-values with **(A)** $r = +1.0$, **(B)** $r = -1.0$, **(C)** $r = 0.84$, and **(D)** $r = 0.33$.

#### 6.4.4.1 *Interpretation*

Considerable care must be taken with the interpretation of the correlation coefficient whenever $|r| \neq 1$ because these values can be produced in multiple non-unique ways. For example, $r = 0.0$ is conventionally interpreted to infer no correlation between $\hat{b}_i$ and $b_i$ - i.e., one of the datasets is flat-lined, showing no variations with changes of the other dataset. However, the zero correlation coefficient also results when half of the datasets are positively correlated and the other half are negatively correlated.

Indeed, as described more fully in **Chapter 11.4.2.3**, any value of the correlation coefficient $r \neq \pm 1$ may be interpreted in terms of the percentages of the two datasets that are positively and negatively correlated with each other. This non-unique behavior makes using any particular correlation coefficient value $r \neq \pm 1$ as the threshold for accepting or rejecting datasets for analysis a somewhat dubious proposition.

It also is incorrect to conclude that $r = 0.6$ indicates a linear relationship that is twice as good as that with $r = 0.3$. More properly, given that $r^2 = 1 - [(SS_{\mathbf{b}} - m \times SS_{\mathbf{Ab}})/SS_{\mathbf{b}}]$, then $r^2 \times 100\%$ of the variation in $b_i$ values may be accounted for by the linear relationship with the $a_i$ variables. Accordingly, a correlation of 0.6 means that 36% of the variation in $b_i$ is accounted for by differences in $a_i$.

As ***Example 06.4.4.1a***, consider the daily amount of rainfall and the pollution particulates removed from the air shown in **Table 6.2**. The sample correlation is $r = -0.979$, which indicates a strong inverse relationship between the $b_i$ and $z_i$ values. Because $r^2 = 0.958$, it may be concluded that the linear inverse relationship with the $z_i$ differences accounts for about 96% of the variation in $b_i$.

#### 6.4.4.2 *Significance*

In evaluating the goodness of fit coefficient $r$, consideration must also be given to the number $n$ of data involved. Low numbers of poorly correlated data may have a large magnitude $r$, whereas for large datasets a very small magnitude $r$ value can indicate significant correlation. In **Figure 6.4**, for example, the $[b_i, a_i]$-data have the correlation coefficient $r = 0.74$ that would

**Table 6.2**

**Experimental observations of daily rainfall $z_i$ in 0.01 inches and air pollution particulates $b_i$ in $\mu$gm/m$^3$.**

| i | 1 | 2 | 3 | 4 | 5 | 6 | 7 | 8 | 9 |
|---|---|---|---|---|---|---|---|---|---|
| $z_i$ | 4.3 | 4.5 | 5.9 | 5.6 | 6.1 | 5.2 | 3.8 | 2.1 | 7.5 |
| $b_i$ | 126 | 121 | 116 | 118 | 114 | 118 | 132 | 141 | 108 |

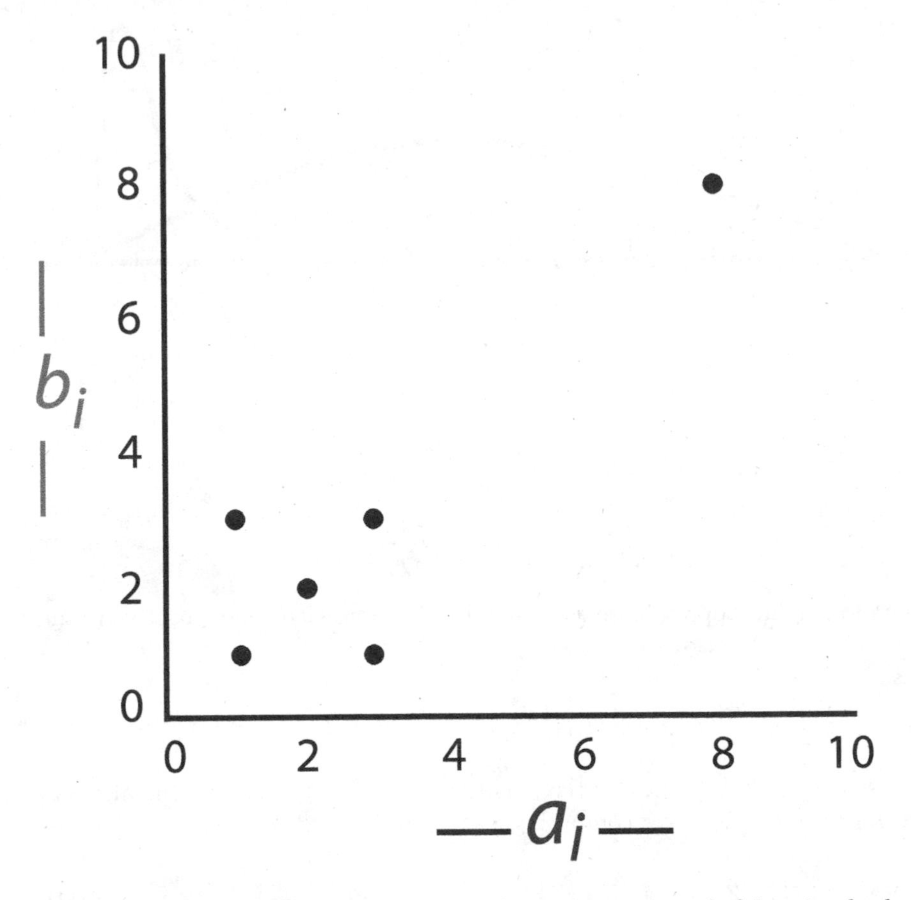


**Figure 6.4** Dataset showing the sensitivity of $r$ to the number of of data involved.

become $r = 0.0$ if the data point with coordinates $[8, 8]$ is eliminated. For small $n$, $r$ will almost always be near $\pm 1$, but the correlation may not be meaningful.

However, a simple test of the significance of $r$ is available. It turns out that the distribution of $r$ depends only on the population values $r_0$ and $n$ [e.g., [45]] as the examples in **Figure 6.5** suggest. From these examples, it is clear that the $r$ distribution is decidedly non-normal for large values of $r_0$. Thus, it will not suffice to obtain the standard deviation of $r$ and use it to determine the statistical significance of $r$ as an accurate estimate of $r_0$.

Fortunately, there is a simple change of variable that transforms the complicated distribution of $r$ into an approximately normal distribution. The transformation can be used to determine the accuracy of the sample $r$ as an estimate of the population $r_0$ in the same way that the normal distribution of $\bar{x}$ was used to determine the accuracy of $\bar{x}$ as an estimate of $\mu_x$.

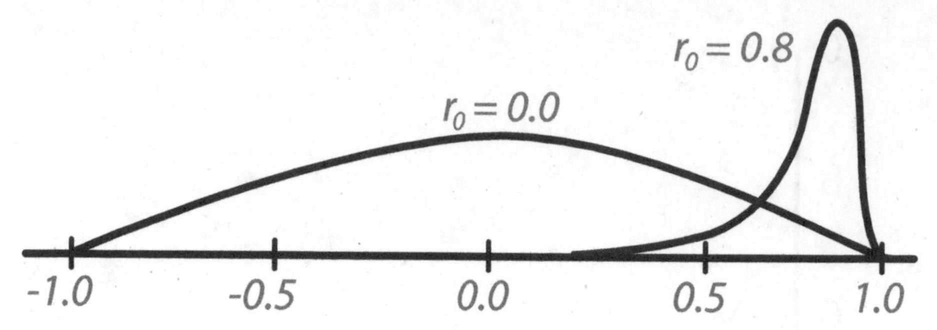


**Figure 6.5** Distribution of $r$ for $r_0 = 0.0$ and $r_0 = 0.8$ when $n = 9$. [Adapted from [45].]

The transformation from $r$ to $Z[r]$ is

$$Z[r] = 0.5\ln[\frac{(1+r)}{(1-r)}], \tag{6.42}$$

which follows an approximately normal distribution with the respective mean and standard deviation given by

$$< Z[r] >= 0.5\ln[\frac{(1+r_0)}{(1-r_0)}] \text{ and } \sigma_{Z[r]} = \frac{1}{\sqrt{n-3}}. \tag{6.43}$$

To test $r$ against $r_0$ - i.e., test the hypothesis $H_0 : r = r_0$ against the alternative $H_1 : r \neq r_0$ - involves computing the statistic

$$Z[r] = [\frac{\sqrt{n-3}}{2}] \times [\ln(\frac{1+r}{1-r}) - \ln(\frac{1+r_0}{1-r_0})], \tag{6.44}$$

which simplifies to

$$Z[r] = [0.5\sqrt{n-3}] \times \ln[\frac{(1+r)(1-r_0)}{(1-r)(1+r_0)}], \tag{6.45}$$

for comparison with the critical points of the standard normal distribution.

As ***Example 06.4.4.2a***, for the results from ***Example 06.4.4.1a***, test the hypothesis at the 5% significance level that the $b_i$ and $z_i$ variables are not linearly associated - i.e., test the hypothesis $H_0 : r = r_0 = 0$. The test is two-sided, and thus to reject the null hypothesis, either the test value $Z[r] < -Z'[r] = -1.96$ or $> +Z'[r] = 1.96$, where $Z'[r]$ is the critical value of the normal distribution at the 5% significance level. For $r = -0.979$ with $n = 9$, $Z[r] = -5.55$ so that $|Z[r]| > |Z'[r]|$ and thus the notion of no linear relationship between the $b_i$ and $z_i$ variables [i.e., $r_0 = 0$] is rejected.

The above test is completely general and can be used to compare sample correlations for any $r_0$ value, not just $r_0 = 0.0$. Confidence limits on sample $r$

| ν | q | | | |
|---|---|---|---|---|
| | 95% | 97.5% | 99% | 99.5% |
| 1 | 98.8 | 99.7 | 99.95 | 99.99 |
| 2 | 90.0 | 95.0 | 98.0 | 99.0 |
| 3 | 80.5 | 87.8 | 93.4 | 95.9 |
| 4 | 72.9 | 81.1 | 88.2 | 91.7 |
| 5 | 66.9 | 75.4 | 83.3 | 87.4 |
| 6 | 62.2 | 70.7 | 78.9 | 83.4 |
| 7 | 58.2 | 66.6 | 75.0 | 79.8 |
| 8 | 54.9 | 63.2 | 71.6 | 76.5 |
| 9 | 52.1 | 60.2 | 68.5 | 73.5 |
| 10 | 49.7 | 57.6 | 65.8 | 70.8 |
| 11 | 47.6 | 55.3 | 63.4 | 68.4 |
| 12 | 45.8 | 53.2 | 61.2 | 66.1 |
| 13 | 44.1 | 51.4 | 59.2 | 64.1 |
| 14 | 42.6 | 49.7 | 57.4 | 62.3 |
| 15 | 41.2 | 48.2 | 55.8 | 60.6 |
| 16 | 40.0 | 46.8 | 54.2 | 59.0 |
| 17 | 38.9 | 45.6 | 52.8 | 57.5 |
| 18 | 37.8 | 44.4 | 51.6 | 56.1 |
| 19 | 36.9 | 43.3 | 50.3 | 54.9 |
| 20 | 36.0 | 42.3 | 49.2 | 53.7 |
| 21 | 35.2 | 41.3 | 48.2 | 52.6 |
| 22 | 34.4 | 40.4 | 47.2 | 51.5 |
| 23 | 33.7 | 39.6 | 46.2 | 50.5 |
| 24 | 33.0 | 38.8 | 45.3 | 49.6 |
| 25 | 32.3 | 38.1 | 44.5 | 48.7 |
| 26 | 31.7 | 37.4 | 43.7 | 47.9 |
| 27 | 31.1 | 36.7 | 43.0 | 47.1 |
| 28 | 30.6 | 36.1 | 42.3 | 46.3 |
| 29 | 30.1 | 35.5 | 41.6 | 45.6 |
| 30 | 29.6 | 34.9 | 40.9 | 44.9 |
| 35 | 27.5 | 32.5 | 38.1 | 41.8 |
| 40 | 25.7 | 30.4 | 35.8 | 39.3 |
| 45 | 24.3 | 28.8 | 33.8 | 37.2 |
| 50 | 23.1 | 27.3 | 32.2 | 25.4 |
| 60 | 21.1 | 25.0 | 29.5 | 32.5 |
| 70 | 19.5 | 23.2 | 27.4 | 30.2 |
| 80 | 18.3 | 21.7 | 25.6 | 28.3 |
| 90 | 17.3 | 20.5 | 24.2 | 26.7 |
| 100 | 16.4 | 19.5 | 23.0 | 25.4 |

**Figure 6.6** Values of $r \times 10^{-2}$ for degrees of freedom $\nu = [n - 2]$ at confidence levels $q = 0.95, 0.975, 0.99$, and $0.995$ [e.g., **[28]**.]

values also can be established using the Z[r] distribution [e.g., **[28]**]. **Figure 6.6** shows a generalization of the test that lists values of $r$ for degrees of freedom $\nu = n - 2$ at several $\alpha$-significance levels.

As ***Example 06.4.4.2b*** to illustrate the test, which is two-sided for positive and negative $r$, consider the $[z_i, b_i]$-dataset that has a value of $r = 0.77$

for $n = 9$. By the tabulation in **Figure 6.6**, there is only one chance in 20 at the 95% confidence level that a random sample of the $[z_i, b_i]$ data with $\nu = 7$ would show a value of $r \geq 0.582$. Thus, the test $r = 0.77$ is statistically significant. Indeed, it is significant even at the 99% confidence level, where there is only one chance in 100 that a random set of 9 data points will show a value of $r \geq 0.75$.

The above test is just a special case of the more general $Z[r]$-test. To see this, consider the result obtained in ***Example 06.4.4.2a***. Accordingly, $|r| = |-0.979| > r'[\nu = 7; \alpha = 0.05] = 0.582$ whereby the hypothesis $H_0 : r = 0.0$ is rejected.

In general, $r$ values must be very close to $\pm 1$ to indicate a significant fit of $\hat{b}_i$ to $b_i$ for small $n$, whereas if $n$ is large, small values of $r$ are significant. For the 6 data points in **Figure 6.4** with $r = 0.74$, for example, the tabulation in **Figure 6.6** shows that the critical value at the 97.5% confidence level is $r' = 0.811 > r$ so that the hypothesis $H_0 : r = 0.0$ cannot be rejected. On the other hand, for $n = 102$, the tabulation shows that $r = 0.164$ is statistically significant at the 95% confidence level. Clearly, the use of $r$ to, say, accept or reject a dataset for analysis can be problematic without considering the number of data involved.

### 6.4.5 SIGNIFICANCE OF THE FIT

The regression partitions the total variation in the data [i.e., $SS_T$ of **Equation 6.36**] into **1)** the variation due to the regression [i.e., $SS_R$ also in **Equation 6.36**] that is presumably related to the causative natural process, and **2)** the variation due to the deviation of the data from the regression given by

$$SS_D = \sum_{i=1}^{n} [\hat{b}_i - b_i]^2, \tag{6.46}$$

which presumably relates to the random data errors. This partitioning of the variations is illustrated in **Figure 6.7**.

The significance of the regression line depends on the ratio of the variances from the two sources. Greater significance is reflected by higher variance due to regression and lower variance due to error. Hence, the significance of the linear regression can be tested by analysis of variance using **Table 6.3**. Accordingly, if $F > F'$, then the hypothesis $H_0 : m = 0$ [i.e., line has zero gradient] is rejected for the alternative $H_1 : m \neq 0$ [i.e., line has non-zero gradient], where the critical value $F' = f_{(\alpha,1,n-2)}$ from the $f$ distribution [e.g., **Figure 4.14**] with significance level $\alpha$ and degrees of freedom $\nu = 1, [n-2]$. Otherwise [i.e., if $F < F'$] there is no reason to reject $H_0$.

In assessing the significance of the regression, it is always good practice to inspect the straight line model on a graphical presentation of the data [e.g., **Figure 6.8**]. Plots of the residuals [e.g., $\hat{b}_i - b_i$] also should be checked for violations of assumptions, biases, and other problems. For example, if the data

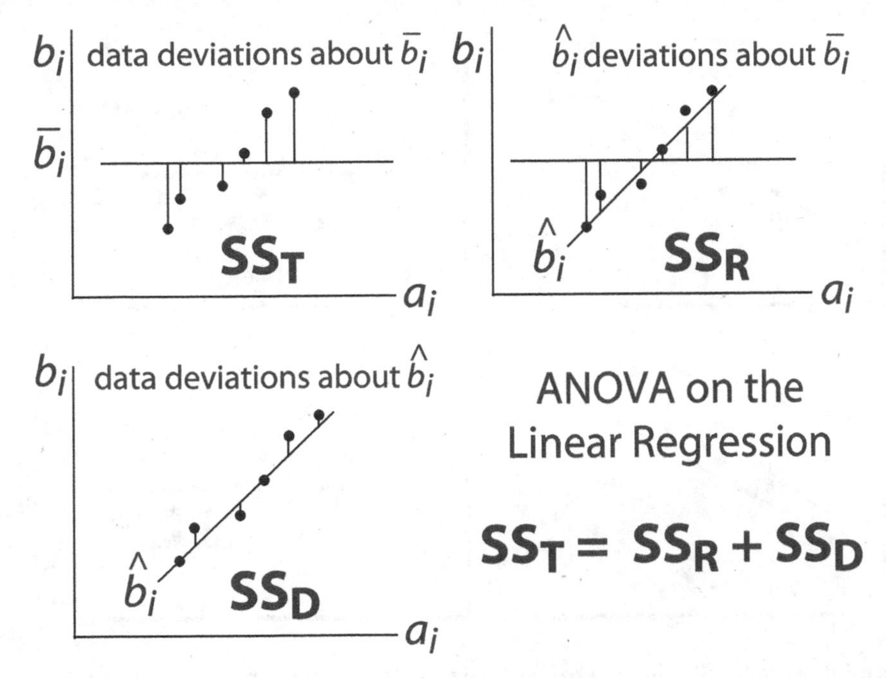


**Figure 6.7** Scatter plots showing the deviations of the sum of squares used in ANOVA to check the significance of the linear regression. The $[a_i, b_i]$-data of independent $a_i$- and dependent $b_i$-variables describe the total variations of the observations $b_i$ about the mean $\bar{b}_i$ involving the total sum of squares $SS_T$, the variations of the regression line predictions $\hat{b}_i$ about the mean $\bar{b}_i$ involving the regression sum of squares $SS_R$, and the deviations of the observations $b_i$ about the regression line predictions $\hat{b}_i$ involving the sum of squared deviations $SS_D$.

errors are random, the histogram of the residuals should look normal. Plotting the residuals against the predictions $\hat{b}_i$ and the independent variables $a_i$ can reveal biases due to data trends not accounted for by the straight line model such as in the bottom two examples of **Figure 6.8**.

**Table 6.3**
**Analysis of variance (ANOVA) table for testing the significance of the regression in terms of the mean variations due to the regression $[MS_R]$ and due to the deviation of the data from the regression $[MS_D]$.**

| VARIATION SOURCE | SUM OF SQUARES | $\nu$ | MEAN SQUARES | F-TEST |
|---|---|---|---|---|
| Regression | $SS_R$ | 1 | $MS_R = \frac{SS_R}{\nu}$ | $F = \frac{MS_R}{MS_D}$ |
| Error | $SS_D$ | $n-2$ | $MS_D = \frac{SS_D}{\nu}$ | |
| Total | $SS_T$ | $n-1$ | $MS_T = \frac{SS_T}{\nu}$ | |

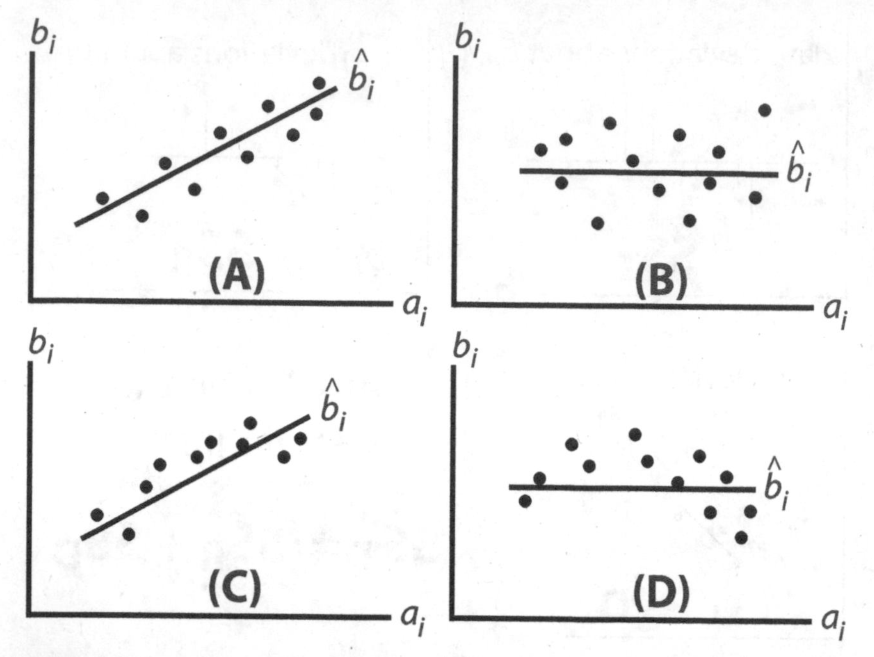


**Figure 6.8** Graphically characterizing the significance of linear regression for various data distributions. Examples include **(A)** significant linear regression with no lack of fit, **(B)** linear regression not significant [e.g., $r = 0.0$] with no lack of fit, **(C)** significant linear regression with signficant lack of fit, and **(D)** linear regression not significant with significant lack of fit. Note also that different scatter plots can be modeled by the same line so that the line is not uniquely indicative of how the data are distributed.

## 6.5 ADDITIONAL LINEAR REGRESSION STRATEGIES

Classical linear regression minimizes the sum of squared deviations $[\hat{b}_i - b_i]^2$ in the $b_i$ direction. The classical regression in the **A**-condition of **Figure 6.9** assumes that the squared deviations are all equal [i.e., each dependent variable $b_i$ has the same uncertainty], each independent variable $a_i$ is errorless, and the squared deviations are independent of each other and the magnitudes of $a_i$ and $b_i$.

Where the squared deviations of the $b_i$ are not equal, the equation for the regression line can be found by *weighted linear regression* that minimizes the sum of weighted squared deviations. As discussed in **Section 6.5.1**, the weighting factor for each squared deviation typically is the inverse variance of the affiliated $b_i$.

Where uncertainty exists in both the variables $b_i$ and $a_i$, several strategies may help to assess an effective linear regression. For example, the *reversed variables* approach in **Section 6.5.2** reverses the classical linear regression

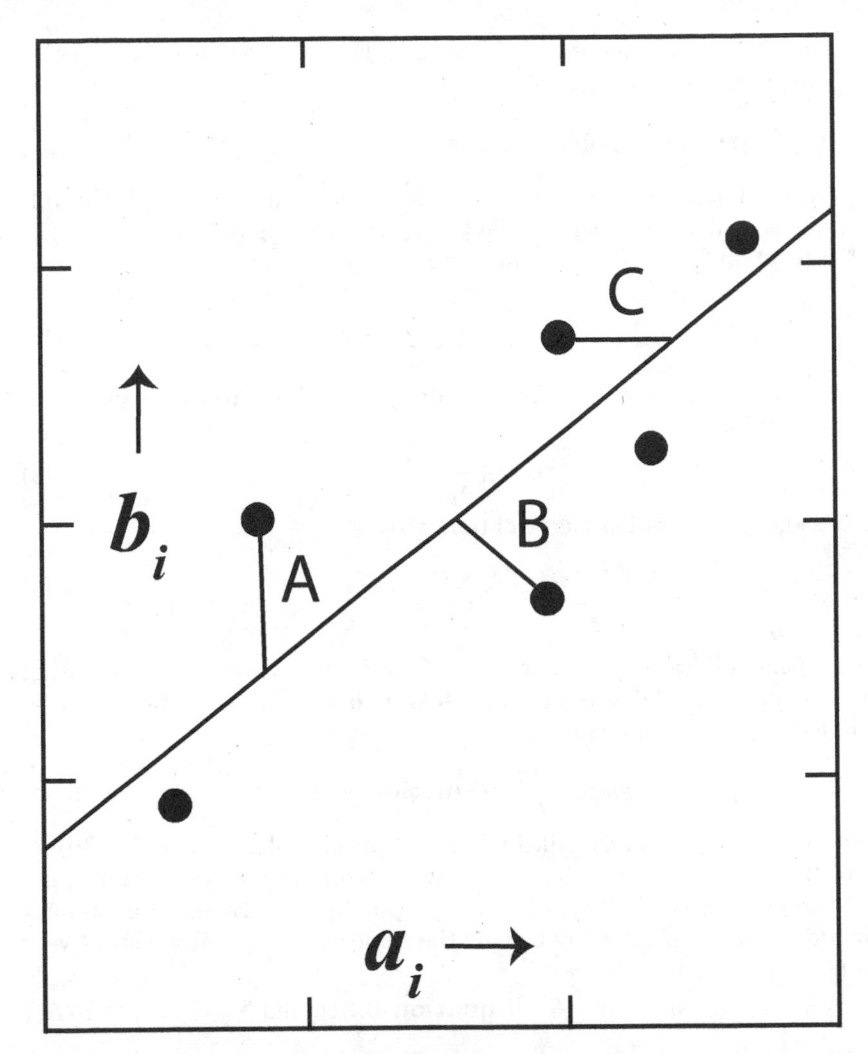


**Figure 6.9** Possible error criteria for minimizing deviations of the $[b_i, a_i]$-data [solid circles] from from the fitted line $\hat{b}_i = c + ma_i$ include **(A)** the classical fitting of $b_i$ to $a_i$, **(B)** minimizing the perpendicular distance between the observations $b_i$ and predictions $\hat{b}_i$ on the *least squares cubic line*, and **(C)** reversing the regression to fit $a_i$ to $b_i$.

variables [i.e., the **C**-condition of **Figure 6.9**] to map out the solution space of lines that may fit the $[b_i, a_i]$ data. **Section 6.5.3** describes another strategy that minimizes the perpendicular distance to the *least squares cubic line* [i.e., the **B**-condition]. A third approach considered in **Section 6.5.4** minimizes the total area of triangles with apexes at the data points and hypotenuses on segments of the *reduced major axis line*. The line's intersections with the vertical and horizontal axes originating from a data point define the end points of the corresponding line segment.

### 6.5.1 WEIGHTED LINEAR REGRESSION

The weighted linear regression is applicable for observations $b_i$ with variable uncertainties and independent variables $a_i$ with no significant errors. The residual function for weighted linear regression [$wlr$] is

$$R_{wlr} = \sum_{i=1}^{n} w_i[b_i - \hat{b_i}]^2 = \sum_{i=1}^{n} w_i[b_i - c - ma_i]^2, \tag{6.47}$$

where the weighting factors $w_i$ are typically proportioned to the inverse variances of the $b_i$ by

$$w_i = (\frac{1}{\sigma_i^2})(\frac{1}{k}) \tag{6.48}$$

and $k$ is the normalization factor obtained from

$$\sum_{i=1}^{n} w_i = \frac{1}{k}[\sum_{i=1}^{n} \frac{1}{\sigma_i^2}] = n. \tag{6.49}$$

The weighted linear regression residual function $R_{wlr}$ may be derived using the maximum likelihood approach in **Section 6.4.1.2** that assumes each $b_i$ is sampled from a normal [Gaussian] distribution so that

$$R_{wlr} = \frac{R_{ml}}{k} \text{ [\textbf{Equation 6.20}]}, \tag{6.50}$$

where $R_{ml}$ is the maximum likelihood [$ml$] residual function from **Equation 6.20**. Other key parameters of the weighted linear regression that may be similarly expressed in terms of the corresponding $k$-normalized maximum likelihood regression parameters include the slope and its variance respectively given by

$$m_{wlr} = \frac{m_{ml}}{k} \text{ [\textbf{Equation 6.21}], and} \tag{6.51}$$

$$\sigma^2_{m(wlr)} = \frac{\sigma^2_{m(ml)}}{k} \text{ [\textbf{Equation 6.27}]}, \tag{6.52}$$

and the intercept and its variance respectively from

$$c_{wlr} = \frac{c_{ml}}{k} \text{ [\textbf{Equation 6.22}], and} \tag{6.53}$$

$$\sigma^2_{c(wlr)} = \frac{\sigma^2_{c(ml)}}{k} \text{ [\textbf{Equation 6.25}]}. \tag{6.54}$$

In addition, the confidence interval for the the $i$-th weighted linear regression prediction is

$$\hat{b}_i \pm t_{(\alpha/2;n-2)}\sqrt{[\frac{\sum_{i=1}^{n}(b_i - \hat{b}_i)^2}{n-2}][\frac{1}{n} + \frac{(a_i - \bar{a}_{wt})^2}{\sum_{i=1}^{n} w_i(a_i - \bar{a}_{wt})^2}]}, \tag{6.55}$$

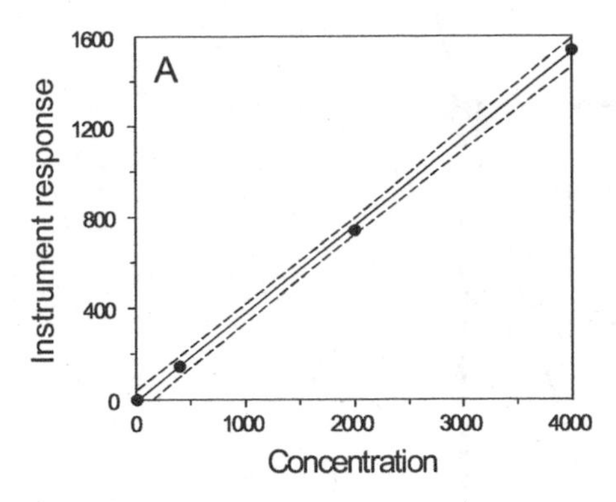


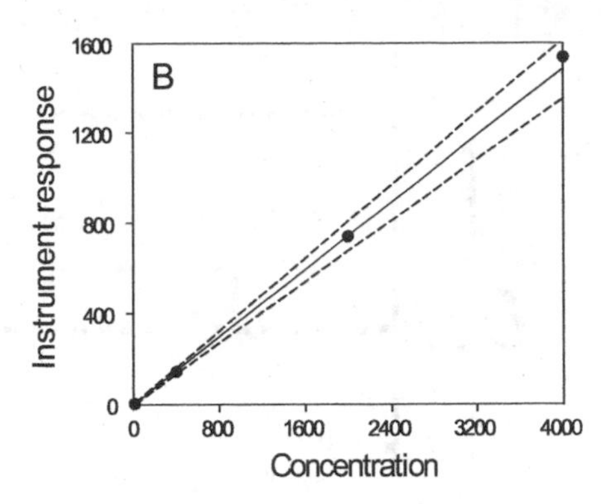


**Figure 6.10** The 90% confidence band for **(A)** the classical linear regression predictions $[\hat{b}_i]$ of the instrument responses $[b_i]$ on concentration $[a_i]$, and **(B)** the weighted linear regression predictions $[\hat{b}_{wlr}]$ of the instrument responses $[b_i]$ on concentration $[a_i]$.

where $\bar{a}_{wt}$ is the weighted mean of the $a_i$ coefficients from

$$\bar{a}_{wt} = \frac{\sum_{i=1}^{n} w_i a_i}{n}. \tag{6.56}$$

In practice, the variances of the $b_i$ coefficients are either known, estimated, or assumed. Replicate measurements of the $b_i$ values for each $a_i$ can be made to estimate the variances. In some cases, the relative standard deviation of the $b_i$ is known to be [approximately] constant so that the variances increase with growing $b_i^2$ [or $a_i^2$]. As ***Example 06.5.1a***, **Figure 6.10** compares the regression lines and confidence intervals for a dataset subjected to **(A)** classical linear regression where the $b_i$-errors are constant as opposed to **(B)** weighted linear regression where the $b_i$-errors increase with increasing $b_i$ [and $a_i$]. To facilitate the comparison, the 4 solid dots in the inserts correspond to the classical linear regression predictions $\hat{b}_i$. The confidence interval for the weighted linear regression line increases with increasing $b_i$ [and $a_i$] because the error is proportional to $b_i$ [and $a_i$].

### 6.5.2 REVERSED VARIABLES

Conventional linear regression assumes that the independent $a_i$-variables are known perfectly so that all deviations of the observations about the model's predictions are due to random errors solely in the dependent $b_i$-variables. Uncertainties in the independent variable can severely affect the stability of the solution because the $\Delta$-term from **Equation 6.14** occurs in denominators for both solution coefficients $m$ and $c$ in **Equations 6.8** and **6.9**, respectively. Thus, as $\Delta \to 0$, the solution coefficients become excessively large, unstable and poorly determined. Additionally, when the slope of the regression is large,

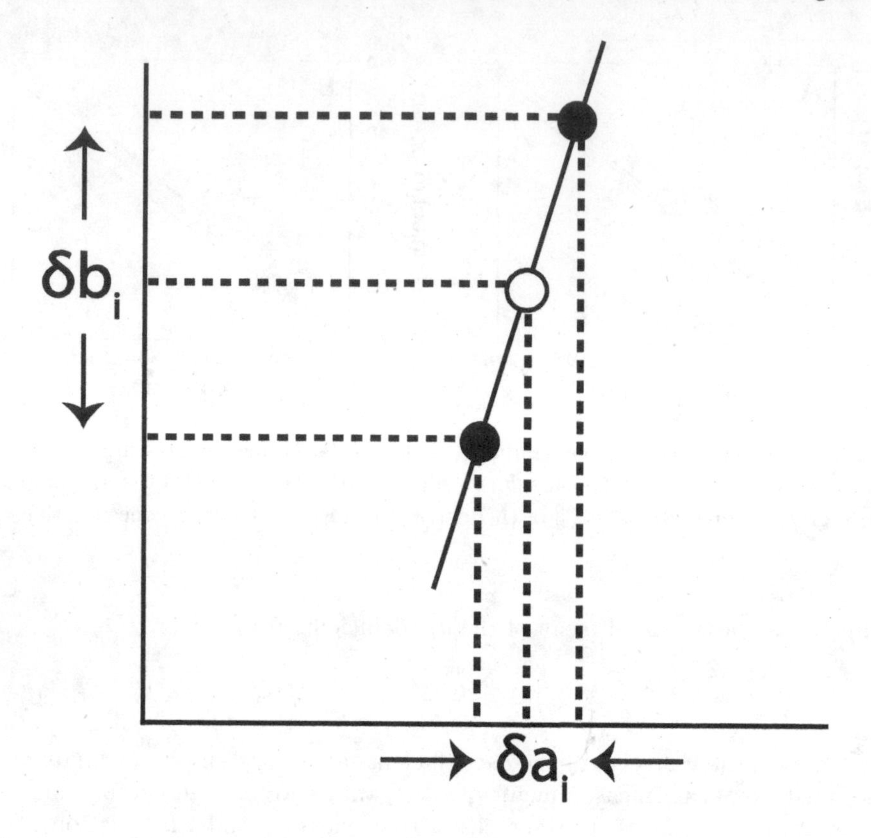


**Figure 6.11** When the slope $m$ is large, small errors $\delta a_i$ in the independent variable produce large errors $\delta b_i$ in the dependent variable.

small errors $\delta a_i$ in the independent variable can result in large deviations $\delta b_i$ of the dependent variable from the fitted line [e.g., **Figure 6.11**].

However, it is not uncommon to encounter situations where both $a_i$ ($\equiv a_{i2}$) and $b_i$ ($\equiv b_{i1}$) may be in error so that the $\hat{b}$-lines will not necessarily be the best fit to the data. In these cases, reversing the regressions of $a_i$ and $b_i$ may help to map out possible ranges of solution space as suggested by the examples in **Figure 6.12**. Reversing the regression yields two sets of line predictions $\hat{b}$ and $\hat{a}$ with the correlation coefficient, $r$, given by the cosine of the angle between the two prediction lines. Thus, where $|r| \neq 1.00$, the two lines map out a finite range over which they may fit the $[b_i, a_i]$-data, although the lines may not be the extremes if the uncertainties in the variables are not equal.

### 6.5.3 LEAST SQUARES CUBIC LINE

Following [1], [**104, 105, 106**] point out that another possible strategy minimizes the perpendicular distance from the fitted line as indicated by the

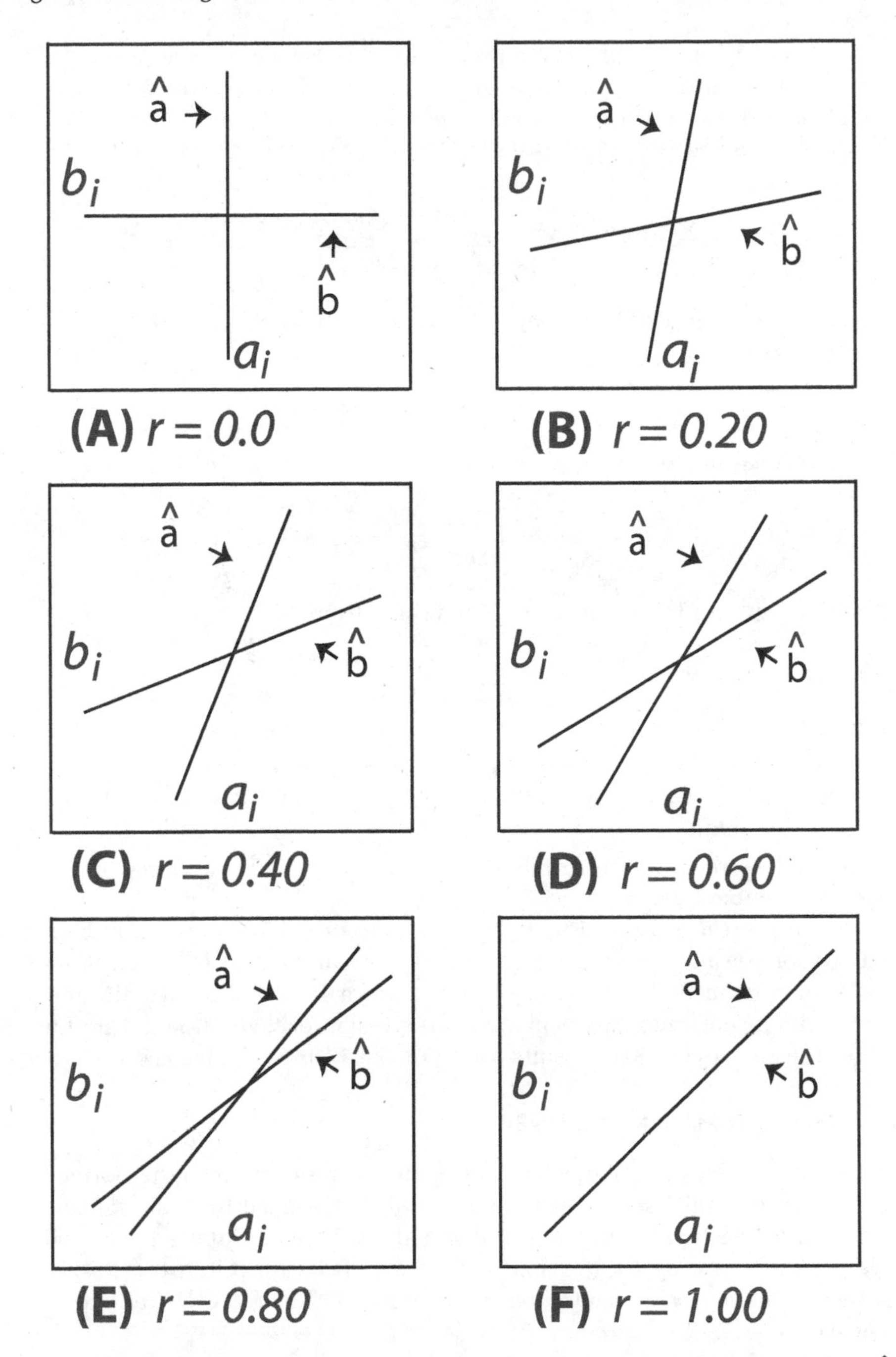


**Figure 6.12** Effects of reversing regressions of $b_i$ and $a_i$ for all $[a_i \in \hat{\mathbf{a}}]$ and $[b_i \in \hat{\mathbf{b}}]$ as a function of various positive values of the correlation coefficient $r$.

**B**-condition in **Figure 6.9**. This approach weights each datum according to the standard deviations in both its $b_i$ and $a_i$ variables. However, the minimized perpendicular distance is not invariant under a change of scale.

Using the *least squares cubic* [lsc] regression, **[55, 104]** estimate the slope by

$$m_{lsc} = \frac{\sum_{i=1}^{n} \Delta b_i^2 - \sum_{i=1}^{n} \Delta a_i^2 + \sqrt{[\sum_{i=1}^{n} \Delta b_i^2 - \sum_{i=1}^{n} \Delta a_i^2]^2 + 4[\sum_{i=1}^{n} \Delta a_i \Delta b_i]^2}}{2\sum_{i=1}^{n} \Delta a_i \Delta b_i}, \tag{6.57}$$

where $\Delta a_i = [a_i - \bar{a}]$ and $\Delta b_i = [b_i - \bar{b}]$. Additionally, the standard deviation of the slope is

$$s_{m(lsc)} = \frac{m_{lsc}}{r}\sqrt{\frac{1-r^2}{n}}, \tag{6.58}$$

with the correlation coefficient given by

$$r = \frac{\sum_{i=1}^{n} \Delta a_i \Delta b_i}{\sqrt{\sum_{i=1}^{n} \Delta a_i^2 \sum_{i=1}^{n} \Delta b_i^2}}. \tag{6.59}$$

The intercept, on the other hand, is obtained from

$$c_{lsc} = \bar{b} - m_{lsc}\bar{a}, \tag{6.60}$$

with the standard deviation

$$s_{c(lsc)} = \sqrt{\frac{1}{n}[s_{\hat{\mathbf{b}}} - s_{\hat{\mathbf{a}}} \cdot m_{lsc}]^2 + [1-r]m_{lsc}(2s_{\hat{\mathbf{a}}}s_{\hat{\mathbf{b}}} + \frac{\bar{a} \cdot m_{lsc}[1+r]}{r^2}}). \tag{6.61}$$

Here, $s_{\hat{\mathbf{a}}}$ and $s_{\hat{\mathbf{b}}}$ refer to the standard deviations for the coordinates on the least squares cubic line.

As ***Example 06.5.3a***, consider the least squares cubic regression in **Figure 6.13** for seismic pressure wave velocity measurements $[P_n]$ against reduced heat flow measurements $[q]$ obtained down bore holes with variable uncertainties in both measurements. This regression excluded the datum for SN, and thus is based on the results for 12 of the 13 listed provinces.

### 6.5.4 REDUCED MAJOR AXIS LINE

When both variables are subject to error, the regression can be performed on the equivalent unitless versions of the variables obtained by standardizing them - i.e., for each value of the variable, subtract the variable's mean and divide the difference by the variable's standard deviation [**Chapter 2.4.1**]. This *reduced major axis* [rma] regression minimizes the area $CDE$ as shown in **Figure 6.14**

A simple way to perform this minimization is to form the slope by

$$m_{rma} = (\frac{r}{|r|})\frac{s_{\mathbf{b}}}{s_{\mathbf{a}}}, \tag{6.62}$$

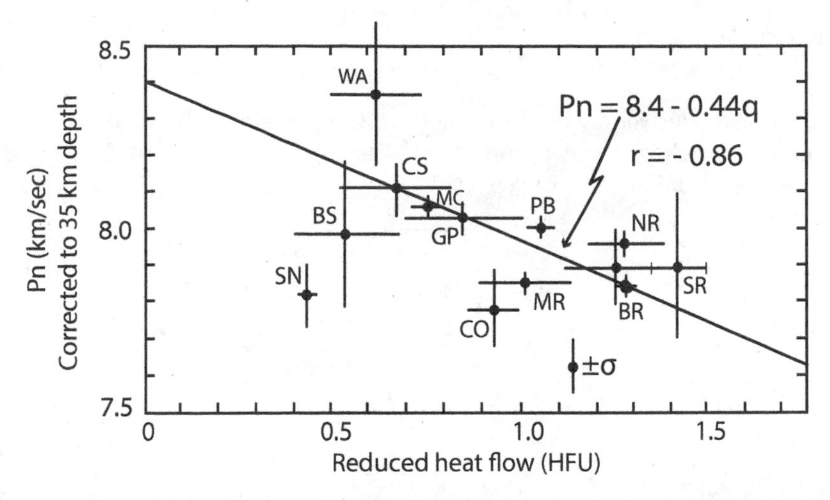


**Figure 6.13** Least squares cubic regression line [[**104, 105**]] for seismic $P$-wave velocities [$P_n$] and reduced heat flow [$q$] weighted for variable uncertainties in both sets of bore hole measurements. The 13 physiographic province codes include BR [Basin and Range], BS [Baltic Shield], CA [Columbia Plateau], CO [Colorado Plateau], CS [Canadian Shield], GP [Great Plains], MC [Mid-Continent], MR [Middle Rocky Mountains], NR [Northern Rocky Mountains], PB [Pacific Border], SN [Sierra Nevada], SR [Southern Rocky Mountains], and WA [Western Australia]. [Adapted from [**7**].]

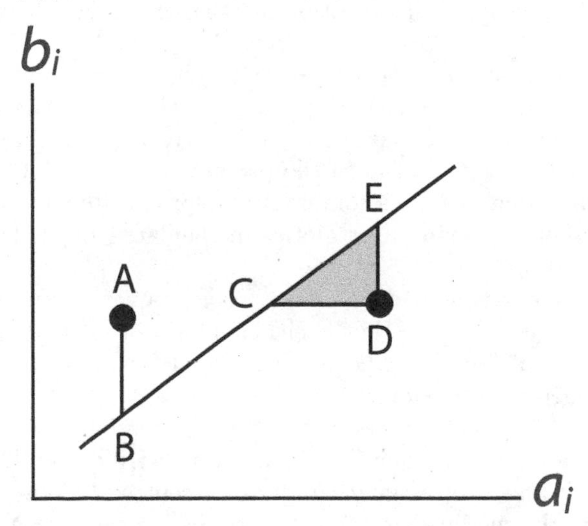


**Figure 6.14** Comparison of the classical regression that minimizes the squared distance [A, B] with the reduced major axis line that minimizes the shaded area of the [C, D, E]-triangle.

and the intercept from

$$c_{rma} = \bar{b} - m_{rma}\bar{a}. \tag{6.63}$$

These results are derived by a similar method to that which obtained the classical regression coefficients [e.g., [55, 104]]. However, the standard deviations for the intercept and slope are respectively

$$s_{c(rma)} = s_{\mathbf{b}}\sqrt{(\frac{1-r^2}{n})(1+\frac{\bar{a}^2}{s_{\mathbf{a}}^2})} \tag{6.64}$$

and

$$s_{m(rma)} = m_{rma}\sqrt{\frac{1-r^2}{n}}. \tag{6.65}$$

## 6.6 KEY CONCEPTS

1. Least squares linear regression estimates intercept and slope parameters for a straight line with minimum summed squared deviations of the data from the straight line.
2. The least squares intercept and slope estimates come from solving the set of normal equations derived by differentiating a residual function involving the sum of squared differences between the data and predictions of the straight-line model. The normal equations are set to zero to obtain intercept and slope estimates that minimize the residual function.
3. Desired goodness-of-fit measures on the regression's performance include minimizing intercept, slope and prediction variances and confidence intervals, and maximizing the correlation coefficient and ANOVA significance test of fit of the predictions to the data.
4. Intercept and slope variances commonly are propagated assuming either uniform or Gaussian uncertainties in the data's dependent variable magnitudes.
5. For uniformly distributed data errors, the parameter variances are based on the data variance that is either known or estimated by the unbiased residual function where the residual function was divided by $[n-2]$-degrees of freedom.
6. For normally distributed data errors, each observation is assumed to be contaminated by a unique Gaussian error distribution. Thus, the maximum likelihood of estimating the intercept and slope parameters requires the modified residual function with each term divided by the variance of the corresponding data point's error distribution. Using the maximum likelihood residual function, maximum likelihood estimates of the intercept and slope and their variances are obtained.
7. Statistical error bars or confidence intervals on the intercept and slope are basically estimated from weighted standard deviations of

the parameters multiplied by the $t$-distribution value $t_{(\alpha/2;\, n-2)}$ at the desired $\alpha$-significance level for $[n - 2]$ degrees of freedom.

8. Confidence intervals on the predictions invoke a comparable product with the desired $t$-value multiplied by the standard deviation of the prediction differences with the data.
9. The correlation coefficient is the square root of the coherency coefficient that ranges between zero and unity and gives the percent of observations fit by the straight-line model. Thus, the coherency coefficient lacks the slope sensitivity that the correlation coefficient possesses, which trends positively to $+1$ as more of the predictions directly follow the observations, and negatively to $-1$ as more of the predictions inversely follow the data.
10. The statistical significance of the correlation coefficient is directly related to the number of data and can be tested using the normal distribution.
11. The linear regression's statistical significance can be tested by the ratio of the mean squared deviations of the predictions about the data mean to the mean squared deviations of the predictions about the data in the ANOVA table. As long as the ratio is less than the critical $f$-distribution value $f_{(\alpha;\, 1;\, n-2)}$ at the desired $\alpha$-significance level and degrees of freedom 1 and $[n - 2]$, there is no reason based on the $F$-test to reject the hypothesis that the data are not fit by the estimated straight line.
12. The common classical linear regression minimizes the sums of squared deviations in the dependent variables of the predictions and observations assuming constant error in the observations.
13. Less common additional error minimizations in earth and environmental applications include using 1) the least squares line weighted by the inverse data variances to minimize the variable data deviations for observations with variable uncertainties. For problems involving variable uncertainties in both dependent and independent variables, 2) reversing the variables in the regression helps to map out possible solution space for an appropriately fitting line, 3) the least squares cubic line minimizes deviations perpendicular to the line, and 4) the reduced major axis line minimizes the areas of right triangles between the line and the data point apexes.

# 7 Matrix Linear Regression

## 7.1 OVERVIEW

*The linear regression problem describes a system of simultaneous equations that is linear in the unknown intercept and slope parameters. This system, in turn, can be elegantly written in matrix notation as* $\mathbf{Ax} = \mathbf{b}$*, where* $\mathbf{b}$ *is the* $n$*-rows by* $1$*-column observation vector,* $\mathbf{x}$ *is the* $2$*-rows by* $1$*-column vector of the unknown intercept and slope parameters, and* $\mathbf{A}$ *is the design matrix with the first column holding* $n$*-ones and the second column containing the* $n$*-independent variable values. The product* $\mathbf{Ax}$ *defines the forward model of the system.*

*This system has the general least squares solution* $\mathbf{x} = [\mathbf{A^tA}]^{-1}\mathbf{A^tb}$ *with variance* $\sigma^2_{\mathbf{x}} = [\mathbf{A^tA}]^{-1}\sigma^2_{\mathbf{b}}$ *as long as* $|\mathbf{A^tA}| \neq 0$*. Here,* $\sigma^2_{\mathbf{b}}$ *is the variance of the observations that, if not known, may be unbiasedly estimated from* $\sigma^2_{\mathbf{b}} = (\mathbf{b^tb} - \mathbf{x^t}[\mathbf{A^tA}]\mathbf{x})/(n-2)\ \forall\ n \geq 3$.

*Rather than computing the inverse* $[\mathbf{A^tA}]^{-1}$ *directly, the solution in practice is more efficiently calculated using elementary row operations to partly or fully diagonalize the system* $[\mathbf{A^tA}]\mathbf{x} = \mathbf{A^tb}$ *in place by the Gaussian or Gauss-Jordan elimination methods, respectively.*

*Statistically probable error bars or confidence limits on the intercept and slope estimates are given by the product of the respective parameter's standard deviation from* $\sigma^2_{\mathbf{x}}$ *times the t-distribution value* $t_{(\alpha/2;\ n-2)}$ *at the desired* $\alpha$*-significance level for* $[n-2]$ *degrees of freedom.*

*The confidence interval on the* $i$*th prediction involves the product of the desired t-value and the standard error or deviation of the* $i$*th prediction given by* $(s_{\mathbf{b}} \cdot \sqrt{\mathbf{a_i^t}[\mathbf{A^tA}]^{-1}\mathbf{a_i}})$*, where* $s_{\mathbf{b}} \approx \sigma_{\mathbf{b}}$ *and* $\mathbf{a_i^t}$ *is the* $i$*th row vector of* $\mathbf{A}$.

*The coherency and its square root, the correlation coefficient, are predominantly functions of the ratio of* $\mathbf{x^tA^tb}$ *to* $\mathbf{b^tb}$*. The first matrix product also defines the mean squares regression error to within a scalar, and the difference in the two matrix products defines the mean squares residual deviation in the ANOVA table. Thus, the ratio of these mean squares matrix products is the basis of the* $F$*-test on the significance of the linear regression.*

*The classical or ordinary linear regression solution* $\mathbf{x} = [\mathbf{A^tA}]^{-1}\mathbf{A^tb}$ *minimizes the sum of the squared dependent variable magnitude differences in the predictions* $\mathbf{Ax}$ *and observations* $\mathbf{b}$*. For linear regressions involving unequal errors in the dependent observations, the weighted least squares line is applicable with* $\mathbf{x_{wls}} = [\mathbf{A^tWA}]^{-1}\mathbf{A^tWb}$*. Here,* $\mathbf{W}$ *is the* $[n \times n]$ *diagonal weight matrix holding the inverse variances of the observations.*

*Linear regression strategies for problems involving data with errors in both the dependent and independent variables of the observations include 1) reversing the variables in the regression to investigate the range of acceptable line fits to the data, and estimating 2) the least squares cubic line minimizing the total data deviations perpendicular to the line, or 3) the reduced major axis line that minimizes the total right triangle areas between the data and the line.*

*The three matrix equations of ordinary linear regression describe the general problem* $\mathbf{Ax} = \mathbf{b}$, *and its least squares solution* $\mathbf{x} = [\mathbf{A}^t\mathbf{A}]^{-1}\mathbf{A}^t\mathbf{b}$, *with solution and prediction errors from* $\sigma^2_{\mathbf{x}} = [\mathbf{A}^t\mathbf{A}]^{-1}\sigma^2_{\mathbf{b}}$. *Although developed here specifically for classical least squares linear regression in n observations with* $n \geq [m = 2$ *unknowns], this matrix formulation more generally describes any least squares linear system with* $n \geq [m \geq 2$ *unknowns].*

## 7.2 INTRODUCTION

The complex looking algebra of least squares is greatly simplified in the algebra of the *matrix* or *array*, which is the fundamental format for analysis in digital computing [**Chapter 2.3**]. For the linear regression problem in particular, the $n$ observations may be expressed as a system of *simultaneous linear equations* [SLE] given by

$$\begin{aligned}
b_1 = &\quad c + m \cdot a_{12} = x_{11} \cdot a_{11} + x_{21} \cdot a_{12} \equiv b_{11} \\
b_2 = &\quad c + m \cdot a_{22} = x_{11} \cdot a_{21} + x_{21} \cdot a_{22} \equiv b_{21} \\
\vdots &\qquad \ddots \qquad\qquad \ddots \qquad\qquad \vdots \\
b_n = &\quad c + m \cdot a_{n2} = x_{11} \cdot a_{n1} + x_{21} \cdot a_{n2} \equiv b_{n1},
\end{aligned} \tag{7.1}$$

which is linear in the 2-unknown variables [i.e., $c = x_{11}$ and $m = x_{21} \in \mathbf{x}$] and the $[n \times 2]$-known variables $[a_{ij} \in \mathbf{A} \ \forall \ i = 1, 2, \ldots, n$, and $j = 1, 2]$ of the straight line forward model [$\mathbf{Ax}$] that the investigator assumes will account for the observations.

Note that the column index is suppressed in the left hand side of the above SLE, whereas the right hand side expresses the SLE with full indices. The abbreviated notation is used where convenient in this book to simplify expressions, although both conventions are equivalent.

In matrix notation, the above linear equation system becomes

$$\mathbf{Ax} = \mathbf{b}, \tag{7.2}$$

where $\mathbf{b}$ is the $n$-rows by 1-column matix or vector containing the $n$ observations, and $\mathbf{x}$ is the 2-rows by 1-column vector of the 2 unknowns. The design matrix $\mathbf{A}$ is the $n$-rows by 2-columns matrix containing the $2n$ known coefficients of the assumed forward model $\mathbf{Ax}$ - i.e., $a_{i1} = 1$, and $a_{i2}$ is the value of the $i$th independent variable [**Figure 7.1**].

The use of matrices and the elementary matrix properties in **Chapter 2.3.1** allows for particularly elegant and efficient descriptions of the linear regression problem. For example, the straight line relationship [e.g., **Equation 6.3**] between the observations $[b_{i1} \equiv b_i \in \mathbf{b}]$ and the independent variables

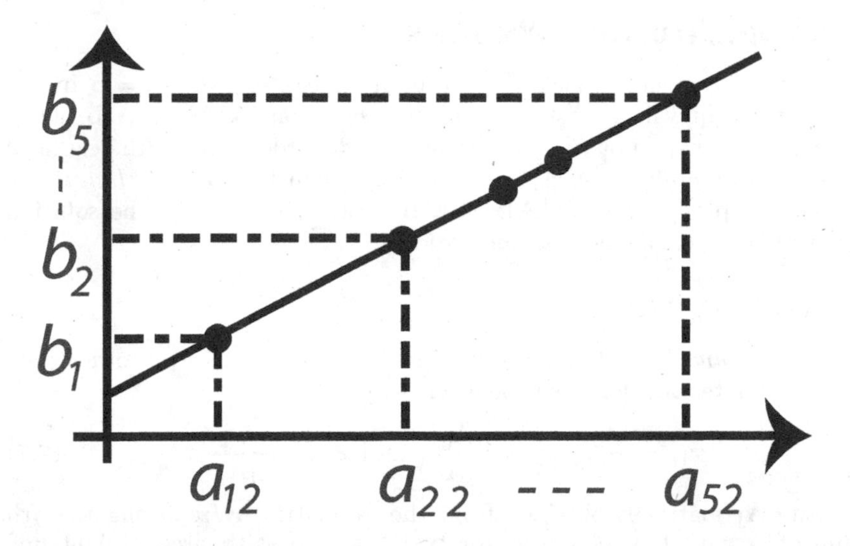


**Figure 7.1** Array parameters for the regression line of 5 data points.

$[a_{i2} \equiv a_i \in \mathbf{A}]$ in terms of the unknown intercept and slope variables $[c = x_{11} \equiv x_1$ and $m = x_{21} \equiv x_2 \in \mathbf{x}]$ is just $\mathbf{b} = \mathbf{A}\mathbf{x}$.

Accordingly, the residual function in **Equation 6.3** in matrix notation becomes simply

$$R = [\mathbf{b} - \mathbf{A}\mathbf{x}]^t[\mathbf{b} - \mathbf{A}\mathbf{x}], \tag{7.3}$$

so that by the definition of matrix multiplication [**Chapter 2.3.1**], the normal equations in **Equation 6.7** are

$$\begin{pmatrix} n & \sum_{i=1}^{n} a_{i2} \\ \sum_{i=1}^{n} a_{i2} & \sum_{i=1}^{n} a_{i2}^2 \end{pmatrix} \begin{pmatrix} x_{11} \\ x_{21} \end{pmatrix} = \begin{pmatrix} \sum_{i=1}^{n} b_{i1} \\ \sum_{i=1}^{n} a_{i2}b_{i1} \end{pmatrix}, \tag{7.4}$$

which, in turn, reduce simply to

$$[\mathbf{A}^t\mathbf{A}]\mathbf{x} = \mathbf{A}^t\mathbf{b}. \tag{7.5}$$

In matrix notation, then, the linear regression problem has the least squares solution given by

$$\mathbf{x} = [\mathbf{A}^t\mathbf{A}]^{-1}\mathbf{A}^t\mathbf{b}. \tag{7.6}$$

This elegant solution is not limited just to solving the straight line model for least squares estimates of the two unknown intercept and slope coefficients. Indeed, the solution is quite general as described in **Chapter 9**, and capable of solving in the least squares sense any linear system of $n$-observations in $m$-unknowns, where $n \geq m$ and $n$ and $m$ are positive integers.

## 7.3 SOLVING FOR THE UNKNOWNS

Several approaches are available to solve the linear system $\mathbf{Ax} = \mathbf{b}$ in the context of its equivalent least squares representation $[\mathbf{A^tA}]\mathbf{x} = \mathbf{A^tb}$. They include the application of *Cramer's Rule* for even order sytems, the classical *inverse determination* of $[\mathbf{A^tA}]^{-1}$ to multiply against $\mathbf{A^tb}$, and *elimination methods* that process the $[(\mathbf{A^tA})\mathbf{x} = \mathbf{A^tb}]$-system in place for the solution coefficients using elementary row operations.

### 7.3.1 CRAMER'S RULE

In general, *Cramer's Rule* for the system $\mathbf{b} = \mathbf{Ax}$ that is even order [i.e., $n = m$], estimates the unknown coefficients by

$$x_{11} = \frac{|\mathbf{A_1}|}{|\mathbf{A}|},\ x_{21} = \frac{|\mathbf{A_2}|}{|\mathbf{A}|},\ \text{and } x_{m1} = \frac{|\mathbf{A_m}|}{|\mathbf{A}|}, \tag{7.7}$$

where the $\mathbf{A_j}$ matrix is obtained from the $\mathbf{A}$ matrix by replacing the $j$th column of $\mathbf{A}$ with the column vector $\mathbf{b}$. For the least squares straight line adaption of **Equation 7.5** with $m = 2$ unknowns, in particular, the Cramer's Rule estimate of the intercept $[c \equiv x_{11}]$ is

$$x_{11} = \frac{1}{\Delta}\begin{vmatrix} \sum_{i=1}^n b_{i1} & \sum_{i=1}^n a_{i2} \\ \sum_{i=1}^n a_{i2}b_{i1} & \sum_{i=1}^n a_{i2}^2 \end{vmatrix} = \frac{[\sum_{i=1}^n a_{i2}^2 \sum_{i=1}^n b_{i1} - \sum_{i=1}^n a_{i2} \sum_{i=1}^n a_{i2}b_{i1}]}{\Delta}, \tag{7.8}$$

and its estimate of the slope $[m \equiv x_{21}]$ is

$$x_{21} = \frac{1}{\Delta}\begin{vmatrix} n & \sum_{i=1}^n b_{i1} \\ \sum_{i=1}^n a_{i2} & \sum_{i=1}^n a_{i2}b_{i1} \end{vmatrix} = \frac{[n\sum_{i=1}^n a_{i2}b_{i1} - \sum_{i=1}^n a_{i2} \sum_{i=1}^n b_{i1}]}{\Delta}, \tag{7.9}$$

where

$$\Delta = \begin{vmatrix} n & \sum_{i=1}^n a_{i2} \\ \sum_{i=1}^n a_{i2} & \sum_{i=1}^n a_{i2}^2 \end{vmatrix} = n\sum_{i=1}^{n} a_{i2}^2 - [\sum_{i=1}^{n} a_{i2}]^2. \tag{7.10}$$

As ***Example 07.3.1a***, consider the $\mathbf{b} = \mathbf{Ax}$ system given by

$$\begin{aligned} 12.220 &= x_{11} + 4.440 \cdot x_{21} \\ 47.737 &= x_{11} + 75.474 \cdot x_{21}, \end{aligned} \tag{7.11}$$

where

$$\mathbf{b} = \begin{pmatrix} 12.220 \\ 47.737 \end{pmatrix},\ \mathbf{A} = \begin{pmatrix} 1.000 & 4.440 \\ 1.000 & 75.474 \end{pmatrix},\ \text{and } \mathbf{x} = \begin{pmatrix} x_{11} \\ x_{21} \end{pmatrix}. \tag{7.12}$$

Thus, the intercept by Cramer's Rule is

$$x_{11} = \begin{vmatrix} 12.220 & 4.440 \\ 47.737 & 75.474 \end{vmatrix} /|\mathbf{A}| = \frac{710.340}{71.034} = 10.000 = c, \tag{7.13}$$

and the corresponding slope is

$$x_{21} = \begin{vmatrix} 1.000 & 12.220 \\ 1.000 & 47.737 \end{vmatrix} /|\mathbf{A}| = \frac{35.517}{71.034} = 0.500 = m. \tag{7.14}$$

### 7.3.2 INVERSE DETERMINATION

Another classical approach to solving the even order system $\mathbf{Ax} = \mathbf{b}$ is to compute the inverse $\mathbf{A}^{-1}$ directly so that the solution is obtained from $\mathbf{x} = \mathbf{A}^{-1}\mathbf{b}$. Where the number of observations $[n = 2]$ equals the number of unknowns $[m = 2]$, and $|\mathbf{A}| \neq 0$, the inverse $\mathbf{A}^{-1}$ can be readily determined by elementary row operations [e.g., **Equation 2.9**], or equivalently by augmented matrix determinants [e.g., **Equation 2.14**].

As ***Example 07.3.2a***, the inverse of the design matrix in the previous ***Example 07.3.1a*** is

$$\mathbf{A}^{-1} = \begin{pmatrix} 1.063 & -0.063 \\ -0.014 & 0.014 \end{pmatrix} \ni \mathbf{A}^{-1}\mathbf{b} = \mathbf{x} = \begin{pmatrix} 10.000 = c \\ 0.500 = m \end{pmatrix}. \quad (7.15)$$

Where $n > m = 2$, the inverse $\mathbf{A}^{-1}$ [and determinant $|\mathbf{A}|$] cannot be determined because $\mathbf{A}$ is not a square matrix. In these cases, however, the inverse $[\mathbf{A}^t\mathbf{A}]^{-1}$ for the alternate least squares solution $\mathbf{x} = [\mathbf{A}^t\mathbf{A}]^{-1}\mathbf{A}^t\mathbf{b}$ can be established as long as $|\mathbf{A}^t\mathbf{A}| \neq 0$.

***Example 07.3.2b*** shows the equivalence of the solutions obtained using the direct inverse $\mathbf{A}^{-1}$ in **Equation 7.15** and the *natural* inverse $[\mathbf{A}^t\mathbf{A}]^{-1}\mathbf{A}^t$. Thus, from the $\mathbf{A}$ matrix in **Equation 7.12**,

$$\mathbf{A}^t\mathbf{A} = \begin{pmatrix} 0.002 & 0.080 \\ 0.080 & 5.716 \end{pmatrix} \times 10^3 \ni |\mathbf{A}^t\mathbf{A}| = 5.046 \times 10^3 \neq 0. \quad (7.16)$$

The inverse of $\mathbf{A}^t\mathbf{A}$ from elementary row operations [or augmented matrix determinants] and the weighted observation matrix are, respectively,

$$[\mathbf{A}^t\mathbf{A}]^{-1} = \begin{pmatrix} 1.133 & -0.016 \\ -0.016 & 0.000 \end{pmatrix}, \text{ and } \mathbf{A}^t\mathbf{b} = \begin{pmatrix} 0.060 \\ 3.657 \end{pmatrix}. \quad (7.17)$$

The least squares solution accordingly is

$$\mathbf{x} = [\mathbf{A}^t\mathbf{A}]^{-1}\mathbf{A}^t\mathbf{b} = \begin{pmatrix} 10.000 = c \\ 0.500 = m \end{pmatrix}, \quad (7.18)$$

which is the same as the solution in **Equation 7.15** obtained directly from the inverse $\mathbf{A}^{-1}$.

### 7.3.3 ELIMINATION METHODS

In general, it is numerically inefficient to solve the linear systems by directly computing an inverse and then multiplying it by the column vector of observations. A more efficient approach is to solve the system using *Gaussian elimination* or the *sweep-out method* where the system is diagonalized by elementary row operations [e.g., **Equation 2.9**].

As ***Example 07.3.3a***, the system in **Equation 7.11** can be solved by multiplying the top row by (75.474/4.440) and subtracting the modified top row from the bottom row. Thus, these elementary row operations reduce the system as

$$\left[\begin{array}{lcl} 12.220 & = & 4.440x_{11} + x_{21} \\ 47.737 & = & 75.474x_{11} + x_{21} \end{array}\right] \sim \left[\begin{array}{lcl} 12.220 & = & 4.440x_{11} + x_{21} \\ -160.000 & = & 0.000 - 16.000x_{21} \end{array}\right],$$

from which the solution coefficients clearly are again

$$x_{21} = \frac{-160.000}{-16.000} = 10.000 \text{ and } x_{11} = \frac{(12.220 - 10.000)}{4.440} = 0.500, \tag{7.19}$$

where $x_{11}$ was obtained by back substituting the solution for $x_{21}$. The salient feature of Gaussian elimination is that the system is reduced by elementary row operations to a triangular system that can be essentially solved in place.

Instead of back substitution, the row operations can be continued to also sweep out the upper triangular part of the system. This procedure called *Gauss-Jordan elimination* obtains the solution coefficients explicitly along the main diagonal as

$$\left[\begin{array}{lcl} 12.220 & = & 4.440x_{11} + x_{21} \\ -160.000 & = & 0.000 - 16.000x_{21} \end{array}\right] \sim \left[\begin{array}{lcl} 2.220 & = & 4.440x_{11} + 0.000 \\ -160.000 & = & 0.000 - 16.000x_{21} \end{array}\right] \sim$$

$$\left[\begin{array}{lcl} 0.500 & = & x_{11} + 0.000 \\ 10.000 & = & 0.000 + x_{21} \end{array}\right].$$

These elimination methods solve the **system 7.5** in place without the need for additional computations and manipulations of its determinant or inverse. Thus, they are computationally more efficient than Cramer's Rule in solving the classical linear regression problem, even though it involves only a $2 \times 2$ equation system no matter how large $n$ gets. However, as **Chapter 9** shows, the significance of this computing advantage increases substantially for the more general least squares problem where $n \geq m \geq 3$.

## 7.4 SENSITIVITY ANALYSIS

The various measures of fit considered in **Chapter 6.4** may also be expressed with matrices to describe the performance characteristics of linear regression in the array format.

### 7.4.1 INTERCEPT AND SLOPE VARIANCES

Because the solution coefficients are related to the linear equation system in **Equation 7.1**, the variance propagation rule [e.g., **Equation 4.13**] can be applied to the system to obtain the solution's variance [e.g., **[6]**] given by

$$s^2_{\mathbf{x}} = [\mathbf{A}^t\mathbf{A}]^{-1}s^2_{\mathbf{b}} = \begin{pmatrix} s^2_c & s^2_{cm} \\ s^2_{mc} & s^2_m \end{pmatrix}, \tag{7.20}$$

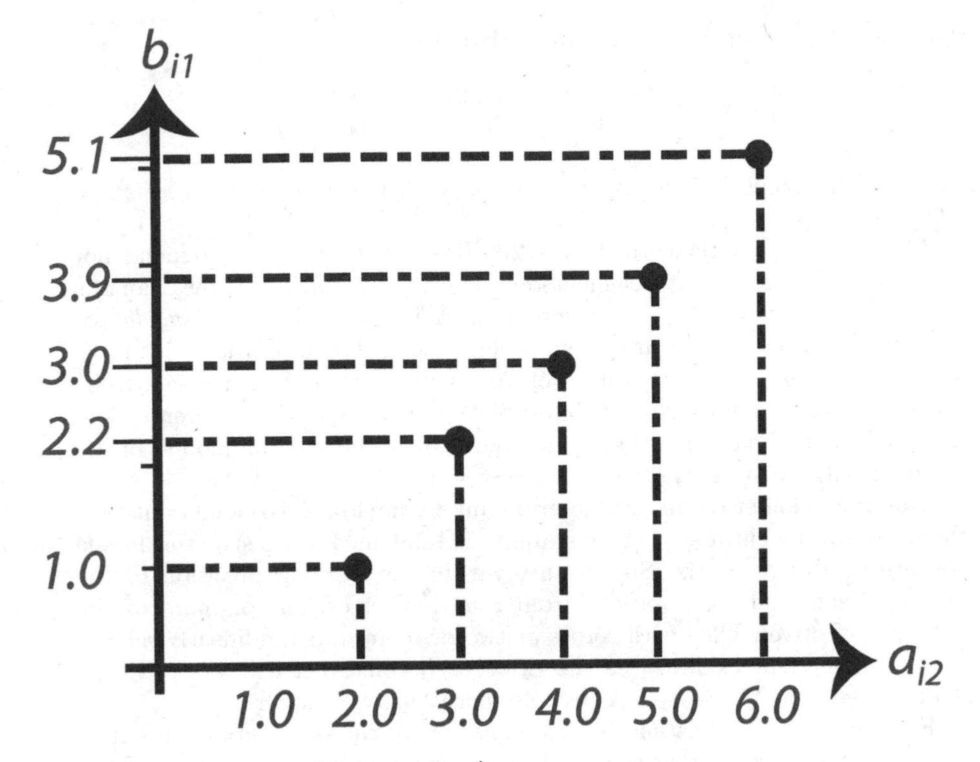


**Figure 7.2** Data for the regression line $\hat{b}_{i1} = -0.92 + 0.99a_{i2}$.

where the diagonal and off-diagonal elements of the matrix are the respective variances and co-variances of the **x**-solution coefficients for the intercept $[c \equiv x_{11}]$ and slope $[m \equiv x_{21}]$. The variance $s_{\mathbf{b}}^2$ of the observations, if not known, can be obtained from the unbiased estimate

$$s_{\mathbf{b}}^2 \approx \frac{\mathbf{b}^t\mathbf{b} - \mathbf{x}^t[\mathbf{A}^t\mathbf{A}]\mathbf{x}}{n-2}, \tag{7.21}$$

which is the matrix equivalent of **Equation 6.13**.

As ***Example 07.4.1a***, consider the data in **Figure 7.2** with

$$\mathbf{b} = \begin{pmatrix} 1.0 \\ 2.2 \\ 3.0 \\ 3.9 \\ 5.1 \end{pmatrix}, \ \mathbf{A} = \begin{pmatrix} 1.0 & 2.0 \\ 1.0 & 3.0 \\ 1.0 & 4.0 \\ 1.0 & 5.0 \\ 1.0 & 6.0 \end{pmatrix}, \ \mathbf{x} = \begin{pmatrix} x_{11} = -0.92 = c \\ x_{21} = 0.99 = m \end{pmatrix}, \tag{7.22}$$

whoro

$$\mathbf{A}^t\mathbf{b} = \begin{pmatrix} 15.2 \\ 70.7 \end{pmatrix}, \ \mathbf{A}^t\mathbf{A} = \begin{pmatrix} 5.0 & 20.0 \\ 20.0 & 90.0 \end{pmatrix}, \ \ni \ [\mathbf{A}^t\mathbf{A}]^{-1} = \begin{pmatrix} 1.8 & -0.4 \\ -0.4 & 0.1 \end{pmatrix}. \tag{7.23}$$

Thus, for this example, $s_{\mathbf{b}}^2 = 0.0170$ so that

$$s_{\mathbf{x}}^2 = [\mathbf{A}^t\mathbf{A}]^{-1}s_{\mathbf{b}}^2 = \begin{pmatrix} 0.0306 & -0.0068 \\ -0.0068 & 0.0017 \end{pmatrix}. \tag{7.24}$$

Here, the intercept variance is $s_c^2 = 0.0306$ with the slope variance of $s_m^2 = 0.0017$.

The variance/covariance matrix **7.20** shows that $s_c^2$ and $s_m^2$ depend not only on $s_{\mathbf{b}}^2$, but also on the elements of $[\mathbf{A}^t\mathbf{A}]^{-1}$. In particular, by Cramer's Rule, $s_{\mathbf{x}}^2 \to \infty$ as $|\mathbf{A}^t\mathbf{A}| \to 0$, whereupon $\mathbf{A}$ is said to be *near-singular* or *ill conditioned* and the linear system is characterized by a solution with large and erratic values [i.e., large variance]. Accordingly, the solution is said to be *unstable* in that its uses [e.g., for interpolation, extrapolation, differentiation, integration, etc.] may be meaningless, even though the solution models or fits the data with great accuracy.

The discussion of the matrix determinant **Equation 2.10** points out that the source of the linear system's instability is fundamentally a structural issue concerning the **A**-matrix. Specifically, within the working precision of the measurements and computations from round-off error, one or more of the rows are effectively filled with zeroes or two or more rows are linearly related to each other. For example, as the slope $m$ becomes vertical [e.g., **Figure 6.11**] $s_{\mathbf{x}}^2 \to \infty$ as the rows of **A** become increasingly co-linear.

**Figure 7.3** illustrates another example where the observations are clustered so close to each other that within working precision, the computations

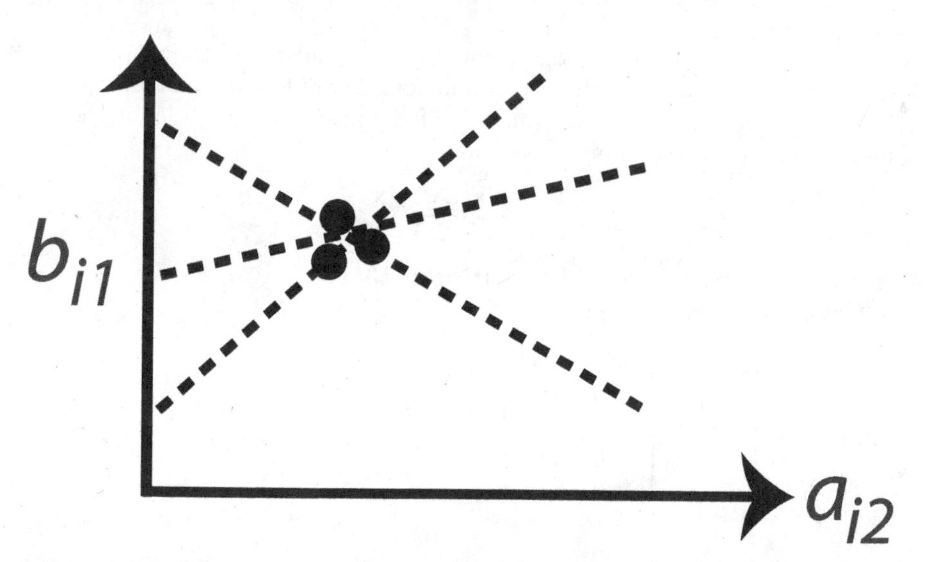


**Figure 7.3** Data cluster with independent variable coordinates that are undifferentiated within working precision, and thus destablizing the use of the straight line for modeling the data - i.e., an effectively infinite number of lines fit the data.

cannot differentiate the differences in the independent variable values from zero. Hence, $|\mathbf{A}^t\mathbf{A}| \rightarrow 0$ so that $s_{\mathbf{x}}^2 \rightarrow \infty$, and any line from the dependent vertical axis through the cluster effectively satisfies the data. This non-discriminating result eliminates the use of the straight line representation of these data for analysis.

The conditioning of the system is enhanced where the absolute magnitudes of the off-diagonal elements of $\mathbf{A}^t\mathbf{A}$ are small relative to those of the diagonal elements. Thus, numerical engineering techniques have been developed to optimize this structure for any $\mathbf{A}^t\mathbf{A}$-matrix. However, further consideration of these techniques is deferred to **Chapter 9** where higher order linear systems involving more than two unknowns are considered.

### 7.4.2 INTERCEPT AND SLOPE ERROR BARS

Following the results presented in **Chapter 6.4.2**, the $q\%$ confidence [or $(1-\alpha)\times 100\%$ significance] intervals or error bars on the intercept $c$ and slope $m$ are respectively given by

$$c \pm s_c \cdot t_{(\alpha/2;n-2)} \text{ and } m \pm s_m \cdot t_{(\alpha/2;n-2)} \tag{7.25}$$

for standard deviations $s_c$ and $s_m$, and the values of the $t$-distribution [e.g., **Figure 4.12**] at the $\alpha$ significance level with $\nu = n-2$ degrees of freedom.

### 7.4.3 PREDICTION VARIANCE AND ERROR BARS

Simplifying the index notation of the matrices [i.e., rewriting $b_{i1}$ more simply as $b_i$ or $b_{i1} \equiv b_i$, $a_{i2} \equiv a_i$, $x_{11} \equiv c$, and $x_{21} \equiv m$] yields linear regression predictions that can be written as $\hat{b}_i = c + m \cdot a_i$. Defining the $i$th row of $\mathbf{A}$ as the row vector $\mathbf{a}_\mathbf{i}^t \equiv [1\ a_i]$, the predictions in the simplified matrix notation become

$$\hat{b}_i = (1\ a_i)\begin{pmatrix} c \\ m \end{pmatrix} = \mathbf{a}_\mathbf{i}^t\mathbf{x} = \mathbf{x}^t\mathbf{a}_\mathbf{i}. \tag{7.26}$$

Applying the variance propagation rule of **Equation 4.14** to $\hat{b}_i$ yields

$$\begin{aligned} s_{\hat{b}_i}^2 &\approx s_c^2 + 2a_i \cdot s_{cm}^2 + a_i^2 \cdot s_m^2 \\ &\approx \begin{pmatrix} 1 & a_i \end{pmatrix}\begin{pmatrix} s_c^2 & s_{cm}^2 \\ s_{mc}^2 & s_m^2 \end{pmatrix}\begin{pmatrix} 1 \\ a_i \end{pmatrix} \\ &\approx \mathbf{a}_\mathbf{i}^t\left([\mathbf{A}^t\mathbf{A}]^{-1} \cdot s_\mathbf{b}^2\right)\mathbf{a}_\mathbf{i}, \end{aligned} \tag{7.27}$$

where an unbiased estimate of $s_\mathbf{b}^2$ again may be obtained from **Equation 7.21** if it is not known.

As ***Example 07.4.3a***, for the data in **Figure 7.2**, the variance of the prediction at the 4th observation point is

$$s_{\hat{b}_4}^2 \approx \begin{pmatrix} 1 & 5 \end{pmatrix}\begin{pmatrix} 0.0306 & -0.0068 \\ -0.0068 & 0.0017 \end{pmatrix}\begin{pmatrix} 1 \\ 5 \end{pmatrix} = 0.0051. \tag{7.28}$$

The $q\%$ confidence [$= (1-\alpha) \times 100\%$ significance] limits for the mean value of $b_i$ at $a_i$ [i.e., $\hat{b}_i$] are obtained from

$$\hat{b}_i \pm s_{\hat{b}_i} \cdot t_{(\alpha/2;n-2)} = \hat{b}_i \pm s_{\mathbf{b}} \sqrt{\mathbf{a_i}[\mathbf{A}^{\mathbf{t}}\mathbf{A}]^{-1}\mathbf{a}_{\mathbf{i}}^{\mathbf{t}}} \cdot t_{(\alpha/2;n-2)}. \tag{7.29}$$

### 7.4.4 CORRELATION COEFFICIENT

In matrix notation, the coherency coefficient [**Equation 6.36**] is given by

$$r^2 = \frac{\mathbf{x}^{\mathbf{t}}\mathbf{A}^{\mathbf{t}}\mathbf{b} - n\bar{b}^2}{\mathbf{b}^{\mathbf{t}}\mathbf{b} - n\bar{b}^2}, \tag{7.30}$$

so that the correlation coefficient is

$$r = \frac{m}{|m|}\sqrt{\frac{\mathbf{x}^{\mathbf{t}}\mathbf{A}^{\mathbf{t}}\mathbf{b} - n\bar{b}^2}{\mathbf{b}^{\mathbf{t}}\mathbf{b} - n\bar{b}^2}}. \tag{7.31}$$

As ***Example 07.4.4a***, the correlation coefficient for the data in **Figure 7.2** is $r = 0.9974$.

### 7.4.5 ANOVA SIGNIFICANCE

**Table 7.1** gives the analysis of variance table expressed in matrices to test the significance of the linear regression fit of the data observations. As before, if $F > F'$, then the hypothesis that no straight line fits the data as suggested by $\hat{\mathbf{b}}$ [i.e., $H_0 : m = 0$] is rejected for the critical value $F' = f_{(\alpha,1,n-2)}$ from the $f$ distribution [e.g., **Figure 4.14**] with significance level $\alpha$ and degrees of freedom $\nu = 1, [n-2]$. Otherwise, there is no reason to reject $H_0$ based on the $F$-test.

**Table 7.1**

**Matrix formatted ANOVA table for testing the significance of a linear regression model's fit to data.**

| ERROR SOURCE | $\nu$ | CORRECTED SUM OF SQUARES (CSS) | MEAN SQUARES | F-TEST |
|---|---|---|---|---|
| Regression Error | 1 | $\mathbf{x}^{\mathbf{t}}\mathbf{A}^{\mathbf{t}}\mathbf{b} - (\sum_{i=1}^{n} b_i)^2/n$ | $MS_R = CSS/\nu$ | $F =$ |
| Residual Deviation | $n-2$ | $\mathbf{b}^{\mathbf{t}}\mathbf{b} - \mathbf{x}^{\mathbf{t}}\mathbf{A}^{\mathbf{t}}\mathbf{b}$ | $MS_D = CSS/\nu$ | $MS_R/MS_D$ |
| Total Error | $n-1$ | $\mathbf{b}^{\mathbf{t}}\mathbf{b} - (\sum_{i=1}^{n} b_i)^2/n$ | $MS_T = CSS/\nu$ | |

## 7.5 ADDITIONAL LINEAR REGRESSION STRATEGIES

As pointed out in **Chapter** 6.5, several minimization criteria for linear regression are available for implementation depending on the affiliated data errors [**Figure 6.9**]. The matrix formulations of these regressions are particularly elegant and amenable to electronic computing compared with their equivalent algebraic formulations. The *classical* or *ordinary linear regression* [*olr*], for example, has the generalized matrix solution from **Equation 7.6** of the complex algebraic formulations in **Equations 6.8** and **6.9**. In general, the *olr* is applicable where all observations $b_i$ are measured with a constant uncertainty or variance $\sigma_{\mathbf{b}}^2$.

However, where some of the observations have smaller uncertainties than the rest, the *weighted linear regression* [*wlr*] provides the straight line estimate that is biased to the more reliable observations by

$$\mathbf{x}_{\mathbf{wlr}} = [\mathbf{A}^{\mathbf{t}}\mathbf{W}\mathbf{A}]^{-1}\mathbf{A}^{\mathbf{t}}\mathbf{W}\mathbf{b}, \tag{7.32}$$

where $\mathbf{W}$ is the $[n \times n]$ diagonal matrix with weights $w_{ii} = \sigma_{b_i}^{-2}$. This solution obtains the $k$-normalized maximum likelihood estimates of the slope and intercept in **Equations 6.51** and **6.53**, respectively, with the related variances given by **Equations 6.52** and **6.54**, and the confidence interval on the predictions from **Equation 6.55**.

When both $b_i$ and $a_i$ variables of the data contain errors, *reversing the variables* can map out possible solution space to help assess an effective line estimate [**Chapter** 6.5.2]. Further appropriate strategies may include estimating the *least squares cubic line* that minimizes the variance of the line-perpendicular distances to the data [**Chapter** 6.5.3], or the *reduced major axis line* that minimizes the total area of right triangles with hypotenuses on the line and apexes at the data points [**Chapter** 6.5.4]. Converting the algebraic formulations of these strategies in **Chapter** 6 into their equivalent matrix expressions are exercises left for the reader.

## 7.6 SUMMARY

The use of matrices greatly simplifies the description of linear regression in terms of three principal equations. In particular, every linear regression problem can be universally formulated as the matrix system $\mathbf{Ax} = \mathbf{b}$ [i.e., **Equation 7.2**], which relates the observations $b_i \in \mathbf{b}\ \forall\ i = 1, 2, \ldots, n$ to the *forward model* $\mathbf{Ax}$ involving known coefficients $a_{ij} \in \mathbf{A}\ \forall\ j = 1, 2$ and the unknown coefficients $x_j \in \mathbf{x}$ that include the intercept $c \equiv x_1$ and slope $m \equiv x_2$.

The second principal equation of the system is its universal least squares solution $\mathbf{x} = [\mathbf{A}^{\mathbf{t}}\mathbf{A}]^{-1}\mathbf{A}^{\mathbf{t}}\mathbf{b}$ [i.e., **Equation 7.6**] that exists $\forall\ |\mathbf{A}^{\mathbf{t}}\mathbf{A}| \neq 0$. Clearly, once the coefficients of $\mathbf{b}$ and $\mathbf{A}$ have been established, the unknown coefficients of $\mathbf{x}$ have also been established. The third principal equation of the system given by $s_{\mathbf{x}}^2 = [\mathbf{A}^{\mathbf{t}}\mathbf{A}]^{-1}s_{\mathbf{b}}^2$ [i.e., **Equation 7.20**] universally constrains the errors of the least squares solution $\mathbf{x}$ and the related $\hat{b}_i$-predictions [i.e., **Equation 7.29**].

The straight line is perhaps the most used forward model in representing environmental science and engineering data. Of course, regressions involving more than two unknowns [i.e., $m \geq 3$] are also possible as illustrated in the profile-, map-, and volume-dimensioned examples of **Figure 7.4**. The profile [i.e., one dimensional or $1D$] examples also include the equations for the rows of the $1D$-forward model $\mathbf{Ax}$ that are presumed to account for the observations $b_i$, along with the $[n \times m]$ dimensions of the applicable $\mathbf{A}$ matrix. Here,

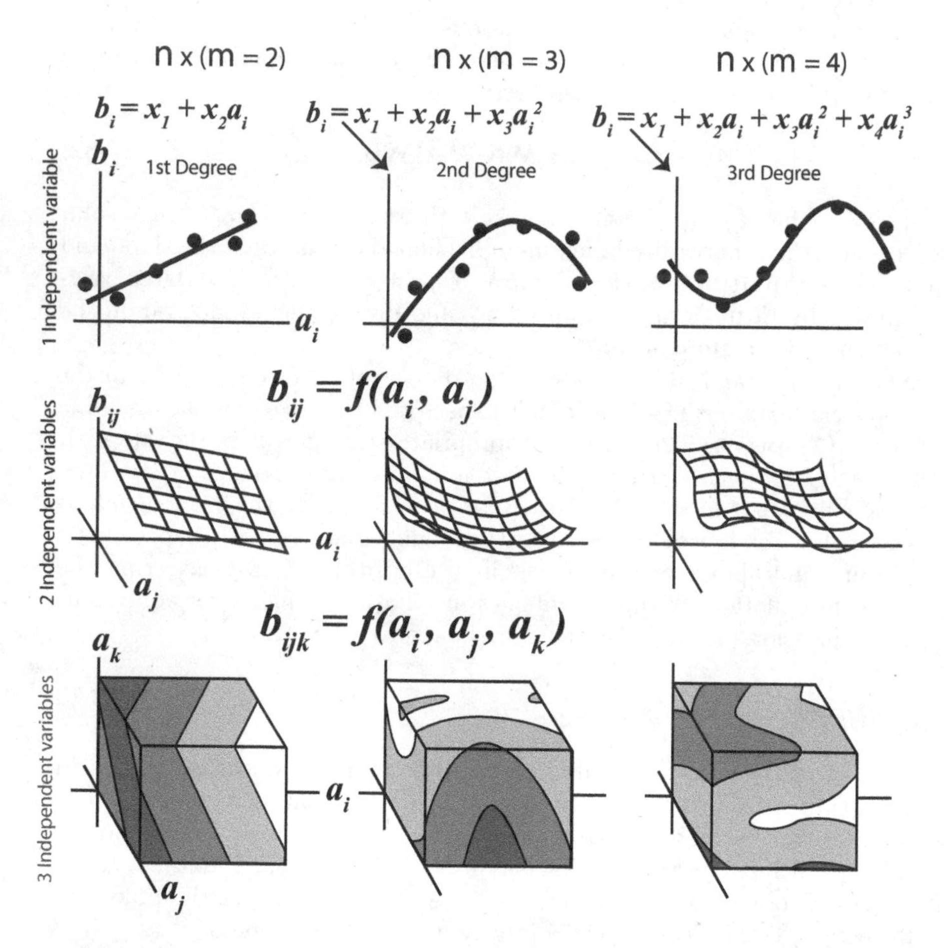


**Figure 7.4** Examples of regression problems for one [profile], two [map], and three [volume] independent variables for polynomial equations of the first, second, and third degree. The top row includes examples of the forward modeling equations along with the $[n \times m]$-dimensions of the related **A**-matrix for $n$-observations in $m$-unknowns. [Adapted from **[20]** after **[40]**.]

the presumed $1D$-polynomial's *degree* controls the number of unknown coefficients according to $m[1D] = degree + 1$. Note, however, that subtracting the means $\bar{b}_i$ and $\bar{a}_i$ from the respective $b_i$- and $a_i$-values makes $x_1 = 0$ and commensurately reduces the number of unknowns.

Clearly, the forward modeling equations become increasingly complex as the degree of variation in the observations increases and the number of independent variables grows. **Part II**, however, shows that the three principal matrix equations of linear regression can be extended to also accommodate these more complex regression problems involving larger numbers of unknowns.

Thus, the reader who is confortable with the matrix formulation of the linear regression problem involving $n$ observations in $m = 2$ unknowns is also prepared in principle to take on regression problems involving $m \geq 3$ unknowns. The reader's main issue in practice is to discern from observations with unavoidable measurement and data processing errors the applicable forward model. The top row of **Figure 7.4** shows some examples, where signals [solid lines] with determinable forward modeling parameters [x's] must be inferred from discrete point observations [dots] containing errors or noise.

Data errors also guarantee that the forward model chosen for the data cannot be unique, so that the selected model is only one of a set of models that can satisfy the data within their error limits. Thus, resolving which of the possible models is optimum for achiving the objectives of the data analysis is now a principal focus of modern practice. **Part II** considers these and other practical issues further for representing and analyzing data.

## 7.7 KEY CONCEPTS

1. The linear regression problem relates deterministic data to the unknown intercept and slope parameters of the straight-line equation by the system of simultaneous linear equations $\mathbf{Ax} = \mathbf{b}$, where the column vectors $\mathbf{b}_{(\mathbf{n}\times\mathbf{1})}$ and $\mathbf{x}_{(\mathbf{2}\times\mathbf{1})}$ hold respectively the $n$-observations and the 2-unkowns, and the first and second columns of the design matrix $\mathbf{A}_{(\mathbf{n}\times\mathbf{2})}$ hold respectively only $n$-ones and the $n$-independent variable values.
2. The least squares solution from $\mathbf{x} = [\mathbf{A}^t\mathbf{A}]^{-1}\mathbf{A}^t\mathbf{b}\ \forall\ |\mathbf{A}^t\mathbf{A}| \neq 0$ yields predictions $\hat{\mathbf{b}}$ $(= \mathbf{Ax})$ that match the observations $\mathbf{b}$ so that the sum of the squared differences $\sum(\hat{b}_i - b_i)^2$ is minimum.
3. Errors in the data characterized by the variance $\sigma^2_{\mathbf{b}}$ propagate into the least squares solution from $\sigma^2_{\mathbf{x}} = [\mathbf{A}^t\mathbf{A}]^{-1}\sigma^2_{\mathbf{b}}$, where $\sigma^2_{\mathbf{b}}$ is either known or unbiasedly estimated from $\sigma^2_{\mathbf{b}} = (\mathbf{b}^t\mathbf{b} - \mathbf{x}^t[\mathbf{A}^t\mathbf{A}]\mathbf{x})/(n - 2)\ \forall\ n \geq 3$.
4. Statistical confidence limits or error bars on the intercept $c$ and slope $m$ estimates are given by the $\sigma^2_{\mathbf{x}}$-derived values of $\sigma_c$ and $\sigma_m$ times $t_{(\alpha/2;\,n-2)}$ from the $t$-distribution at the desired $\alpha$-significance level for $[n - 2]$ degrees of freedom.

5. The confidence interval on the $i$th prediction is $(\pm s_{\mathbf{b}} \cdot \sqrt{\mathbf{a}_{\mathrm{i}}^{\mathrm{t}}[\mathbf{A}^{\mathrm{t}}\mathbf{A}]^{-1}\mathbf{a}_{\mathrm{i}}} \cdot t_{(\alpha/2;\, n-2)})$, where $s_{\mathbf{b}} \approx \sigma_{\mathbf{b}}$ and $\mathbf{a}_{\mathrm{i}}^{\mathrm{t}}$ is the $i$th row vector of $\mathbf{A}$.
6. Further regression performance parameters include the coherency and its square root, the correlation coefficient, as well as the mean squares regression error and mean squares residual deviation that constrain regression significance via the $F$-test in the ANOVA table. These parameters are all basically functions of the matrix products $\mathbf{x}^{\mathrm{t}}\mathbf{A}^{\mathrm{t}}\mathbf{b}$ and $\mathbf{b}^{\mathrm{t}}\mathbf{b}$.
7. Ordinary linear regression as described above is applicable where the dependent observations are observed with the same uncertainty. However, where the observations have unequal errors, the line biased to the minimal error observations may be obtained using weighted least squares regression with the solution $\mathbf{x}_{\mathbf{wlr}} = [\mathbf{A}^{\mathrm{t}}\mathbf{W}\mathbf{A}]^{-1}\mathbf{A}^{\mathrm{t}}\mathbf{W}\mathbf{b}$, where $\mathbf{W}$ is the $[n \times n]$ diagonal weight matrix holding the inverse variances of the observations.
8. Additional strategies for linear regressions involving uncertainties in both the dependent and independent variables of the observations include 1) reversing the regression variables to map out solution space holding effective line estimates, 2) estimating the least squares cubic line that minimizes the total data deviations perpendicular to it, and 3) estimating the reduced major axis line that minimizes the total area of right triangles between it and the data.
9. The elegant matrix expressions developed for ordinary least squares linear regression of $n$-observations in $m = 2$-unknowns where $n \geq m$ are also applicable for the more general least squares linear system with $n \geq m \geq 2$.

# *Part II*

## *Digital Data Inversion, Spectral Analysis, Interrogation, and Graphics*

# 8 Basic Digital Data Analysis

## 8.1 *OVERVIEW*

*A voluminous literature full of application-specialized jargon describes numerous analytical procedures for processing and interpreting deterministic data. However, when considered from the digital perspective of electronic computing, these procedures simplify into the core problem of manipulating a digital forward model of the data to achieve the data analysis objectives.*

*The forward model consists of a set of coefficients specified by the investigator and a set of unknown coefficients that must be determined by inversion from the input dataset and the specified forward model coefficients. The inversion typically establishes a least squares solution, as well as errors on the estimated coefficients and predictions of the solution in terms of the data and specified model coefficients.*

*The inversion solution is never unique because of the errors in the data and specified model coefficients, the truncated calculation errors, and any ambiguities of the theory that may be involved [e.g., the source ambiguity of potential fields]. Thus, a sensitivity analysis is commonly required to establish an 'optimal' set or range of solutions that conforms to the error constraints. Sensitivity analysis assesses solution performance in achieving data analysis objectives including the determination of the range of environmentally reasonable parameters that satisfy the observed data.*

*Basic inversion procedures include relatively simple trial-and-error testing, array methods, and extremely fast and accurate spectral analysis. The spectral model represents gridded data by the superposition of sine and cosine waves with amplitudes derived from the data, whereas the array model represents the* $[n \times 1]$ *observation vector by the product of the* $[n \times m]$ *design matrix of investigator-supplied coefficients and the* $[m \times 1]$ *solution vector of unknown coefficients obtained from the data.*

*Trial-and-error inversion with relatively small numbers of observations and unknown parameters is numerically manageable and effective. It is also useful for training inexperienced investigators on the interpretational limits of the data. However, this approach becomes increasingly laborious and unmanageable as more complicated models involving larger numbers of unknowns must be considered. For more complex inversions, array or matrix and spectral analysis methods are generally implemented as described in the next chapters.*

## 8.2 INTRODUCTION

The remaining chapters of this book predominantly focus on digital deterministic data analysis by inversion [e.g., **Figure 2.1**], and extend the inverse problem from the simple least squares straight-line equation of **Chapters 6 and 7** involving $n$-observations in $m = 2$-unknowns with $n \geq m$ to the general least squares problem with $n \geq m \geq 2$. The profiles in **Figure 1.2** illustrate some of the practical issues in setting up and solving inverse problems. For example, measurements of the signal $f(t)$ in practice always involve a finite set of discrete point observations containing errors. Thus, the observations can never completely replicate the signal from which they originate, but only approximate a portion of it as illustrated in the middle-right insert of **Figure 1.2**.

To infer the function or signal giving rise to data observations, the patterns in the data are fit with a hypothesized function [i.e., a likely story] as described in the sections below. The effective explanation of the point observations in the middle-right insert of **Figure 1.2**, for example, requires us to intuit that the analytical expression for the appropriate function (solid line) describing the data is $f(t) = a(\frac{t^2}{2})$, where $f(t)$ is the measured dependent variable value, $t$ is the measured independent variable value, and $a$ is the unknown variable value. In general, the procedure for determining the unknown parameter of the inferred function is fundamentally straightforward and the same for any application.

This story-telling process is an artform, however, and thus goes by a sometimes baffling variety of application-dependent labels even though the labels all essentially cover the same procedural details. For example, *factor analysis* commonly describes the data-fitting process in psychology, the social sciences or economics, whereas it may be labeled *blackbox theory* in physics, or *parameter estimation*, *deconvolution*, *inverse modeling*, or simply *modeling* in other applications. In this book, the process will be referred to as *inversion*, which is the term most commonly used in mathematics, engineering, and the natural sciences.

The primary objective of inversion is to establish an effective analytical model for the data that can be exploited for new insights on the processes causing the data variations. This exploitation phase is commonly called *systems processing* or *data interrogation* and may range from simple data interpolations or extrapolations to assessing the integral and differential properties of the data. Indeed, as described in **Chapter 1.3**, the two fundamental theorems of calculus guarantee that the mere act of plotting or mapping data fully reveals their analytical properties.

The ultimate objective in mapping data is to identify an analytical function or model that can effectively explain or *forward model* the data variations. Graphical differentiations and integrations of the data can be insightful when choosing an appropriate fitting function or forward model for data inversion. When the number of model unknowns is small [e.g., $m \leq 5 - 10$] like it was for the $f(t)$-model in the middle-right insert of **Figure 1.2**, data inversion by *trial-and-error* forward modeling to test various values for the unknown parameters [e.g., $a$] can be effective. However, this approach is not practical for many modern inversions that routinely solve for vastly greater numbers of unknowns. For these larger scale inversions, computer adaptations of the early 19th century least squares and array processing procedures of Gauss efficiently obtain the solution that minimizes the sum of the squared residuals between the model's predictions and the data.

For electronic processing, data are always pixilated or rendered into gridded datasets. In gridded format, the data can be modeled by the *fast Fourier transform* or simply the *FFT*, which is a very elegant version of the early 19th century Fourier series. The FFT is a *spectral model* that closely represents the data as the summation or integration of a series of cosine and sine waves at discrete wavelength/frequency arguments which are completely fixed by the interval and number of data points. The data inversion estimates only the amplitudes of these cosine and sine waves, and thus the FFT is extremely fast to implement. Indeed, the spectral model is the fastest known procedure for fitting mapped data and invokes no assumptions about the dataset other than that it is uniformly discretized in each of its 1, 2, 3, or more orthogonal dimensions.

The generic spectral model is commensurately efficient and comprehensive in accessing the differential and integral properties of data. In **Figure 8.1**, for example, the integral of $A\cos(t)$ is simply $A\sin(t)$ so that the inverse

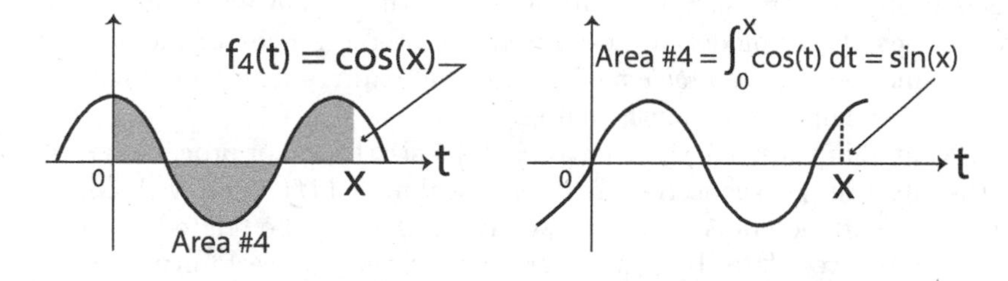


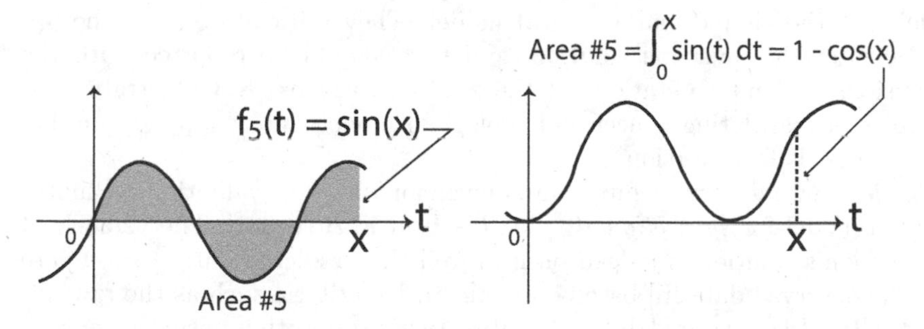


**Figure 8.1** Extensions of the two fundamental theorems of calculus from **Figure 1.2** include the integral of the cosine $[f_4(t)]$ from $t = 0$ to $x$ or Area #4 which is the cosine phase-shifted by $90^o$ or the sine, whereas the integral of the sine $[f_5(t)]$ from $t = 0$ to $x$ is Area #5 or one minus the cosine [e.g., **[2]**].

differential or the derivative of $A\sin(t)$ is $A\cos(t)$, where $A$ and $t$ are the respective inversion-derived amplitude and investigator-specified argument of the wave functions. Similarly, the integral of $A\sin(t)$ is $-A\cos(t)$ so that the derivative of $A\cos(t)$ is $-A\sin(t)$. Thus, to model the data's analytical properties, the spectral model's cosine and sine components are simply exchanged for the appropriate derivative- or integral-producing waves and the modified series re-evaluated. The entire process only involves summing the products of the wave amplitudes, which were determined from the data, with the related waves that the interval and number of data points specified. In summary, spectral modeling dominates modern digital data analysis because of the computational efficiency, accuracy, and minimal number of assumptions involved.

The remaining chapters outline further practical details concerning the inversion and systems processing of digital data. The role of modern digital graphics is also considered for developing insightful presentations of the data and their interpretations. The next section outlines the *process* of data inversion or modeling that invokes the investigator's insight to set up and relatively routine mathematical procedures to solve.

## 8.3 DATA INVERSION OR MODELING

A principal objective in mapping data is to plausibly quantify or model the data variations. These models are produced from the inversion of the data. It is the fundamental *inverse problem* of science that expresses measured natural processes in the language of mathematics.

As illustrated in **Figure 8.2**, inversion is a quasi-art form or process that requires the interpreter to conceive a mathematical model $[f(\mathbf{A}, \mathbf{x})]$ with known and unknown parameters $\mathbf{A}$ and $\mathbf{x}$, respectively, that can be related to variations in the observed data $[\mathbf{b}]$ to an appropriate level of uncertainty or error. The mathematical model is a simple usually linear approximation of the conceptual model that the interpreter proposes to account for the data variations. Assumptions arise both implicitly and explicitly to convert the conceptual model into the simple mathematical model. They critically qualify the applicability of the inversion solution, and thus should be reported with the conclusions and interpretations of the analysis. **Figure 8.3** illustrates two examples of translating conceptual geological models into simplified mathematical models for inversion.

The *forward modeling* component of inversion involves evaluating the mathematical model for *synthetic data* [i.e., $\hat{b} \in \hat{\mathbf{b}}$ in **Figure 8.2**]. The veracity of an inversion's solution is judged on how well the predicted data $[\hat{\mathbf{b}}]$ compare with the observed data $[\mathbf{b}]$ based on quantitative criteria such as the raw differences $[\hat{\mathbf{b}} - \mathbf{b}]$ or squared differences $[\hat{\mathbf{b}} - \mathbf{b}]^2$ or some other norm. In general, a perfect match between the predictions and observations is not necessarily expected or desired because the observational data $[\mathbf{b}]$ and the coefficients $[\mathbf{A}]$ assumed for the forward model are subject to inaccuracy, insufficiency, interdependency, and other distorting effects [**50**].

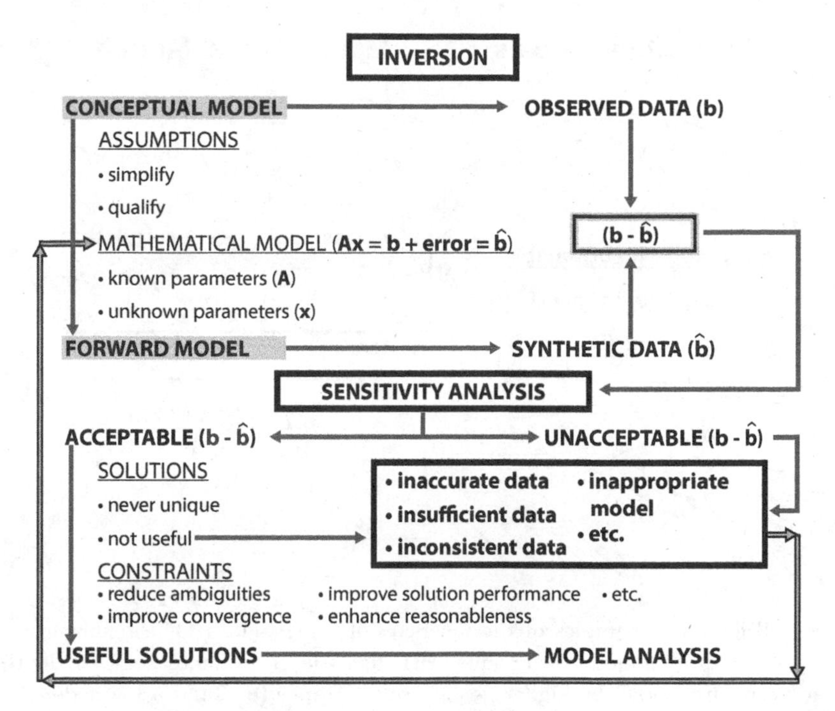


**Figure 8.2** The process of inversion [solid arrows] establishes a forward model [**Ax**] of the observations [**b**] and a sensitivity analysis to find an optimal subset of possible solutions for obtaining new insights on the processes producing the data variations. The feed-back loop [black-bordered gray arrows] from unacceptable residuals and useless solutions to the mathematical model reformulates the forward model until a useful solution is produced [**43**].

Data *inaccuracies* result from mapping errors, random and systematic, related to imprecision in the measurement equipment or techniques, inexact data reduction procedures, truncation [round-off] errors, and other factors. They scatter the data about a general trend or line [e.g., middle-right graph in **Figure 8.3**] so that a range of model solutions yields predictions within the scattered observations. However, as described in **Chapters 7.4 and 9.3.1**, the statistical properties of the column vector of observational data [**b**] can be propagated to characterize uncertainties in the column vector of solution variables [**x**].

*Insufficient* data do not contain enough information to describe the model parameters completely. Examples include the lack of data observations on the signal's peak, trough, flank, or other critical regions. The data coverage also may not be sufficiently comprehensive to resolve critical longer wavelength properties of data, or the station interval may be too large to resolve important shorter wavelength features in the data. Another source of data insufficiency is the imprecision in the independent variables [**A**] that promotes solution

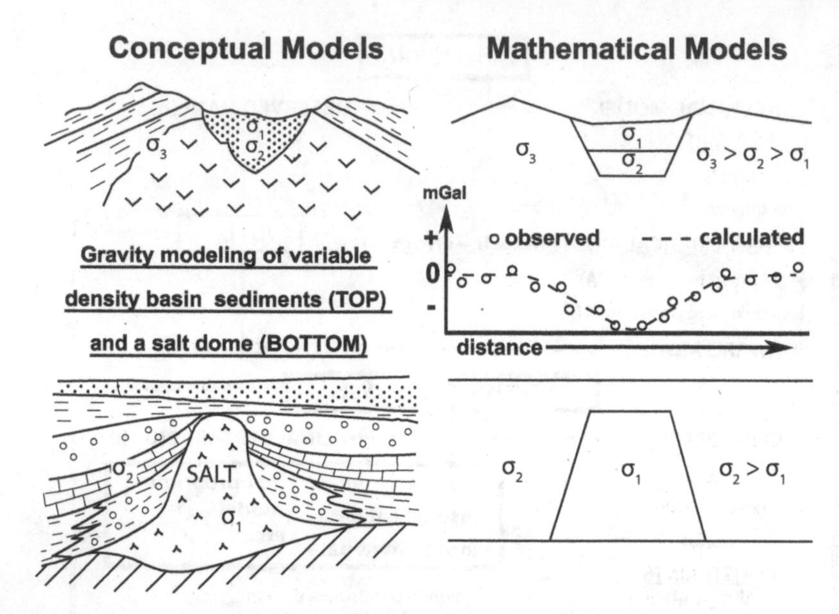


**Figure 8.3** Schematic examples of the inversion of gravity data for the subsurface distributions of lower density sediments with densities that increase with depth within a basin surrounded by higher density rocks [upper row] and a lower density salt dome displacing higher density sedimentary rocks [bottom row]. Both conceptual models can be mathematically represented by the simplified density models on the right that predict a relative gravity minimum over the center of the sources as illustrated by the generic gravity signal calculated in the middle panel of the right column [**43**].

biases. In general, data insufficiency issues are difficult to resolve without new data observations or other supplemental information.

*Interdependency* in datasets is where the independent variables for two or more data points cannot be distinguished from each other within the working precision of the mathematical model. Examples include data points that are too close to each other for the mathematical model to discriminate, or where perfect symmetry exists so that only half of the data observations need to be considered in the inversion [e.g., **Figure 9.1**]. Incorporating dependent data into the computations provides no new information and results in considerable computational inefficiency and solutions with erratic and often unreasonable performance characteristics. However, as described in **Chapter 9.3.2**, this problem can be mitigated somewhat using random noise in the **A**-coefficients to optimize the solution for acceptable performance attributes with only marginal growth in the deviation between the observed and predicted data.

The use of an *inappropriate* model for inversion results in significant negative consequences ranging from invalid solutions to valid solutions produced from excessive or unnecessary effort. Compromised solutions occur, for example, from attempts to fit a two-dimensional model to data that exhibit three-dimensional effects, or a homogeneous model where the effects of anisotropy are evident. In addition, the use of a complex model where a simpler one can achieve the objectives of the inversion leads to unnecessary and often unacceptable expenditures of effort. Furthermore, data such as potential field observations [e.g., gravity, magnetic, electrical, thermal, etc.] vary inversely with distance to their sources so that in theory at least, an infinite number of equivalent source distributions can be fit to these fields. Thus, the inherent nonuniqueness of these solutions allows considerable flexibility in developing appropriate models for effective data inversions.

In general, the solutions of inverse problems are not unique and require the application of *constraints* to identify the subset of the more effective solutions for a given objective. Constraints are additional information about the model parameters that when imposed on the inverse problem help reduce interpretational ambiguities, improve convergence to a solution, enhance the reasonableness of the solution and its predictions, and otherwise improve the effectiveness of the solution.

The identification of a useful solution or set of solutions has only established a set of numbers for $\mathbf{x}$ that map the assumed numbers in $\mathbf{A}$ into the observed numbers of $\mathbf{b}$. These three sets of numbers constitute the core of the inversion and the basis for other investigators to test the modeling efforts. The final stage of the process interprets these numbers for a story and graphics in a *report* that addresses the veracity of the conceptual hypothesis, the consequences of accepting or rejecting the hypothesis, alternate interpretations, and other issues for applying the inversion results effectively.

In summary, the creative, artful elements of data modeling or inversion involve developing the conceptual and forward models and the final interpretation. The analytical efforts to obtain solutions [$\mathbf{x}$], by contrast, are routine and well known since Gauss and Fourier established the respective least squares and spectral methods at the beginning of the $19^{th}$ century. In practice, the greatest analytical challenge confronting modern inversion efforts is the sensitivity analysis [see **Chapter 9**] required to identify an optimal set of solution coefficients for the given modeling objective.

Inversion methods range from relatively simple trial-and-error forward modeling to numerically more extensive and complex inverse modeling using array methods. The former approach usually involves a relatively small number [e.g., $m \leq 5 - 10$] of unknowns, where trial-and-error adjustments of the unknown parameters $\mathbf{x}$ are made until the predictions of the forward model reproduce the observed data within a specified margin of error. For greater numbers of unknowns, on the other hand, inverse modeling is commonly applied that determines the unknown parameters directly from the observed data using an optimization process comparing the observed and predicted data. However, inverse modeling is applicable on any number of unknowns, and thus the more general approach.

## 8.4 TRIAL AND ERROR METHODS

Historically, the trial-and-error method has been perhaps the most widely used and successful approach for data inversion. In this approach, the forward problem is solved for predicted data which are compared to the observed data. If the two datasets differ substantially, the model parameters are modified and the forward problem is solved again for another set of predictions. This process is repeated until a *satisfactory* match is obtained between the predicted and observed data.

This approach offers several advantages for data interpretation. For example, it requires only a moderate amount of experience to implement. In addition, the forward modeling exercise quickly provides insights on the range of data that the given model satisfies, as well as the sensitivity of the model's parameters to the data. In general, moderately experienced investigators can often apply trial-and-error inversion to obtain considerable practical insight or training on the scientific implications of the data. However, difficulties with the approach include the lack of uniqueness and reliable error statistics for the final solution. Furthermore, for more complicated models involving large numbers [e.g., $m > 5 - 10$] of unknown parameters, the convergence to the final solution by trial-and-error inversion becomes increasingly laborious, slow and unmanageable. Thus, to process the more complex inversion, array or matrix methods are commonly applied as considered in the next chapter.

## 8.5 KEY CONCEPTS

1. Digital analysis of deterministic data is invariably based on completing a mathematical forward model with partly known or investigator-specified coefficients and unknown coefficients that are determined from the data by inversion. Subsequent interrogation of the complete set of model coefficients reveals the data's differential and integral properties to within the inversion's error limits.
2. The process of data inversion begins with subjective conceptualization of a model for the observed data variations. The conceptual model is transformed via assumptions into a simplified mathematical model of known and unknown parameters.
3. Solving for the unknown parameters by data inversion yields forward modeling predictions [i.e., synthetic data] that are compared with the observed data in a sensitivity test of the solution's veracity.
4. The data residuals between observations and predictions may be unacceptable to analysis objectives due to data inaccuracies, insufficiencies, and inconsistencies, model inappropriateness, and other errors. Here, the investigator may consider updating the mathematical model for an improved set of predictions.
5. However, even if the data residuals are deemed acceptable, data and modeling errors still may conspire to make the solution useless for

achieving analysis objectives. In this case, updating the mathematical model may again be warranted.

6. Inversion never yields a unique solution due to the inevitable presence of data and modeling errors. Thus, constraints are invoked to identify a subset of plausible solutions with reduced ambiguities, improved numerical performance characteristics, enhanced reasonableness, and other desired properties.
7. Finally, the accepted model is transformed into text and graphics in a report that addresses the conceptual model's veracity, the consequences of accepting or rejecting it, alternate interpretations, and other issues for effectively applying the inversion results.
8. Popular data inversion methods include the trial-and-error testing of possible solutions, which is most effective on relatively small numbers of unknowns [e.g., $m \leq 5 - 10$] and observations. It is also useful for training inexperienced investigators on the data's interpretational limits.
9. For more complex inversions involving larger numbers of unknowns and data, the matrix and spectral analysis methods are commonly implemented as described in the next chapters.

# 9 Array Methods

## 9.1 OVERVIEW

*The general regression problem describes a system of simultaneous linear equations in m-unknown parameters that accounts for the n-observations, where* $n \geq m > 2$. *In matrix notation, this system can be expressed by* $\mathbf{A}_{(n\times m)}\mathbf{x}_{(m\times 1)} = \mathbf{b}_{(n\times 1)}$, *where* $\mathbf{Ax}$ *is the system's forward model.*

*As it is with least squares linear regression, once the investigator has specified the design matrix elements, the least squares general regression solution is taken in the known elements of the problem by* $\mathbf{x} = [\mathbf{A}^t\mathbf{A}]^{-1}\mathbf{A}^t\mathbf{b}$ *with variance* $\sigma^2_{\mathbf{x}} = [\mathbf{A}^t\mathbf{A}]^{-1}\sigma^2_{\mathbf{b}}$ *as long as* $|\mathbf{A}^t\mathbf{A}| \neq 0$. *Again,* $\sigma^2_{\mathbf{b}}$ *is the variance of the observations that, if not known, may be unbiasedly estimated from* $\sigma^2_{\mathbf{b}} = (\mathbf{b}^t\mathbf{b} - \mathbf{x}^t[\mathbf{A}^t\mathbf{A}]\mathbf{x})/(n-2)\ \forall\ n \geq 3$.

*Methods for finding the solution range from the relatively cumbersome Cramer's Rule that uses determinants to find by the inverse* $[\mathbf{A}^t\mathbf{A}]^{-1}$ *directly to multiply against the weighted observation vector* $\mathbf{A}^t\mathbf{b}$, *to the numerically more efficient elementary row operations that partially or fully diagonalize the system* $\mathbf{A}^t\mathbf{Ax} = \mathbf{A}^t\mathbf{b}$ *in place by the Gaussian or Gauss-Jordan elimination methods, respectively. However, even more efficient matrix procedures evaluate the solution coefficients recursively from data series set up with the Cholesky factors of* $\mathbf{A}^t\mathbf{A}$.

*The simple least squares linear and general regressions have the same statistically probable error bars or confidence intervals on the solution coefficients and predictions that are based on the solution's variance* $\sigma^2_{\mathbf{x}}$. *Common matrix expressions for the coherency and correlation coefficients, and the mean squares errors of the ANOVA test are also applicable between simple and general regressions.*

*Even though a solution is found that provides essentially a 100% match of the observations, it can be useless for other applications if, within working precision, any row [or column] of* $\mathbf{A}^t\mathbf{A}$ *is effectively filled with near-zero values or two of the rows are nearly linear multiples of each other. This near-singular or unstable condition of the system increases as* $|\mathbf{A}^t\mathbf{A}| \to 0$ *so that by Cramer's Rule the solution variance* $\sigma^2_{\mathbf{x}} \to \infty$. *In addition, the system's condition number* $COND \to \infty$ *[or equivalently its reciporcal condition number* $RCOND \to 0$*], where COND is the ratio of the maximum eigenvalue to the minimum eigenvalue of* $\mathbf{A}^t\mathbf{A}$.

*The optimally performing well-conditioned system, by contrast, is characterized by* $|\mathbf{A}^t\mathbf{A}| \neq 0$, *minimum* $\sigma^2_{\mathbf{x}}$, *and* $COND = RCOND = 1$ *so that each of the m elements of* $\mathbf{A}^t\mathbf{b}$ *is represented by one of the p fully independent rows of* $\mathbf{A}^t\mathbf{A}$ *- i.e.,* $p = m$.

*In practice, all regressions tend to show some degree of ill conditioning [i.e., $RCOND < COND > 1$ or $p < m$] due to errors in the computed coefficients of the assumed forward model. However, numerical adjustments of the ill conditioned* **A***-coefficients that push the condition number closer to unity may improve solution stability and effectiveness. These adjustments range from the relatively laborious effort of developing a new forward model [i.e., reparameterizing the system] to simpler efforts of numerically tweaking the existing* $\mathbf{A}^t\mathbf{A}$ *matrix to reduce its ill conditioned effects.*

*An example of the latter approach is ridge regression, which contaminates each element of* **A** *by random noise drawn from a distribution with error variance EV and zero mean. Ridge regression enhances the conditioning of the full* $\mathbf{A}^t\mathbf{A}$ *matrix by adding EV to the diagonal elements of all rows [including the [m−p] near-singular rows] that is just large enough to stabilize the solution, yet small enough to maximize its predictive power. The optimal EV-value may be obtained from the trade-off diagram over EV-values that compares the predicted data with the observed data against the difference behavior in the model estimates of the objective of the analysis.*

*Another numerical tweaking approach to stabilize the solution for improved prediction accuracy uses singular value decomposition to trim* $\mathbf{A}^t\mathbf{A}$ *of its [m−p] offending rows. This more computationally intensive approach involves the eigenvalue-eigenvector decomposition of* $\mathbf{A}^t\mathbf{A}$ *to identify the* $p \leq m$ *linearly independent rows that correspond within working precision to the effectively non-zero eigenvalues. The improved solution is obtained from the generalized linear inverse that is constructed from the maximum subset of the p linearly independent rows of the system. The threshold p-value can be established from the gradients in the plotted eigenvalues or their inversely proportional solution variances.*

*The generalized linear inverse always exists and thus solves all systems - i.e., systems that are even [$p = m = n$], overconstrained [$p = m < n$], strictly underdetermined [$p \leq n < m$], and simultaneously underdetermined and overconstrained [$p < m < n$].*

*The information density matrix of any system is obtained from multiplying the inverse on the left by* **A** *to identify the subset of observations in* **b** *that are most linearly independent in the context of the presumed forward model. Implementing this subset of observations offers the possibility of improving solutions for overconstrained, simultaneously underdetermined and overconstrained, and the strictly underdetermined systems where* $p < n < m$.

*The model resolution matrix of any system is obtained from multiplying the inverse on the right by* **A** *to identify the subset of unknowns in* **x** *that are most linearly independent in the context of the presumed forward model. Implementing this subset of unknowns offers the possibility of improving solutions for all strictly underdetermined, and simultaneously underdetermined and overconstrained systems.*

*The ridge regression and generalized linear inverses allow the investigator to modify the conditioning of the* $\mathbf{A}^t\mathbf{A}$ *matrix to obtain solutions with im-*

*proved variances. Because the ridge regression inverse is most amenable to uses of fast linear equation solvers, it is computationally much more efficient to obtain than the generalized linear inverse. However, the most efficient implementation of matrix inversion is described in the next chapter where the spectral forward model is used to represent orthogonally gridded data by inverting a fully diagonalized* $\mathbf{A}^t\mathbf{A}$ *matrix with minimal conditioning issues.*

## 9.2 INTRODUCTION

Electronic computations are digital and carried out by linear series manipulations on linear digital models of the input data. The ultimate objective of modeling data or inversion is not simply to replicate the data, but rather to establish a model that can be analytically interrogated for new insights on the processes causing the data variations. The digital data model is the basis for achieving any data processing objective by electronic computing. It can be demonstrated with the problem of relating $n$-data values $(b_{i1} \equiv b_i \in \mathbf{b},\ \forall\, i = 1, 2, \ldots, n)$ to the simultaneous system of $n$-equations

$$\begin{aligned} a_{11}x_1 + a_{12}x_2 + \cdots + a_{1m}x_m &= b_1 \\ a_{21}x_1 + a_{22}x_2 + \cdots + a_{2m}x_m &= b_2 \\ \ddots \qquad\qquad \ddots \qquad &= \vdots \\ a_{n1}x_1 + a_{n2}x_2 + \cdots + a_{nm}x_m &= b_n, \end{aligned} \tag{9.1}$$

which is linear in the $m$-unknown variables $(x_{j1} \equiv x_j \in \mathbf{x},\ \forall\, j = 1, 2, \ldots, m)$ and the $(n \times m)$-known variables $(a_{ij} \in \mathbf{A})$ of the forward model $(\mathbf{Ax})$ that the investigator assumes will account for the observations. In matrix notation, this linear equation system becomes

$$\mathbf{Ax} = \mathbf{b}, \tag{9.2}$$

where $\mathbf{b}$ is the $n$-row by 1-column matrix or vector containing the $n$-observations, $\mathbf{x}$ is the $m$-row by 1-column vector of the $m$-unknowns, and the design matrix $\mathbf{A}$ is the $n$-row by $m$-column matrix containing the known coefficients of the assumed model $\mathbf{Ax}$. Note that the fundamental linear equation system **Eq. 9.2** was previously considered in **Chapter 7** in the context of linear regression involving $n \geq 2$ observations in $m = 2$ unknowns.

The forward model is the heart of any analysis and recognizing its attributes is essential to productive data analysis. Linear models dominate modern analysis, but non-linear forward models also can be invoked and were popular in inversion before the advent of electronic computing because of their relative ease in manual application. For computer processing, the non-linear model is linearized either by the application of the Taylor series [e.g., **Eq. 3.13**] or chopped up numerically into array computations.

In general, any linear or non-linear inversion involves only three sets of coefficients with one set taken from the measured data (i.e., $\mathbf{b}$), another set imposed by the interpreter's assumptions (i.e., **A**), and the final set of unknowns (i.e., **x**) determined by solving **Eq. 9.2**. Electronic computers process these coefficients as data arrays with basic matrix properties so that matrix inversion is extensively used for obtaining the solution (**x**).

## 9.3 MATRIX INVERSION

Using the elementary matrix properties described in **Chapter 2.3.1**, the common array methods for solving linear systems of simultaneous equations are readily developed. To illustrate these methods, consider ***Example 9.3a*** of the simple gravity profile in **Figure 9.1** that crosses perpendicularly over a two-dimensional [i.e., a strike-infinite] horizontal cylinder source in the presence of a constant regional gravity field. Each observation along the profile is the gravity effect of the cylindrical mass plus the regional field computed from

$$g_{(2D)z} = 41.93\Delta\sigma(\frac{R^2}{z})(\frac{1.00}{(d^2/z^2)+1.00}) + C, \tag{9.3}$$

where $g_{(2D)z}$ is the vertical component of gravity in milligal [mGal] and $d$ is the distance along the profile in km crossing the horizontal cylinder with radius $R = 3.00$ km, depth to the central axis $z = 5.00$ km, and density (or

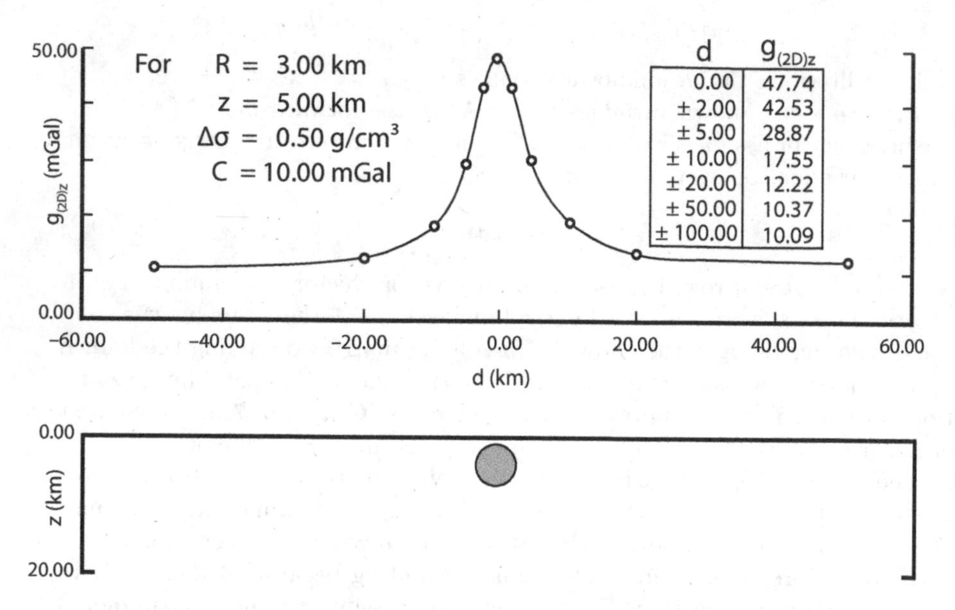


**Figure 9.1** Profile of the gravity effect ($g_{(2D)z}$) across a buried horizontal $2D$ cylinder striking perpendicular to the profile superimposed on a constant regional field $C$. The 11 point observations of the profile (open circles) are used to illustrate the data inversion process as described in the text. [Adapted from [**43**].]

density contrast) $\Delta\sigma = 0.50$ g/cm$^3$, in the regional gravity field of amplitude $C = 10.00$ mGal.

**Equation 9.3** involves two linear terms so that an interpreter can estimate a variable in each of the terms from the remaining variables and the observations. For example, consider determining $\Delta\sigma$ and $C$ assuming that $R$, $z$, and $d$ are known and the model [**Equation 9.3**] is appropriate for explaining the variations in the gravity profile observations. To evaluate the two unknowns, at least two observations are needed which may be modeled by the linear system of equations

$$\begin{aligned} b_1 &= a_{11}x_1 + a_{12}x_2 \\ b_2 &= a_{21}x_1 + a_{22}x_2, \end{aligned} \tag{9.4}$$

where the design matrix coefficients are

$$\begin{aligned} a_{11} &= [41.93(R^2/z)]/[(d_1^2/z^2) + 1.00], \quad a_{12} = 1.00, \\ a_{21} &= [41.93(R^2/z)]/[(d_2^2/z^2) + 1.00], \quad a_{22} = 1.00. \end{aligned} \tag{9.5}$$

The design matrix coefficients represent the forward model in **Equation 9.3** evaluated with the known parameters [$R$, $z$, $d$] set to their respective values and the unknown parameters [$\Delta\sigma$, $C$] numerically set to equal unity.

In mathematical theory, the two equations in two unknowns can be solved *uniquely* from any two observations. However, if the two observations are located symmetrically about the signal's peak, then the two equations in the above system are co-linear and $|\mathbf{A}| = 0$ so that no solution is possible. Thus, in the context of the assumed model, the solution can be completely obtained from the peak value and the observations either to the right or left of the peak. In other words, the interpreter's choice of the model [**Equation 9.3**] established the interdependency that made roughly 50% of the data in **Figure 9.1** redundant for subsurface analysis.

Consider now the signal's values on the peak at $d_1 = 0.00$ km and the flank at $d_2 = 20.00$ km so that the system becomes

$$\begin{aligned} 47.74 &= 75.47x_1 + x_2 \\ 12.22 &= 4.44x_1 + x_2. \end{aligned} \tag{9.6}$$

The design matrix and its determinant and inverse for this system are, respectively,

$$\mathbf{A} = \begin{pmatrix} 75.47 & 1.00 \\ 4.44 & 1.00 \end{pmatrix}, \; |\mathbf{A}| = 71.03, \text{ and } \mathbf{A}^{-1} = \begin{pmatrix} 0.01 & -0.01 \\ -0.06 & 1.06 \end{pmatrix}. \tag{9.7}$$

*Cramer's Rule* is a relatively simple method for obtaining the solution $\mathbf{x}$ whereby $x_{j1} = |\mathbf{D_j}|/|\mathbf{A}|$. Here, the *augmented matrix* $\mathbf{D_j}$ is obtained from $\mathbf{A}$ by replacing the $j$-column of $\mathbf{A}$ with the column vector $\mathbf{b}$. Thus, by Cramer's Rule,

$$x_1 = \frac{35.52}{71.03} = 0.50 \text{ and } x_2 = \frac{710.34}{71.03} = 10.00, \tag{9.8}$$

which are the correct values. Cramer's Rule, however, is practical only for simple systems with relatively few unknowns because of the computational labor of computing the determinants. For larger systems, a more efficient approach uses elementary row operations to process **A** for its inverse so that the solution to **Equation 9.6** can be obtained from

$$\mathbf{x} = \mathbf{A}^{-1}\mathbf{b} = \begin{pmatrix} 0.50 \\ 10.00 \end{pmatrix}. \tag{9.9}$$

Even more efficient approaches directly diagonalize the linear equations in place for the solution coefficients. *Gaussian elimination* [**Chapter 7.3.3**] uses elementary row operations to sweep out the lower triangular part of the equations. The **system 9.6**, for example, reduces to

$$\begin{bmatrix} 47.74 & = & 75.47x_1 + x_2 \\ 12.22 & = & 4.44x_1 + x_2 \end{bmatrix} \sim \begin{bmatrix} 47.74 & = & 75.47x_1 + x_2 \\ 9.41 & = & 0.00 + 0.94x_2 \end{bmatrix},$$

from which the solution coefficients to the nearest hundredths are

$$x_2 = \frac{9.41}{0.94} = 10.01 \text{ and } x_1 = \frac{(47.74 - 10.01)}{75.47} = 0.50, \tag{9.10}$$

where $x_1$ was obtained by back substituting the solution for $x_2$. Instead of back substitution, the row operations can be continued to also sweep out the upper triangular part of the system. This procedure called *Gauss-Jordan elimination* obtains the solution coefficients explicitly along the main diagonal as

$$\begin{bmatrix} 47.74 & = & 75.47x_1 + x_2 \\ 9.41 & = & 0.00 + 0.94x_2 \end{bmatrix} \sim \begin{bmatrix} 37.74 & = & 75.47x_1 + 0.00 \\ 9.41 & = & 0.00 + 0.94x_2 \end{bmatrix} \sim$$

$$\begin{bmatrix} 0.50 & = & x_1 + 0.00 \\ 10.01 & = & 0.00 + x_2 \end{bmatrix}.$$

These matrix inversion methods are applicable for systems where the number of unknowns $m$ equals the number of input data values $n$. However, for most data inversions, $m < n$ and the system of equations is always related to an incomplete set of input data containing errors, as well as uncertainties in estimating the coefficients of the design matrix **A** and other errors. Thus, the *least squares* solution is commonly adopted that minimizes the sum of squared differences between the predicted and observed data. This early 19th century solution due to Gauss involves solving the system of *normal equations* derived by differentiating the sum of squared differences with respect to the model parameters being estimated in the inversion.

To obtain a least squares solution, the simple raw error **e** is used to define the *residual function*

$$R = \mathbf{e}^{\mathbf{t}}\mathbf{e} = (\mathbf{b} - \mathbf{A}\mathbf{x})^{\mathbf{t}}(\mathbf{b} - \mathbf{A}\mathbf{x}) = \sum_{i=1}^{n}(b_i - \sum_{j=1}^{m} a_{ij}x_j)^2.$$

Next, the partial derivatives of $R$ with respect to the solution coefficients are taken and set equal to zero to obtain the normal equations. For the $q$th coefficient $x_q$, for example,

$$\frac{\partial R}{\partial x_q} = 2\sum_{j=1}^{m} x_j \sum_{i=1}^{n} a_{iq}a_{ij} - 2\sum_{i=1}^{n} a_{iq}b_i \equiv 0,$$

which by the definition for matrix multiplication reduces to

$$\mathbf{A^t A x} - \mathbf{A^t b} \equiv 0.$$

Thus, the system $\mathbf{Ax} = \mathbf{b}$ is equivalent to the $\mathbf{A^t}$-weighted system given by $\mathbf{A^t Ax} = \mathbf{A^t b}$, which has the least squares solution

$$\mathbf{x} = (\mathbf{A^t A})^{-1}\mathbf{A^t b} \tag{9.11}$$

if $|\mathbf{A^t A}| \neq 0$.

Of course, the solution is found more efficiently by processing the weighted system in place with Gaussian elimination than by taking the inverse $(\mathbf{A^t A})^{-1}$ directly and multiplying it against the weighted observation vector $(\mathbf{A^t b})$. This method works for any set of equations which can be put into linear form. No derivatives need be explicitly taken for the least squares solution, which contains only the known dependent and independent variable coefficients from $\mathbf{b}$ and $\mathbf{A}$, respectively.

As ***Example 9.3b***, for the simple (2 × 2) system in **Equation 9.6** with the design matrix $\mathbf{A}$ in **Equation 9.7**,

$$(\mathbf{A^t A})^{-1} = \begin{pmatrix} 0.04 & -1.58 \\ -1.58 & 113.28 \end{pmatrix} \times 10^{-2}$$

so that the least squares or *natural* inverse is given by

$$(\mathbf{A^t A})^{-1}\mathbf{A^t} = \begin{pmatrix} 1.41 & -1.41 \\ -6.25 & 106.25 \end{pmatrix} \times 10^{-2},$$

which is identical to the inverse $\mathbf{A}^{-1}$ that was found previously using elementary row operations. Thus, for simple systems where $(n = m)$, the natural inverse reduces to the elementary inverse $\mathbf{A}^{-1}$.

However, the natural inverse also applies for the more usual case where $(n > m)$ and $\mathbf{A}^{-1}$ does not exist. As ***Example 9.3c***, consider the 7 linearly independent gravity observations in **Figure 9.1** that extend from the peak along the right flank [RF] given by

$$\mathbf{b_{RF}} = (47.74\ 42.53\ 28.87\ 17.55\ 12.22\ 10.37\ 10.10)^t$$

for which

$$\mathbf{A^t} = \begin{pmatrix} 75.47 & 65.06 & 37.74 & 15.10 & 4.44 & 0.75 & 0.19 \\ 1.00 & 1.00 & 1.00 & 1.00 & 1.00 & 1.00 & 1.00 \end{pmatrix}.$$

In this case, the natural inverse is

$$(\mathbf{A^tA})^{-1}\mathbf{A^t} = \begin{pmatrix} 0.79 & 0.62 & 0.16 & -0.22 & -0.40 & -0.46 & -0.47 \\ -8.15 & -3.19 & 9.83 & 20.62 & 25.70 & 27.46 & 27.73 \end{pmatrix} \times 10^{-2},$$

which again correctly estimates the unknown parameters from

$$(\mathbf{A^tA})^{-1}\mathbf{A^t b_{RF}} = \mathbf{x} = (0.50\ 10.00)^t.$$

The matrix $\mathbf{A^tA}$ is symmetric and positive definite with $|\mathbf{A^tA}| > 0$, and thus can be factored into an explicit series solution that is much faster and more efficient to evaluate than the conventional Gaussian elimination solution. In particular, the matrix $\mathbf{A^tA}$ can be decomposed into $(m \times m)$ lower and upper triangular *Cholesky factors* $\mathbf{L}$ and $\mathbf{L^t}$, respectively, so that $\mathbf{LL^t} = \mathbf{A^tA}$. The coefficients of $\mathbf{L}$ are obtained from

$$\begin{aligned}
l_{11} &= \sqrt{a_{11}}, && \forall i = j = 1 \\
l_{j1} &= a_{j1}/l_{11}, && \forall j = 2, 3, \cdots, m \\
l_{ii} &= \sqrt{a_{ii} - \sum_{k=1}^{i-1} l_{ik}^2}, && \forall i = 2, 3, \cdots, (m-1) \\
l_{ji} &= 0 && \forall i > j \\
&= \left(\frac{1}{l_{ii}}\right)\left(a_{ji} - \sum_{k=1}^{i-1} l_{ik} l_{jk}\right), && \forall i = 2, 3, \cdots, (m-1) \ \mathit{and} \\
& && j = (i+1), (i+2), \cdots, (m), \ \mathit{and} \\
l_{mm} &= \sqrt{a_{mm} - \sum_{k=1}^{m-1} l_{mk}^2}.
\end{aligned} \tag{9.12}$$

Note that in the above Cholesky elements, the summations were effectively taken to be zero for $i = 1$.

To see how the Cholesky decomposition obtains the faster series solution, consider the previous example of the 7 observations in **Figure 9.1**, where $\mathbf{A^tA} = \begin{pmatrix} 1.1601 & 0.0199 \\ 0.0199 & 0.0007 \end{pmatrix} \times 10^4$ and $\mathbf{A^tb} = \begin{pmatrix} 7.7885 \\ 0.1694 \end{pmatrix} \times 10^3$. Now, since $|\mathbf{A^tA}| = 4.1707 \times 10^4 > 0$, $\mathbf{A^tA}$ can be decomposed into Cholesky factors $\mathbf{L} = \begin{pmatrix} 107.7121 & 0 \\ 1.8452 & 1.8962 \end{pmatrix}$ and $\mathbf{L^t}$ such that $\mathbf{LL^t} = \mathbf{A^tA}$. Also, let the weighted observation column vector $\mathbf{A^tb} = \mathbf{k} = \mathbf{LL^tx}$. Thus, setting $\mathbf{p} = \mathbf{L^tx}$ obtains

$$\begin{aligned}
p_1 &= l_{11}x_1 + l_{21}x_2 \\
p_2 &= 0.00 + l_{22}x_2,
\end{aligned} \tag{9.13}$$

as well as $\mathbf{Lp} = \mathbf{k}$ with

$$\begin{aligned} p_1 &= k_1/l_{11} \\ p_2 &= k_2/l_{22} - (l_{21}/l_{22})p_1, \end{aligned} \tag{9.14}$$

which can be combined for the solution coefficient estimates

$$\begin{aligned} x_2 = \quad & (k_2/l_{22}^2) - (k_1/l_{22}^2)/(l_{21}l_{11}) &= \quad & 10.00 \\ x_1 = \quad & (k_1/l_{11}^2) - (l_{21}/l_{11})x_2 &= \quad & 0.50 \end{aligned} \tag{9.15}$$

In general, the Cholesky factorization allows $x_m$ to be estimated from a series expressed completely in the known coefficients of $\mathbf{L}$, $\mathbf{A}$ and $\mathbf{b}$, which is back-substituted into the series expression for $x_{m-1}$, which in turn is back-substituted into the series for $x_{m-2}$, etc. The Cholesky solution is much faster than Gaussian elimination approaches that require twice as many numerical operations to complete. The Cholesky solution also has minimal in-core memory requirements because the symmetric $\mathbf{A^tA}$ matrix can be packed into a singly dimensioned array of length $m(m+1)/2$. For systems that are too large to hold in active memory, *updating Cholesky algorithms* are available that process the system on external storage devices [e.g., **[59]**].

The Cholesky factorization is problematic, however, in applications where the coefficients $l_{i,j} \longrightarrow 0$ within working precision. In these instances, the products and powers of the coefficients are even smaller so that solution coefficient estimates become either indeterminant or blow up wildly and unrealistically. *Indeterminant* or *unstable* solutions respectively signal *singular* or *near-singular* $\mathbf{A^tA}$ matrices. However, these *ill conditioned* matrices can still be processed for effective solutions using the error statistics of the inversions to help suppress their singular or near-singular properties.

### 9.3.1 ERROR STATISTICS

In general, reporting of data inversion results routinely requires quantifying the statistical uncertainties of the analysis. These assessments commonly focus on the variances and confidence intervals on the solution coefficients and predictions of the model, and the coherency and statistical significance of the fit of the model's predictions to the data observations.

For example, applying *variance propagation* [e.g., **[6]**] to the linear function $(\mathbf{A^tA})^{-1}\mathbf{A^tb}$ shows that the solution variance can be obtained from

$$\sigma_{\mathbf{x}}^2 = [(\mathbf{A^tA})^{-1}\mathbf{A^t}]^2\sigma_{\mathbf{b}}^2 = (\mathbf{A^tA})^{-1}\sigma_{\mathbf{b}}^2, \tag{9.16}$$

where the statistical variance in the observations $\sigma_{\mathbf{b}}^2$ is specified either *a priori* or from

$$\sigma_{\mathbf{b}}^2 \simeq [\mathbf{b^tb} - \mathbf{x^t}(\mathbf{A^tA})\mathbf{x}]/(n-m) \ \forall \ n > m, \tag{9.17}$$

which is an unbiased estimate if the model $\mathbf{Ax}$ is correct. **Equation 9.16** is the symmetric *variance-covariance matrix* of $\mathbf{x}$ with elements that are the products $\sigma_{x_k}\sigma_{x_j}$ of the standard deviations of $x_k$ and $x_j$. Thus, the elements along the diagonal where $k = j$ give the variances $\sigma^2_{x_k} = \sigma_{x_k}\sigma_{x_j}$ and the off-diagonal elements give the covariances that approach zero if $x_k$ and $x_j$ are not correlated.

The $100(1-\alpha)\%$ *confidence interval* [or equivalently the $\alpha\%$ *significance interval*] on each $x_k \in \mathbf{x}$ is given by

$$x_k \pm (\sigma_{x_k})t(1-\alpha/2)_{n-m}, \tag{9.18}$$

where $t(1-\alpha/2)_{n-m}$ is the value of Student's $t$-distribution for the confidence level $(1-\alpha/2)$ and degrees of freedom $\nu = (n-m)$. The individual confidence intervals for $x_k$ can also be readily combined into joint confidence regions [e.g., **[51, 27]**], but they are difficult to visualize for $m > 3$.

Application of the variance propagation rule to $\mathbf{Ax}$ shows that the variance on each prediction $\hat{b_i} \in \hat{\mathbf{b}}(= \mathbf{Ax} + \mathbf{error} \simeq \mathbf{b})$ can be estimated from

$$\sigma^2_{\hat{b_i}} = \mathbf{a}_i^t(\mathbf{A}^t\mathbf{A})^{-1}\mathbf{a}_i\sigma^2_{\mathbf{b}}, \tag{9.19}$$

where the $i$-th row vector of the design matrix is $\mathbf{a}_i^t = (a_{i1}\ a_{i2}\ \cdots\ a_{im})$. Thus, the $100(1-\alpha)\%$ *confidence limits* or *error bars* for the $\hat{b_i}$ at $\mathbf{a}_i^t$ can be obtained from

$$\hat{b_i} \pm \sigma_{\mathbf{b}}[\sqrt{\mathbf{a}_i^t(\mathbf{A}^t\mathbf{A})^{-1}\mathbf{a}_i}]t(1-\alpha/2)_{n-m}. \tag{9.20}$$

Another measure of the fit is the *correlation coefficient* $r$, which when squared gives the *coherency* that indicates the percent of the observations $b_i$ fitted by the predictions $\hat{b_i}$. The coherency in matrix form is given by

$$r^2 = (\mathbf{x}^t\mathbf{A}^t\mathbf{b} - n\bar{b}^2)/(\mathbf{b}^t\mathbf{b} - n\bar{b}^2), \tag{9.21}$$

where $\bar{b}$ is the mean value of the $b_i \in \mathbf{b}$. In general, our confidence in the model $\mathbf{Ax}$ tends to increase as $r^2 \longrightarrow 1$.

The coherency (**Equation 9.21**) reduces algebraically to

$$r^2 = [\sum_{i=1}^{n}(\hat{b_i} - \bar{b})^2]/[\sum_{i=1}^{n}(b_i - \bar{b})^2], \tag{9.22}$$

where $\sum_{i=1}^{n}(\hat{b_i} - \bar{b})^2$ is the sum of squares due to regression, and $\sum_{i=1}^{n}(b_i - \bar{b})^2$ is the sum of squares about the mean. In general, the sum of squares about the regression is $\sum_{i=1}^{n}(b_i - \hat{b_i})^2 = \sum_{i=1}^{n}(\hat{b_i} - \bar{b})^2 + \sum_{i=1}^{n}(b_i - \hat{b_i})^2$ so that if $\mathbf{Ax}$ is correct, then $[\sum_{i=1}^{n}(\hat{b_i} - \bar{b})^2] >> [\sum_{i=1}^{n}(b_i - \hat{b_i})^2] \ni r^2 \longrightarrow 1$. Thus, an *analysis of variance* or *ANOVA* table [e.g., **Chapter 6.4**] can be constructed to test the statistical significance of the fit between $\hat{b_i}$ and $b_i$ as shown in **Table 9.1**. The *null hypothesis* that the model's predictions ($\hat{\mathbf{b}}$) do not significantly fit the observations ($\mathbf{b}$) is tested by comparing the estimate $F = MS_R/MS_D$ from the ANOVA table with the *critical value* ($F'$) from the Fisher distribution with degrees of freedom $\nu = (m-1), (n-m)$ at the desired level $\alpha$ of significance [or alternatively $(1-\alpha)$ of confidence]. The hypothesis is rejected if $F > F'$; otherwise there is no reason to reject the hypothesis based on the $F$-test.

**Table 9.1**
**The analysis of variance (ANOVA) table for testing the significance of a model's fit to data.**

| ERROR SOURCE | $\nu$ | CORRECTED SUM OF SQUARES (CSS) | MEAN SQUARES | F-TEST |
|---|---|---|---|---|
| Regression Error | $m-1$ | $\mathbf{x}^t\mathbf{A}^t\mathbf{b} - (\sum_{i=1}^{n} b_i)^2/n$ | $MS_R = CSS/\nu$ | $F =$ |
| Residual Deviation | $n-m$ | $\mathbf{b}^t\mathbf{b} - \mathbf{x}^t\mathbf{A}^t\mathbf{b}$ | $MS_D = CSS/\nu$ | $MS_R/MS_D$ |
| Total Error | $n-1$ | $\mathbf{b}^t\mathbf{b} - (\sum_{i=1}^{n} b_i)^2/n$ | | |

### 9.3.2 SENSITIVITY ANALYSIS

The above sections have illustrated how inversion determines an unknown set of discrete numbers ($\mathbf{x}$) that relates a set of discrete numbers ($\mathbf{b}$) observed from nature to a given set of discrete numbers ($\mathbf{A}$) from the model ($\mathbf{Ax}$) that the interpreter presumes can account for the observations ($\mathbf{b}$). Of course, once the forward model has been postulated, the solution ($\mathbf{x}$) and its error statistics are routinely established in the least squares sense from **Equations 9.11** and **9.16**, respectively.

The objective of inversion, however, is not simply to model the observations, but to develop a model that is effective in estimating additional key properties of the observations such as the related derivatives, integrals, values at unsurveyed coordinates, physical property variations, and other attributes. The solution in practice is never unique, but one of a range of solutions that is compatible with the ever-present error bars on the **A**- and **b**-coefficients and the computational round-off or truncation errors. Thus, a sensitivity analysis is commonly implemented to sort out the more effective solutions for a particular processing objective.

The assumed **A**-coefficient and computing errors can be particularly troublesome because an important requirement for obtaining the least squares solution is that ($\mathbf{A}^t\mathbf{A}$) is non-singular or equivalently that $|\mathbf{A}^t\mathbf{A}| \neq 0$. Singularity of the system occurs if a row (or column) contains only zeros, or if two or more rows (or columns) are linear multiples of each other and thus co-linear. The more common situation in practice, however, is for the system to be near-singular or ill conditioned, where in the computer's working precision the elements of a row (or column) are nearly zero, or two or more rows (or columns) are nearly linear multiples of each other and thus have a high degree of co-linearity.

For poorly conditioned systems, $|\mathbf{A}^t\mathbf{A}| \longrightarrow 0$ so that by Cramer's rule $\mathbf{x} \longrightarrow \infty$. Thus, near-singularity of the linear system is characterized by a solution with large and erratic values or large variance. In this instance, the solution is said to be unstable and its linear transformations beyond predicting the observations will be largely meaningless, even though it models the observed data with great accuracy.

An effective numerical approach for stabilizing solutions is to add a small positive constant $(EV)$ to the diagonal elements of $(\mathbf{A^tA})$ to dampen the effects of nearly co-linear or zero rows in the system. This approach is equivalent to obtaining the damped *mean squared error* solution [[**47, 46**]] given by

$$\mathbf{x_D} \equiv [\mathbf{A^tA} + (EV) \times \mathbf{I}]^{-1}\mathbf{A^tb}, \tag{9.23}$$

with the variance [**63**]

$$\sigma^2_{\mathbf{x_D}} = [\mathbf{A^tA} + (EV)\mathbf{I}]^{-1}(\mathbf{A^tA})[\mathbf{A^tA} + (EV)\mathbf{I}]^{-1}. \tag{9.24}$$

This approach is frequently called *ridge regression* and the constant $(EV)$ is known by several names including *damping factor*, *Marquardt parameter*, and *error variance*. The last term reflects the fact that **Equation 9.23** is equivalent to adding into each **A**-coefficient random noise from the normal distribution with zero mean and standard deviation $\sqrt{EV}$. Thus, the noise products in the off-diagonal elements of $(\mathbf{A^tA})$ average out to zero, whereas along the diagonal they equal the noise or random error variance $EV$.

To reduce solution variance and improve the mean-squared error of prediction for ill-conditioned systems, a value for the error variance must be found that is just large enough to stabilize the solution, but small enough to maximize the predictive properties of the solution. A straightforward approach to quantifying the *optimal* value $EV_{opt}$ is to develop *trade-off diagrams* like the example in **Figure 9.2**. Here, Curve #1 compares the sum of squared residuals $[SSR_j[\mathbf{b}, \hat{\mathbf{b}}(EV_j)] = \sum_{i=1}^{n}[b_i - \hat{b}_i(EV_j)]^2 = C]$ between the observed and modeled data for a range of error variances $[EV_j]$ that usually is established by trial-and-error. For any application of the model, solutions with predictions that do not deviate significantly from the observations are clearly desired such as those $\leq EV_{opt}$.

Curve #2, on the other hand, is developed to test the relative behavior of the solution in estimating derivatives, integrals, interpolations, continuations, vector components or other attributes $\hat{b}'(EV_j)$ of the observations as a function of $EV_j$. This performance curve may be evaluated from norms such as $SSR_j[\hat{\mathbf{b}}'(EV_j), \hat{\mathbf{b}}'(EV_{max})] = D$, where $\hat{\mathbf{b}}'(EV_{max})$ are the predictions at an excessively damping and dominant error variance $EV_{max}$. Another useful form of this curve tests the behavior of the differences between successive predictions by $SSR_j[\hat{\mathbf{b}}'(EV_j), \hat{\mathbf{b}}'(EV_{j-1})]$. Curve #2 targets the range $EV_j \leq EV_{lo}$ where small changes in the **A**-coefficients yield radically changing and thus highly unstable and suspect predictions $\hat{\mathbf{b}}'(EV_j)$. The more stable and robust predictions over the range $EV_j \geq EV_{hi}$, on the other hand, are increasingly limited because they are increasingly less constrained by the observations as shown in Curve #1.

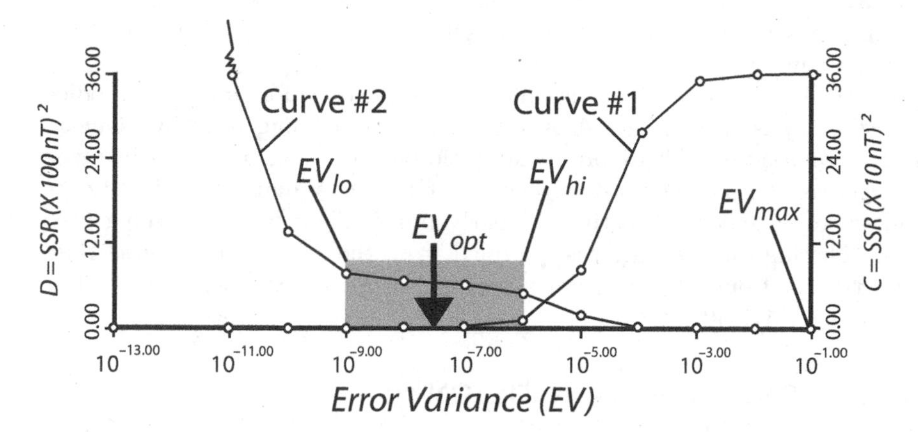


**Figure 9.2** Using performance curves to estimate the 'optimal' value of error variance ($EV_{opt}$) for effective data inversion. In this example, curves #1 and #2 with respective $C$- and $D$-ordinates were established to determine the EV-range (shaded) over which regional magnetic anomalies of southern China and India could be effectively modeled for their radial components. [Adapted from [**43**].]

Thus, the trade-off diagram in **Figure 9.2** marks the range of error variances $EV_{lo} \leq EV_j \leq EV_{hi}$ for obtaining reliable stable predictions, where the 'optimal' predictions $\hat{\mathbf{b}}'(EV_{opt})$ are taken for the error variance at the range's midpoint. In addition, constraints on the errors of $\hat{\mathbf{b}}'(EV_{opt})$ may be obtained by considering the difference map $[\hat{\mathbf{b}}'(EV_{lo}) - \hat{\mathbf{b}}'(EV_{hi})]$ [e.g., [**98**]]. Note, however, that the optimal error variance for one application may not necessarily hold for other applications. Thus, the most successful inversion requires constructing the performance curve that is specific to the application [e.g., Curve #2] for comparison with Curve #1.

The range of acceptable error variances for an application also establishes error bars on the solution coefficients via **Equation 9.24**. Thus, the sensitivity analysis can be extended to investigate the range of environmentally reasonable solution coefficients that satisfy the observed data.

In summary, the ridge regression approach enhances the conditioning of the complete $\mathbf{A}^t\mathbf{A}$ matrix by adding $EV$ to the diagonal elements of the nearly co-linear or zero rows that is just large enough to stabilize the solution, yet small enough to maximize its predictive power. This approach substantially improves the accuracy of model predictions and optimizes available computer time and storage by implementing fast, efficient matrix decomposition algorithms like Cholesky factorization. Roughly 90% of the computational labor of inversion is taken up in evaluating the forward problem for the $\mathbf{A}$-coefficients, whereas the Cholesky factorization of $(\mathbf{A}^t\mathbf{A})$ and subsequent determination of $\mathbf{x}$ account for the remaining 10%. Thus, for the cost of computing and storing the **A**-coefficients, roughly nine sets of solutions can be prepared for sensitivity analysis.

A contrasting approach for stabilizing the solution for improved prediction accuracy is to trim the $\mathbf{A^tA}$ matrix of its offending rows by *singular value decomposition*. This more computationaly intensive approach involves the *eigenvalue-eigenvector decomposition* of the $\mathbf{A^tA}$ matrix to identify its non-zero rows that are effectively linearly independent within working precision. The improved solution is obtained from the *generalized inverse* that is constructed from the maximum subset of linearly independent rows in the system [e.g., [**57, 50, 103, 36**]].

## 9.4 GENERALIZED LINEAR INVERSION

Another measure of an ill-conditioned $\mathbf{A^tA}$ matrix is its *condition number* defined as the ratio $(\lambda_{max}/\lambda_{min})$ or its inverse $[RCOND]$, the *reciprocal condition number*, where $\lambda_{max}$ and $\lambda_{min}$ are respectively the largest and smallest eigenvalues of $\mathbf{A^tA}$ [e.g., [**59**]]. Well-conditioned systems are characterized by $RCOND \approx 1$, where each **x**-coefficient accounts for roughly an equal measure of variation in the **b**-observations. Poorly conditioned systems, on the other hand, exhibit $RCOND \approx 10^{-K}$, where the solution coefficients can usually be expected to have $K$ fewer significant figures of accuracy than the elements of $\mathbf{A^tA}$. Thus, **x** may have no significant figures if RCOND is so small that in floating point arithmetic it is negligible when compared to unity - i.e., where $RCOND + 1.0 = 1.0$ in the computer's working precision, the derived solution may be useless for application no matter how well the $\hat{\mathbf{b}}$-predictions match the **b**-observations.

In modern practice, the computationally efficient Cholesky factorization of $\mathbf{A^tA}$ is used to estimate $RCOND$ as a measure of the sensitivity of the solution **x** to errors in $\mathbf{A^tA}$ and the weighted observation vector $\mathbf{A^tb}$ [e.g., [**15, 25**]]. However, to design the generalized inverse, the complete collection of eigenvalues and related eigenvectors is usually considered, where the commitment of computational labor is several times greater than required for a ridge regression solution of the full matrix system with fast linear equation solvers like Cholesky factorization.

### 9.4.1 EIGENVALUES AND EIGENVECTORS

Eigenvalues and eigenvectors are characteristic or unique scalars and vectors, respectively, associated with the system $\mathbf{A^tAx} = \mathbf{A^tb}$ [or more generally the equivalent system $\mathbf{Ax} = \mathbf{b}$], which linearly transform the vector **x** into the vector $\mathbf{A^tb}$ [or more generally the vector **b**] according to the linear model or relationship defined by the coefficients of $\mathbf{A^tA}$ [or more generally the coefficients of **A**]. Formally for the least squares system, the vector **x** is an eigenvector of $\mathbf{A^tA}$ if **x** is a non-zero vector and $\lambda$ is a scalar [which may be zero] so that $\lambda\mathbf{x} = \mathbf{A^tb}$. The scalar $\lambda$ is an eigenvalue of $\mathbf{A^tA}$ only if a non-zero vector $\mathbf{x}$ exists so that $\lambda\mathbf{x} = \mathbf{A^t b}$. At most $m$ not necessarily distinct values of $\lambda$ and associated eigenvectors can be found for the $\mathbf{A^t A}$ matrix that satisfy $\mathbf{A^t A x} = \mathbf{A^t b}$.

In consideration of the above remarks, the system

$$\mathbf{A^t A x} = \mathbf{A^t b} = \lambda\mathbf{x}, \tag{9.25}$$

can be rewritten as

$$(\mathbf{A^t A} - \lambda\mathbf{I})\mathbf{x} = 0, \tag{9.26}$$

where $\mathbf{I}$ is the identity matrix [**Equation 2.7**]. Using Cramer's Rule [e.g., **Equation 9.8**], each $x_j \in \mathbf{x}$ can be found from

$$x_j = \frac{0}{|\mathbf{A^t A} - \lambda\mathbf{I}|}, \tag{9.27}$$

since the determinant of any matrix formed with a zero column vector inserted from the right-hand side of **Equation 9.26** will be zero. Rewriting the above equation gives

$$|\mathbf{A^t A} - \lambda\mathbf{I}|x_j = 0, \tag{9.28}$$

which, excepting for the trivial case of $x_j = 0 \ \forall \ j$, can hold only if

$$|\mathbf{A^t A} - \lambda\mathbf{I}| = 0. \tag{9.29}$$

This determinant can be expanded in a polynomial of degree $m$ [e.g., **Figure 2.3**] and solved for the $m$-eigenvalues, $\lambda_j \ \forall \ j = 1, 2, \ldots, m$. From this determinant, it is also clear that the matrix $\mathbf{A^t A}$ is singular if any eigenvalue is zero. The eigenvector, $\mathbf{x_j}$, corresponding to each eigenvalue, $\lambda_j$, is obtained by inserting the eigenvalue into **Equation 9.26** and solving the resulting system of simultaneous equations for the eigenvector's coefficients.

In summary, to extract the $m$ eigenvalues and associated eigenvectors, the determinant $|\mathbf{A^t A}|$ and the $m$-roots of its characteristic polynomial equation must be found, and the corresponding $m$-sets of $m$-simultaneous equations must be solved, respectively. These computations can be numerically arduous to perform, but computer codes for efficient eigensystem calculations are available from *MATLAB*, *Mathematica*, *Mathcad* and other modern linear equation solving packages. [20] also provides elegant numerical examples for conducting these eigensystem calculations.

### 9.4.2 SINGULAR VALUE DECOMPOSITION

From the above discussion, the relative magnitudes of the eigenvalues clearly provide a measure of the degree of linear dependency in the design matrix and uncertainty in the solution coefficients. Specifically, $RCOND = 1$ indicates complete linear independence of the rows of $\mathbf{A^t A}$ and minimum solution variance $\sigma_{\mathbf{x}}$, whereas $RCOND = 0$ marks a singular system with no solution, and $RCOND \longrightarrow 0$ flags a near-singular, unstable system where $\sigma^2_{\mathbf{x}} \longrightarrow \infty$. The quantitative effect of the eigenalues on the solution variance $\sigma^2_{\mathbf{x}}$ can be studied using the *singular value decomposition* [SVD] of the design matrix $\mathbf{A}$.

In general, the $[n \times m]$ matrix $\mathbf{A}$ can be factored into

$$\mathbf{A} = \mathbf{U}\mathbf{\Lambda}\mathbf{V}^{\mathbf{t}}, \tag{9.30}$$

where $\mathbf{V}$ is the $[m \times m]$ matrix whose columns are the eigenvectors of $\mathbf{A}^{\mathbf{t}}\mathbf{A}$, $\mathbf{\Lambda}$ is the $[n \times m]$ matrix of zeroes except for the diagonal elements of the top $m$-rows that hold the $m$-eigenvalues of $\mathbf{A}$ [which are the square roots of the non-zero eigenvalues that are common to both $\mathbf{A}^{\mathbf{t}}\mathbf{A}$ and $\mathbf{A}\mathbf{A}^{\mathbf{t}}$], and $\mathbf{U}$ is the $[n \times n]$ matrix with columns of the eigenvectors of $\mathbf{A}\mathbf{A}^{\mathbf{t}}$ that correspond to the $m$-eigenvalues of $\mathbf{A}$ [e.g., **[57]**]. The eigenvectors are orthogonal so that $\mathbf{U}^{\mathbf{t}}\mathbf{U} = \mathbf{I}_{\mathbf{n}\times\mathbf{n}}$ and $\mathbf{V}^{\mathbf{t}}\mathbf{V} = \mathbf{I}_{\mathbf{m}\times\mathbf{m}}$, and thus

$$\mathbf{A}^{\mathbf{t}} = \mathbf{V}\mathbf{\Lambda}^{\mathbf{t}}\mathbf{U}^{\mathbf{t}} \text{ and } \mathbf{U} = \mathbf{A}\mathbf{V}\mathbf{\Lambda}^{-1}. \tag{9.31}$$

Using these results shows that $(\mathbf{A}^{\mathbf{t}}\mathbf{A})^{-1} = \mathbf{V}\mathbf{\Lambda}^{-2}\mathbf{V}^{\mathbf{t}}$, where $\mathbf{\Lambda}^{-2} = (\mathbf{\Lambda}^{\mathbf{t}}\mathbf{\Lambda})^{-1}$. Thus, the solution variance in **Equation 9.16** becomes

$$\sigma^2_{\mathbf{x}} = (\mathbf{V}\mathbf{\Lambda}^{-2}\mathbf{V}^{\mathbf{t}})\sigma^2_{\mathbf{b}}, \tag{9.32}$$

and the presence of very small eigenvalues [i.e., $(\lambda_j) \longrightarrow 0$] causes the elements of $\mathbf{\Lambda}^{-2}$ to blow up [i.e., $(1/\lambda^2_j) \longrightarrow \infty$] and dominate solution variance. Thus, uncertainties in the data observations do not influence the singularity conditions of the system, but these conditions can inordinately amplify data errors in the solution variance $\sigma^2_{\mathbf{x}}$.

The SVD of $\mathbf{A}^{\mathbf{t}}\mathbf{A}$ permits forming the *Lanczos* or *natural* inverse given by

$$(\mathbf{A}^{\mathbf{t}}\mathbf{A})^{-1}\mathbf{A}^{\mathbf{t}} = \mathbf{V}\mathbf{\Lambda}^{-1}\mathbf{U}^{\mathbf{t}} \tag{9.33}$$

to manage the variance of the least squares solution. The procedure involves computing the generalized inverse that incorporates the less singular portion of the linear system by simply deleting eigenvalues [and corresponding eigenvectors] that are smaller than an appropriate threshold value [e.g., **[75, 57]**].

### 9.4.3 GENERALIZED LINEAR INVERSE

Formally, the generalized inverse $\mathbf{H}$ is given by

$$\mathbf{H} = \mathbf{V}\mathbf{\Lambda}^{-1}\mathbf{U}^{\mathbf{t}}. \tag{9.34}$$

Here, $\mathbf{H}$ is $[m \times n]$, $\mathbf{V}$ is $[m \times p \leq m]$, $\mathbf{\Lambda}^{-1}$ is $[p \times p]$, $\mathbf{U}^{\mathbf{t}}$ is $[p \times n]$, and the matrices $\mathbf{V}, \mathbf{\Lambda}^{-1}$, and $\mathbf{U}^{\mathbf{t}}$ originate as in **Equation 9.30** except that

1. the eigenvalues below a specified threshold value [and their related eigenvectors] are deleted, and

2. the inverse eigenvalue coefficients in $\mathbf{\Lambda}^{-1}$ are arranged in increasing order - i.e., $\lambda_{11}^{-1} < \lambda_{22}^{-1} < \lambda_{33}^{-1} < \cdots < \lambda_{pp}^{-1}$.

The rationale for deleting the smaller eigenvalues [and the related eigenvectors in $\mathbf{V}$ and $\mathbf{U}^{\mathbf{t}}$] is that they correspond to elements of $\mathbf{A}$ that are either poorly determined or nearly linear multiples of each other. These elements provide negligible additional information concerning the relationship between $\mathbf{x}$ and $\mathbf{b}$, and thus the eigenvalues and related eigenvectors are simply deleted from the problem.

Using the generalized inverse, which always exists, yields the generalized solution of the linear system $\mathbf{Ax} = \mathbf{b}$ given by

$$\mathbf{x_G} \equiv \mathbf{HAx} = \mathbf{Hb}. \tag{9.35}$$

According to **[50]**, an optimal generalized inverse is marked by the following conditions:

1. $\mathbf{AH} \approx \mathbf{I_{n\times n}}$ and the model fits the data because $\hat{\mathbf{b}} = \mathbf{Ax_G} = \mathbf{b}$;
2. $\mathbf{HA} \approx \mathbf{I_{m\times m}}$, and the model's parameters are well resolved because $\mathbf{x_G} = \mathbf{x}$;
3. and $\sigma^2_{\mathbf{x_G}}$ is minimum.

The degree to which $\mathbf{H}$ satisfies the three criteria above depends on the presence of linear dependency in $\mathbf{A}$, the sufficiency, consistency, and accuracy of the data, and the choice of the threshold level [i.e., the $p$-value] for the eigenvalues retained in the calculation of $\mathbf{H}$.

#### 9.4.3.1 *Information Density*

The independence of the observations in the linear inversion can be estimated from the *information density matrix* given by

$$\mathbf{S} \equiv \mathbf{AH} \approx \mathbf{UU^t}. \tag{9.36}$$

**[103]** showed that for overconstrained systems where $n > m$, the degree to which $\mathbf{S} \approx \mathbf{I_{n\times n}}$ is indicative of the degree of independence of the data. In other words, the model $\mathbf{x_G}$ will fit certain combinations of the data depending on how much $\mathbf{S} \approx \mathbf{I_{n\times n}}$ because $\hat{\mathbf{b}}_\mathbf{G} \equiv \mathbf{Ax_G} = \mathbf{AHb} = \mathbf{Sb}$.

A measure of the degree to which $\mathbf{S} \approx \mathbf{I_{n\times n}}$ is estimated by the *deltaness* of the rows of $\mathbf{S}$ given by

$$\Delta s_k = \sum_{j=1}^{n} [s_{kj} - \delta_{kj}]^2 \ \forall \ k = 1, 2, \ldots, n, \tag{9.37}$$

where the *Dirac delta function* [$\delta_{kj}$] is unity for $k = j$ and zero for $k \neq j$. The generalized inverse $\mathbf{H}$ minimizes $\Delta s_k$ and thus provides the best information density.

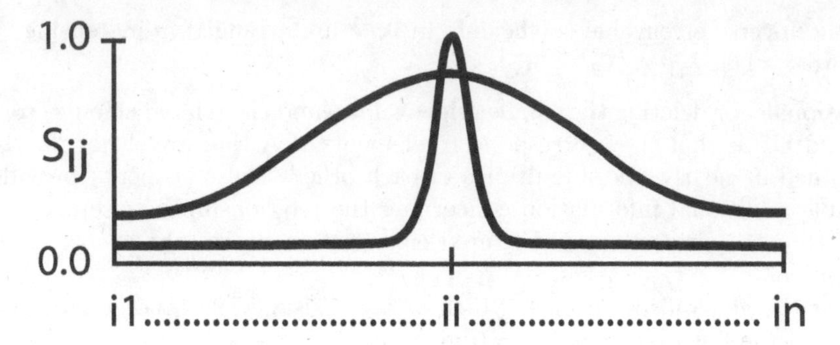


**Figure 9.3** The more the $i$-th row of the information density matrix **S** is a spike function about its diagonal element, the greater is the importance of the corresponding observation point $[b_i]$ to the solution.

The size of the diagonal element $s_{ii} \in S$ is an estimate of the importance of the corresponding observation point $b_i$ to the solution of the model [**Figure 9.3**]. For example, where $s_{ii} \approx 1$, the corresponding data point is nearly independent, and thus is providing information to the problem that is not contained in other observations. Alternatively, if a row of **S** displays only a broad peak centered about $s_{ii}$, then considerable dependency occurs in the observations - i.e., several [usually adjacent] data points are contributing similar information to the solution.

The information density matrix **S** can make important contributions to the design and interpretation of the inverse problem. For example, the **S**-matrix for a solution can be examined to gain insight on the interdependency of the observations, which may be particularly important in joint inversion problems or where large numbers of observations are considered. For these cases, the **S**-matrix brings focus on the observations with the greatest sensitivity to the solution parameters so that the less sensitive observations might be discarded to improve computational efficiency, or additional unknowns might be added to the model to improve the fit of the observations and the information density **S**. Additionally, in the design of experiments, it may be useful to simulate the observed data and various models, where the **S**-matrix identifies the observations which are critical to determining an acceptable model.

#### 9.4.3.2 *Model Resolution*

[4] showed that for underdetermined systems where $m > n$, the *resolution matrix* given by

$$\mathbf{R} \equiv \mathbf{HA} \approx \mathbf{VV}^{\mathrm{t}} \tag{9.38}$$

provides a measure of the uniqueness of the solution's parameters. For linear systems which are not underdetermined, $\mathbf{R} = \mathbf{I}_{\mathbf{m}\times\mathbf{m}}$ always. In general, the model $\mathbf{x_G}$ is related to the solution of $\mathbf{Ax} = \mathbf{b}$ by $\mathbf{x_G} = \mathbf{Rx} = \mathbf{Hb}$ so that if

$\mathbf{R} = \mathbf{I_{m \times m}}$, the solution is mathematically unique in that each parameter of $\mathbf{x_G}$ is perfectly resolved. If $\mathbf{R} \approx \mathbf{I_{m \times m}}$, then $\mathbf{R}$ is a nearly diagonal matrix where each parameter $x_l \in \mathbf{x_G}$ is a weighted average of nearby parameters.

Thus, the degree to which $\mathbf{R} \approx \mathbf{I_{m \times m}}$ is a measure of the resolution of the model's parameters, where rows of $\mathbf{R}$ are called *resolving kernels* with deltaness given by

$$\Delta r_l = \sum_{q=1}^{m} [r_{lq} - \delta_{lq}]^2 \ \forall \ l = 1, 2, \ldots, m. \tag{9.39}$$

Small values of $\Delta r_l$ [i.e., for the $l$-th row of $\mathbf{R}$] imply that the corresponding resolving kernel is approximately a delta function and the related $l$-th variable is well resolved and "unique." The generalized inverse $\mathbf{H}$ minimizes $\Delta r_l$ for all $l$-rows and for all choices of the number of non-zero eigenvalues $p$.

The width of the resolving kernel is a measure of the range of model space that is averaged to obtain $x_l$. If the widths of the resolving kernels are small, then the details of $\mathbf{x_G}$ are well resolved. However, where the width is large, then each $x_l$ [even if it is known to a high degree of accuracy by variance estimates] must be considered an *average* over the parameter space. Thus, the resolving kernels are indicative of the relative trade-offs between model parameters. In particular, large values of the off-diagonal elements $r_{lq} \in \mathbf{R}$ with $l \neq q$ indicate that the parameters $x_l$ and $x_q$ are correlated, either positively or negatively. In either case, the resolution of $x_l$ and $x_q$ separately is degraded and a re-parameterization [i.e., a new formulation of the forward model] may be desireable.

The GLI-literature contains numerous illustrations of the role of resolving kernels in evaluating model resolution. **[103]**, for example, considers the resolving kernels for the inversion of earthquake surface waves and their resonances [i.e., free oscillations] on a simulated surface-to-core density structure of the Earth.

### 9.4.3.3 *Model Variance*

The coefficients of the solution variance from the GLI-adaptation of **Equation 9.32** are estimated from

$$\sigma_G^2(x_k) = [\sum_{j=1}^{p} (v_{kj}/\lambda_j)^2]\sigma_{\mathbf{b}}^2 \ \forall \ v_{kj} \ \in \ \mathbf{V} \text{ and } k = 1, 2, \ldots, p, \tag{9.40}$$

assuming that the observations are statistically independent, and that $\sigma_{\mathbf{b}}^2$ is the variance of the problem [**Chapter 4.2.1**]. Either $\sigma_{\mathbf{b}}^2$ is known or it is estimated from the error of the fit $\mathbf{e} = \mathbf{A}\mathbf{x_G} - \mathbf{b}$ by $\sigma_{\mathbf{b}}^2 = \mathbf{e}^t\mathbf{e}/(n-p)$ for degrees of freedom $(n - p)$.

**[103]** notes that dividing each observation by the standard deviation of the observations $\sigma_{\mathbf{B}}$ yields a dimensionless set of *standardized* observations with the standardized variance $\sigma^2_{S\mathbf{tb}} \approx 1$. Accordingly, if $\sigma^2_{S\mathbf{tb}} << 1$, then the $\sigma^2_{\mathbf{b}}$-estimate is too large compared to the internal consistency of the observations. In this case, $\sigma^2_{S\mathbf{tb}}$ is still applicable, but with a *safety factor* overestimate from the overestimated variance of the observations. On the other hand, if $\sigma^2_{S\mathbf{tb}} >> 1$, then either the variance estimate of the observations is too small, or the model is inadequate to fit the data. Inadequacy of the selected model can generally be seen by examining the error of fit for each of the model's parameters.

### 9.4.4 SENSITIVITY ANALYSIS

The relationship between the information density [**S**] and model resolution [**R**] matrices and solution variances $[\sigma^2_G(x_k)]$ illustrates the classical trade-off problem in assessing inversion sensitivity for maximizing the observational data fit and model detail, while also minimizing errors in the model's parameter estimates. Specifically, large cut-off $p$-values utilizing even very small eigenvalues maximize the fit of the model's predictions to the data, as well as the model's details, but also the errors in the model's parameter estimates. In general, every inverse problem has a resolution limit that depends on **1)** the design matrix **A** for the assumed model, **2)** the sufficiency, accuracy, and consistency of the observations **b**, and **3)** the design of the inverse.

The design of an effective generalized linear inverse **H** requires the choice of the threshold eigenvalue $\lambda_t$ that defines a cut-off $p$-value, which in turn maximizes information density and model resolution and minimzes solution variance [**Chapter 4.2.2**]. A common approach is to arrange the eigenvalues in ascending or descending order and inspect the resulting eigenvalue spectrum for prominent features that may point to an appropriate $p$-value as shown in **Figure 9.4**. Experience suggests that this approach is especially effective for smaller numbers of eigenvalues [e.g., **Figure 9.4**] where the threshold eigenvalue may be more apparent. At large numbers, the corresponding eigenvalue spectra tend to be smoother and increasingly less diagnostic and exploring for $\lambda_t$ requires substantial commitments of computational labor.

## 9.5 SUMMARY

This chapter considered practical solutions for five formulations of the array system $\mathbf{Ax} = \mathbf{b}$, including

1. the *even system* of linearly independent variables where $p = m = n$ and $\mathbf{S} = \mathbf{I_{n \times n}}$ and $\mathbf{R} = \mathbf{I_{m \times m}}$;
2. the *overconstrained system* where $p = m < n$ and $\mathbf{S} \neq \mathbf{I_{n \times n}}$ and $\mathbf{R} = \mathbf{I_{m \times m}}$;
3. the *strictly underdetermined system* where $p = n < m$ and $\mathbf{S} = \mathbf{I_{n \times n}}$ and $\mathbf{R} \neq \mathbf{I_{m \times m}}$;

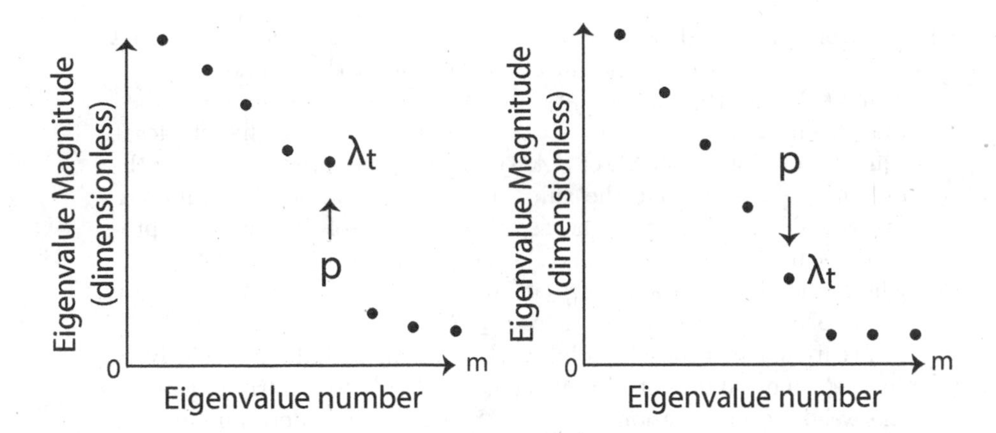


**Figure 9.4** Exploring eigenvalue plots for effective $p$-values. In the left plot, the knee in the curve suggests a threshold value $\lambda_t$ where smaller eigenvalues may be discarded, or equivalently set to zero. In the right plot, the threshold value $\lambda_t$ and corresponding $p$-value may be indicated where the spectrum's gradient flattens.

4. the *simultaneously underdetermined and overconstrained system* where $p < m < n$ and $\mathbf{S} \neq \mathbf{I_{n\times n}}$ and $\mathbf{R} \neq \mathbf{I_{m\times m}}$; and
5. the *strictly underdetermined system* where $p < n < m$ and $\mathbf{S} \neq \mathbf{I_{n\times n}}$ and $\mathbf{R} \neq \mathbf{I_{m\times m}}$.

Accordingly, the information density matrix **S** offers the possibility of improving solutions for systems **#1-**, **#4-**, and **#5-**above, whereas the model resolution matrix **R** may be useful in enhancing solutions for systems **#3-**, **#4-**, and **#5-**above. Note that both the information density [**S**] and model resolution [**R**] matrices may be established for any inverse - i.e., the direct, least squares, ridge regression, and generalized linear inverse.

In general, the most computationally intense element of inversion is evaluating the forward model for the **A**-coefficients. Using fast linear equation solvers, like the Cholesky facorization of $(\mathbf{A^tA})$, to determine $\mathbf{x}$ involves only about 10% of the computational effort that computed **A**. By contrast, the computational labor in determining the eigenvalues and eigenvectors of $(\mathbf{A^tA})$ and implementing the generalized inverse for $\mathbf{x}$ is many times greater and often impractical to apply at the large scales of modern data inverse problems. Thus, modern inversions increasingly invoke fast linear equation solvers for obtaining appropriate solutions. However, the most effective implementation of matrix inversion is based on the spectral forward model, which can be computed with maximum efficiency and minimum assumptions.

## 9.6 KEY CONCEPTS

1. Electronic data analysis is based on analyzing a digital forward model ($\mathbf{Ax}$) of the data ($\mathbf{b}$) using inversion to obtain the unknown coefficients ($\mathbf{x}$) from the known data and $\mathbf{A}$-coefficients of the prescribed model. This fundamental inverse problem (i.e., $\mathbf{Ax} = \mathbf{b}$) has the least squares solution $\mathbf{x} = (\mathbf{A^tA})^{-1}\mathbf{A^tb}$ which yields predictions matching the data such that the sum of squared residuals is minimum. Statistical error bars may be ascribed to the solution and its predictions based on the variance of the solution $\sigma^2_{\mathbf{x}} = (\mathbf{A^tA})^{-1}\sigma^2_{\mathbf{b}}$, where the data variance $\sigma^2_{\mathbf{b}}$, if unknown, may be estimated by $\sigma^2_{\mathbf{b}} = (\mathbf{b^tb} - \mathbf{x^t}[\mathbf{A^tA}]\mathbf{x}/(n-2)\ \forall\ n \geq 3$.
2. Conventional methods of solving the system include the relatively cumbersome taking of the inverse $[\mathbf{A^tA}]^{-1}$ to multiply against the weighted observation vector $\mathbf{A^tb}$ using either determinants by Cramer's Rule or elementary row operations. More efficient solutions use elementary row operations to partially or fully diagonalize the system $\mathbf{A^tAx} = \mathbf{A^tb}$ in place by the respective Gauss or Gauss-Jordan elimination methods. An even more efficient matrix procedure evaluates the solution coefficients recursively from the system's series representation based on the Cholesky factors of $\mathbf{A^tA}$.
3. The inversion solution is never unique due to errors in the data, presumed model coefficients, and truncated calculations. Theoretical ambiguities also may contribute, like the source ambiguity of potential fields, for example, where in theory the data can be matched by an indefinite number of forward models. Thus, a sensitivity analysis based on the solution variance is commonly required to establish an 'optimal' set or range of data solutions that conform to the inversion errors, as well as any additional constraints on the environmental properties of the problem.
4. The utility of a solution to do more than simply match the observations depends on the conditioning of $\mathbf{A^tA}$. The system's conditioning can be tested by the condition number $[COND]$ or its reciporcal [RCOND], where $COND \equiv \lambda_{max}/\lambda_{min}$ for the respective maximum and minimum eigenvalues $\lambda_{max}$ and $\lambda_{min}$ of $\mathbf{A}$, which also are the square roots of the eigenvalues of $\mathbf{A^tA}$.
5. In particular, $COND = RCOND = 1$ marks an optimally performing, well-conditioned system with $|\mathbf{A^tA}| > 0$ and minimum $\sigma^2_{\mathbf{x}}$ so that each element of $\mathbf{A^tb}$ is represented within working precision by a fully independent row in $\mathbf{A^tA}$. On the other hand, $COND \to \infty$ or $RCOND \to 0$ marks a more poorly performing, ill-conditioned system with $|\mathbf{A^tA}| \to 0$ and $\sigma^2_{\mathbf{x}} \to \infty$ because $\mathbf{A^tA}$ holds within working precision a row of nearly zero values or a row that is nearly a linear multiple of another row.

6. In general, every regression is performed with limited working precision, and hence tends to some degree to be ill-conditioned with $COND > 1$. Thus, even though the forward model is non-singular in theory, it is always prudent to investigate the regression's conditioning effects before committing to the results. Insights on the adopted forward model's conditioning are commonly obtained from $\mathbf{A^tA}$ using ridge regression and singular value decomposition.
7. Ridge regression effectively contaminates each element of $\mathbf{A}$ with random noise drawn from the Gaussian distribution $G[0, \sqrt{EV}]$ to obtain the mean squared solution $\mathbf{x_D} \equiv [\mathbf{A^tA} + (EV) \times \mathbf{I}]^{-1}\mathbf{A^tb}$ with variance $\sigma^2_{\mathbf{x_D}} = [\mathbf{A^tA} + (EV) \times \mathbf{I}]^{-1}(\mathbf{A^tA})[\mathbf{A^tA} + (EV) \times \mathbf{I}]^{-1}$.
8. Ridge regression beefs up the diagonal elements of $\mathbf{A^tA}$ with an $EV$-value that is just large enough to make the the $[m - p]$ ill-conditioned rows linearly independent, but also small enough to minimize distortions of the $[p]$ linearly independent rows. This optimal $EV$-value can be resolved from the trade-off diagram that compares over appropriate $EV$-values the data predictions with the observed data against the behavior of the differences in model estimates of the analytical objective [e.g., derivative, integral, interpolation, extrapolation, etc.].
9. Singular value decomposition effectively trims $\mathbf{A^tA}$ of its $[m - p]$ ill-conditioned rows to yield the generalized linear solution $\mathbf{x_G} \equiv \mathbf{V\Lambda^{-1}U^tb}$ with variance $\sigma^2_{\mathbf{x_G}} = (\mathbf{V\Lambda^{-2}V^t})\sigma^2_{\mathbf{b}}$. Here, $\mathbf{V}$ is the $[m \times (p \leq m)]$ matrix with columns that are the $p$ linearly independent eigenvectors of $\mathbf{A^tA}$, $\mathbf{\Lambda}$ is the $[p \times p]$ diagonal matrix of the $p$ effectively non-zero eigenvalues of $\mathbf{A}$ arranged in decreasing order, and $\mathbf{U}$ is the $[n \times p]$ matrix with columns that are the $p$ linearly independent eigenvectors of $\mathbf{AA^t}$.
10. The trade-off diagram comparing the gradients in the eigenvalues against solution variances as a function of $m$ facilitates selecting the 'optimal' or 'threshold' $p$-value. Specifically, the maximum number of eigenvalues at which solution variance is tolerable yields an effective threshold $p$-estimate.
11. The generalized linear inverse $\mathbf{H_{(m\times n)}} \equiv \mathbf{V_{(m\times p)}\Lambda^{-1}_{(p\times p)}U^t_{(p\times n)}}$ and related solution $\mathbf{x_G} = \mathbf{Hb}$ exist for systems that are even $[p = m = n]$, overconstrained $[p = m < n]$, strictly underdetermined $[p \leq n < m]$, and simultaneously underdetermined and overconstrained $[p < m < n]$.
12. In general, the deltaness of the rows of the information density matrix $\mathbf{S} \equiv \mathbf{A[H]} \rightarrow \mathbf{A[(A^tA)^{-1}A^t]}$ identifies the subset of observations in $\mathbf{b}$ that are linearly independent in the context of the presumed forward model. Using this subset of observations, improved lower solution variances result for the overconstrained, simultaneously underdetermined and overconstrained, and the strictly underdetermined systems where $p < n < m$.

13. Additionally, the deltaness of the rows of the model resolution matrix $\mathbf{R} \equiv [\mathbf{H}]\mathbf{A} \rightarrow [(\mathbf{A}^t\mathbf{A})^{-1}\mathbf{A}^t]\mathbf{A}$ identifies the subset of unknowns in $\mathbf{x_G}$ that are linearly independent in the context of the presumed forward model. Using this subset of unknowns, improved lower solution variances result for all strictly underdetermined, and simultaneously underdetermined and overconstrained systems.
14. The generalized linear inversion is several times more computationally laborious to implement than ridge regression where Cholesky factors and other fast linear equation solvers can readily process the damped versions of $\mathbf{A}^t\mathbf{A}$. However, the most efficient matrix inversion is described in the next chapter where the spectral forward model represents orthogonally gridded data by inverting a diagonalized, and thus fully conditioned $\mathbf{A}^t\mathbf{A}$ matrix.

# 10 Spectral Analysis

## 10.1 OVERVIEW

*For electronic analysis, data are typically rendered into a grid that is most efficiently modeled by the fast Fourier transform (FFT) which has largely superseded spatial domain calculations. This spectral model accurately represents the data as the summation of cosine and sine series over a finite sequence of discrete uniformly spaced wavelengths. These wavelengths correspond to an equivalent discrete sequence of frequencies called wavenumbers which are fixed by the number and uniform spacing of the gridded data points. The FFT dominates modern data analysis because of the associated computational efficiency and accuracy, and minimal number of assumptions.*

*Spectral analysis involves the inverse processes of Fourier analysis and synthesis. Fourier analysis decomposes the signal in the data domain into the amplitudes of cosine and sine waves over a fixed number of frequencies that are specified by the number of data and the data interval. Fourier synthesis, on the other hand, converts these respective real and imaginary cosine and sine transforms into data domain models of the data and their derivative and integral attributes.*

*The synthesis can be equivalently achieved using the combination of either cosine and sine transforms, or power and phase spectra. Here, the power spectrum is derived from the frequency-by-frequency sums of the squared cosine and sine transform coefficients or amplitudes, and the phase spectrum results from the frequency-by-frequency arctangents of the ratio of negative sine transform amplitude to the related positive coefficient of the cosine transform.*

*When applied to orthogonally gridded data, the Fourier analysis defines an especially elegant inversion for the sinusoidal amplitudes, where only the diagonal elements of* $\mathbf{A}^t\mathbf{A}$ *are computed because the off-diagonal elements are the zero-valued cross-products of the orthogonal wave functions.*

*Thus, the Fourier transform is applicable to any dataset that can be plotted or mapped onto paper, electronic screens, and other graphical displays. In general, mappable data are single valued over the independent variables, with finite magnitudes, maxima, minima, and discontinuities, and thus also satisfy the Dirichlet conditions on taking the transform.*

*Numerical computations of forward and inverse transforms involve Gibbs' oscillating approximation errors, and data gridding errors that propagate as wavenumber leakage, aliasing, and resolution problems in estimating the data's spectral attributes. However, for nearly any application, these errors can be significantly reduced by numerical modifications [i.e., window carpentry] of the input data.*

*Gibbs' errors mostly occur as edge effects at the data margins and oscillating shadow effects or ringing around data spikes, discontinuities, and other sharp higher frequency features. In addition, wavenumber leakage occurs when the data do not repeat themselves at the data margins, resulting in sharp edge effects in the data domain that introduce significant Gibbs' error in the wavenumber components. Both types of error can be managed by tapering the data along the margins smoothly to zero, or by padding the edges with a synthetic rind of several rows and columns filled in with zeros or possibly the values of the edge-adjacent rows and columns folded out across the margin.*

*Wavenumber aliasing refers to erroneous longer wavelength components of the spectrum introduced by using a grid interval that is too large to capture significant shorter wavelength features of the data. By Shannon's sampling theorem, the Nyquist or shortest wavelength of any grid is two times the grid interval or 3 data points. Establishing the Nyquist interval in practice typically involves gridding and plotting the data at increasingly smaller intervals and selecting the largest grid interval at which the smaller interval data plots no longer change appreciably.*

*Wavenumber resolution errors result from the Fourier transform producing wavenumber amplitudes that are symmetric about the Nyquist frequency. Thus, the Fourier transform of the $N$-point signal with $N$ degrees of freedom in the data domain results in a signal spectrum with only [$N/2$] degrees of freedom in the frequency domain. This loss can seriously limit resolving details in spectral studies involving small numbers [$N \leq 10$] of observations. For analyzing small datasets, however, a statistical maximum entropy-based spectrum with $N$ unique wavenumber coefficients is available.*

## 10.2 INTRODUCTION

The above matrix procedures can be extended to the inversion of gridded data for their representative sine and cosine or spectral models that, in turn, can be evaluated for the differential and integral properties of the data. Spectral analysis requires that the data satisfy the Dirichlet conditions, whereby the signal $\mathbf{b}(t)$ in the interval $(t_1, t_2)$ must be single-valued, and have only a finite number of maxima, minima, and discontinuities, and finite energy (i.e., $\int_{t_1}^{t_2} |\mathbf{b}(t)|\, dt < \infty$). Note that $t$ is a dummy parameter representing space or time or any other independent variable against which the dependent variable $\mathbf{b}$ is mapped. Most mappable datasets satisfy the Dirichlet conditions, and thus can be converted by inversion into representative sine and cosine series for analysis.

Spectral analysis includes the inverse processes of Fourier analysis and Fourier synthesis. Fourier analysis decomposes the $t$-data domain signal $\mathbf{b}(t)$ into representative wave functions in the $f$-frequency domain by taking the

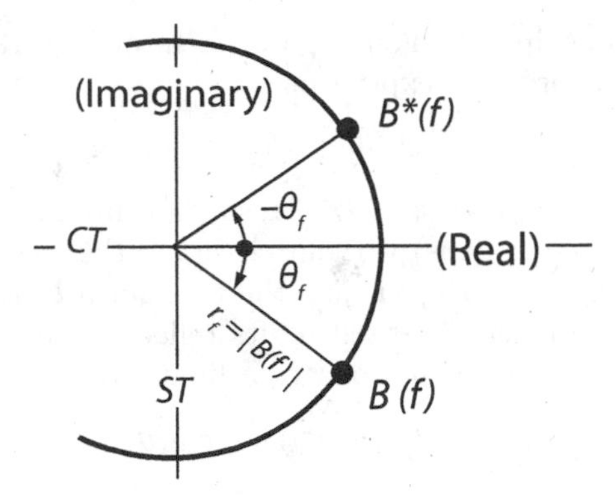


**Figure 10.1** The projection of the Fourier transform $\mathcal{B}(f)$ and its complex conjugate $\mathcal{B}^{\star}(f)$ in the polar complex coordinate $(CT, ST)$-plane. [Adapted from [**43**].]

Fourier transform (FT) of $\mathbf{b}(t)$ given by

$$\mathcal{B}(f) = \int_{-\infty}^{\infty} \mathbf{b}(t)e^{-j(2\pi f)t}\, dt = \int_{-\infty}^{\infty} \mathbf{b}(t)\cos(\omega t)\, dt - j\int_{-\infty}^{\infty} \mathbf{b}(t)\sin(\omega t)\, dt, \tag{10.1}$$

where the imaginary number $j = \sqrt{-1}$. The wave expression of the exponential follows Euler's formula

$$e^{\pm j\omega} = \cos(\omega) \pm j\sin(\omega), \tag{10.2}$$

with the angular frequency $\omega = 2\pi f = 2\pi/\lambda$ for linear frequency $f$ and wavelength $\lambda = 1/f$.

The spectral analysis literature is complicated by the variety of notations used to describe the properties of the Fourier transform. For example, the cosine and sine integrals in **Equation 10.1** are also known as the cosine transform ($CT$) and sine transform ($ST$), respectively, so that the Fourier transform may be expressed as the complex number

$$\mathcal{B}(f) = CT(f) - jST(f). \tag{10.3}$$

As shown in **Figure 10.1**, $\mathcal{B}(f)$ can be uniquely mapped into the $(CT, ST)$-plane using polar coordinates $(r_f, \theta_f)$, where

$$\begin{aligned}
CT(f) &= r_f\cos(\theta_f) &&\Longrightarrow \textbf{Cosine Transform} \text{ or the real part } (\equiv R(f)) \text{ of } \mathcal{B}(f),\\
ST(f) &= r_f\sin(\theta_f) &&\Longrightarrow \textbf{Sine Transform} \text{ or the imaginary part } (\equiv I(f)) \text{ of } \mathcal{B}(f),\\
r_f^2 &= |\mathcal{B}(f)|^2 = CT^2(f) + ST^2(f) &&\Longrightarrow \textbf{Power Spectrum} \text{ of } \mathcal{B}(f),\\
r_f &= |\mathcal{B}(f)| = \sqrt{CT^2(f) + ST^2(f)} &&\Longrightarrow \textbf{Amplitude Spectrum} \text{ of } \mathcal{B}(f), \text{ and}\\
\theta_f &= \tan^{-1}[-ST(f)/CT(f)] &&\Longrightarrow \textbf{Phase Spectrum} \text{ of } \mathcal{B}(f).
\end{aligned} \tag{10.4}$$

From these relationships, the Fourier transform of **Equation 10.1** also has the complex polar coordinate expression

$$\mathcal{B}(f) = r_f[\cos(\theta_f)] - j[\sin(\theta_f)] = r_f e^{-j\theta_f} = |\mathcal{B}(f)|e^{-j\theta_f}. \tag{10.5}$$

The phase angle $\theta_f$ satisfies $-\pi < \theta_f \leq \pi$ so that any integral multiple of $2\pi$ can be added to $\theta_f$ (in radians) without changing the value of $\mathcal{B}(f)$.

The complex conjugate $\mathcal{B}^\star(f)$ is just the transform $\mathcal{B}(f)$ with the polarity of its imaginary component reversed by its reflection about the real axis as shown in **Figure 10.1**. Thus, in polar coordinate form

$$\mathcal{B}^\star(f) = CT(f) + jST(f) = r_f \cos(\theta_f) + j\sin(\theta_f) = r_f e^{j\theta_f} = |\mathcal{B}(f)|e^{j\theta_f}, \tag{10.6}$$

where $|\mathcal{B}^\star(f)| = |\mathcal{B}(f)| = r_f$ and $\theta^\star = \tan^{-1}[ST(f)/CT(f)] = -\theta_f$. In addition, $\mathcal{B}^\star(f)\mathcal{B}(f)$ and $\sqrt{\mathcal{B}^\star(f)\mathcal{B}(f)}$ are the respective power and amplitude spectra.

Fourier synthesis, on the other hand, is the inverse operation that reconstitutes the $t$-data domain signal $\mathbf{b}(t)$ from its wave model $\mathcal{B}(f)$ in the $f$-frequency domain. Fourier synthesis is achieved by taking the inverse Fourier transform (IFT)

$$\mathbf{b}(t) = \int_{-\infty}^{\infty} \mathcal{B}(f)e^{+j(2\pi t)f}\, df. \tag{10.7}$$

Notational complexities in the spectral analysis literature can result, however, from the different forms of normalization that may be applied to the forward and inverse Fourier transforms. For example, the generic transforms $\mathcal{B}(\omega) = a_1 \int_{-\infty}^{\infty} \mathbf{b}(t)e^{\pm j\omega t}dt$ and $\mathbf{b}(t) = a_2 \int_{-\infty}^{\infty} \mathcal{B}(\omega)e^{j\omega t}d\omega$ require that $a_1 \times a_2 = 1/2\pi$. This normalization in the literature is variously achieved by setting $a_1 = 1$ and $a_2 = 1/2\pi$, or $a_1 = a_2 = 1/\sqrt{2\pi}$, or $a_1 = 1/2\pi$ and $a_2 = 1$.

## 10.3 ANALYTICAL FOURIER TRANSFORM

In the previous section, it was shown that the analytical integral transform, and its cosine and sine transforms ($CT, ST$), and amplitude and phase spectra ($|\mathcal{B}(f)|, \theta_f$) provide three equivalent representations of data. To illustrate these and other essential properties of the analytical transform that carry over to data analysis, consider ***Example 10.3a*** after **[11]** that involves the $t$-data domain function in **Figure 10.2(a)** given by

$$\begin{aligned} \mathbf{b}(t) &= \beta e^{-\alpha t} \quad \forall \quad t \geq 0 \\ &= 0 \quad \forall \quad t < 0. \end{aligned} \tag{10.8}$$

Taking the analytical Fourier transform of $\mathbf{b}(t)$ gives

$$\begin{aligned} \mathcal{B}(f) &= \textstyle\int_0^{\infty} \beta e^{-\alpha t} e^{-j2\pi f t}dt = \beta \int_0^{\infty} e^{-(\alpha + j2\pi f)t}dt \\ &= -[(\beta e^{-(\alpha+j2\pi f)t})/(\alpha + j2\pi f)]|_0^{\infty} = \frac{\beta}{(\alpha + j2\pi f)}. \end{aligned} \tag{10.9}$$

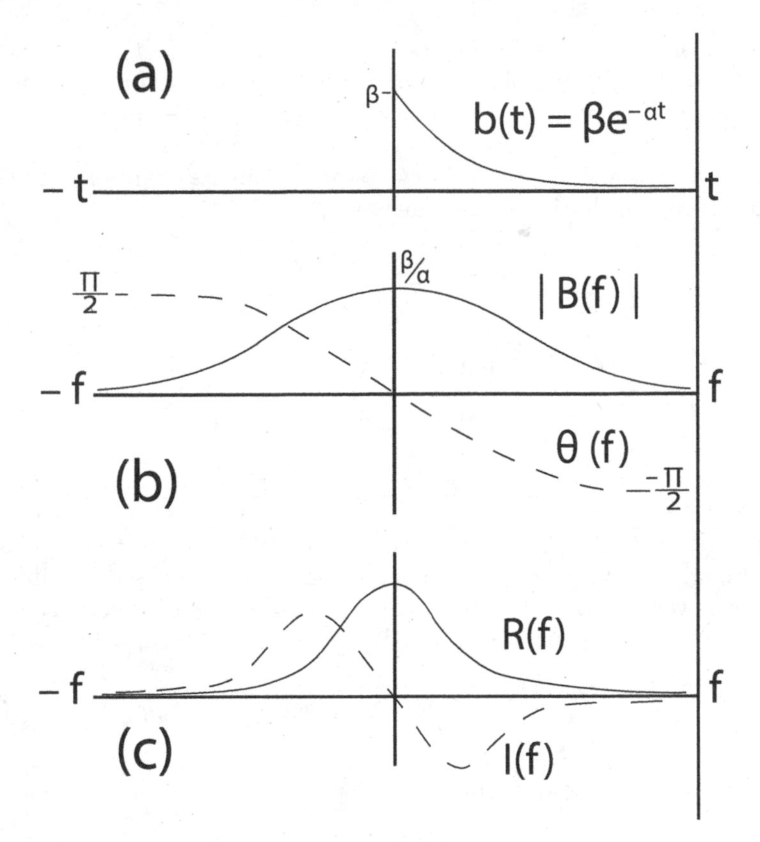


**Figure 10.2** Data **(a)** and frequency **(b and c)** domain representations of **Equation 10.10**. [Adapted from **[43]**.]

The last expression above is the most compact form of the transform and thus the most likely one to be reported in the literature. To expand it into its equivalent $(CT, ST)$ and $(|\mathcal{B}(f)|, \theta_f)$ representations, the integral transform in **Equation 10.9** must be separated into its real and imaginary components $(R(f), I(f))$. This can be done by multiplying the integral transform with the special form of unity given by the ratio of the complex conjugate of the denominator to itself (i.e., $[\alpha - j2\pi f]/[\alpha - j2\pi f]$) so that

$$\begin{aligned} R(f) &= \beta\alpha/(\alpha^2 + (2\pi f)^2) &&= CT(f), \\ I(f) &= -2\pi f\beta/(\alpha^2 + (2\pi f)^2) &&= ST(f), \\ |\mathcal{B}(f)| &= \beta/\sqrt{\alpha^2 + (2\pi f)^2} &&= \textbf{amplitude spectrum}, \text{ and} \\ \theta_f &= \tan^{-1}(-2\pi f/\alpha) &&= \textbf{phase spectrum}. \end{aligned} \tag{10.10}$$

**Figures 10.2(b)** and **10.2(c)** respectively illustrate the equivalent $(CT(f), ST(f))$- and $(|\mathcal{B}(f)|, \theta_f)$-representations of the integral transform. Note also that the cosine transform $(CT(f))$ and amplitude spectrum $(|\mathcal{B}(f)|)$

are both a symmetric or even function with $R(f) = R(-f)$, whereas the sine transform ($ST(f)$) and phase spectrum ($\theta_f$) are an antidiametric or odd function with $I(f) = -I(-f)$ that integrates to zero over symmetrical limits $(-f, f)$.

Consider now the Fourier synthesis of $\mathbf{b}(t)$ from its transform $\mathcal{B}(f)$ in **Equation 10.9**, which can be expanded as

$$\mathcal{B}(f) = \frac{\beta}{(\alpha + j2\pi f)} = \frac{\beta\alpha}{\alpha^2 + (2\pi f)^2} - j\frac{2\pi f\beta}{\alpha^2 + (2\pi f)^2}. \tag{10.11}$$

Taking the IFT of $\mathcal{B}(f)$ and expanding by Euler's formula gives

$$\begin{aligned}\hat{\mathbf{b}}(t) &= \int_{-\infty}^{\infty}[\tfrac{\beta\alpha\cos(2\pi ft)}{\alpha^2+(2\pi f)^2} + \tfrac{2\pi f\beta\sin(2\pi ft)}{\alpha^2+(2\pi f)^2}]df + \\ &+j\int_{-\infty}^{\infty}[\tfrac{\beta\alpha\sin(2\pi ft)}{\alpha^2+(2\pi f)^2} - \tfrac{2\pi f\beta\cos(2\pi ft)}{\alpha^2+(2\pi f)^2}]df.\end{aligned} \tag{10.12}$$

Now, the imaginary integral above goes to zero because the product of an even and odd function or vice versa is an odd function. However, the real integral contains two even functions because the product of two even functions or two odd functions is even. Thus, the IFT reduces to the real term

$$\hat{\mathbf{b}}(t) = \int_{-\infty}^{\infty}[\frac{\beta\alpha\cos(2\pi ft)}{\alpha^2 + (2\pi f)^2} + \frac{2\pi f\beta\sin(2\pi ft)}{\alpha^2 + (2\pi f)^2}]df. \tag{10.13}$$

Standard tables of integrals show that

$$\begin{aligned}\int_{-\infty}^{\infty}\frac{\cos(ax)}{b^2 + x^2}dx &= \tfrac{\pi}{b}e^{-ab} \quad \forall\, a > 0, \text{ and} \\ \int_{-\infty}^{\infty}\frac{x\sin(ax)}{b^2 + x^2}dx &= \pi e^{-ab} \quad \forall\, a > 0,\end{aligned} \tag{10.14}$$

so that the IFT becomes

$$\begin{aligned}\hat{\mathbf{b}}(t) = \frac{\beta\alpha}{(2\pi)^2}[\frac{\pi}{(\alpha/2\pi)}e^{-(2\pi t)(\alpha/2\pi)}] &+ \frac{2\pi\beta}{(2\pi)^2}[\pi e^{-(2\pi t)(\alpha/2\pi)}] \quad \text{or} \\ = \frac{\beta}{2}e^{-\alpha t} + \frac{\beta}{2}e^{-\alpha t} &= \beta e^{-\alpha t} \quad \forall\, t > 0.\end{aligned} \tag{10.15}$$

Accordingly, $\mathbf{b}(t)[\simeq \hat{\mathbf{b}}(t)]$ and $\mathcal{B}(f)$ form a Fourier transform pair that is commonly indicated by the double-arrow notation

$$\begin{aligned}\mathbf{b}(t) &\Longleftrightarrow \mathcal{B}(f) \quad \text{or} \\ \beta e^{-\alpha t} &\Longleftrightarrow \frac{\beta}{\alpha + j2\pi f} = \frac{\beta}{\sqrt{\alpha^2 + (2\pi f)^2}}e^{j[\tan^{-1}(-2\pi f/\alpha)]}.\end{aligned} \tag{10.16}$$

**Table 10.1**
**Analytical properties of the Fourier transform.**

| DATA DOMAIN | $\Longleftrightarrow$ | FREQUENCY DOMAIN |
|---|---|---|
| Addition<br>$\mathbf{b}(t) \pm \mathbf{c}(t)$ | $\Longleftrightarrow$<br>$\Longleftrightarrow$ | Addition<br>$\mathcal{B}(f) \pm \mathcal{C}(f)$ |
| Multiplication<br>$\mathbf{b}(t) \times \mathbf{c}(t)$ | $\Longleftrightarrow$<br>$\Longleftrightarrow$ | Convolution<br>$\mathcal{B}(f) \otimes \mathcal{C}(f)$ |
| Convolution<br>$\mathbf{b}(t) \otimes \mathbf{c}(t)$ | $\Longleftrightarrow$<br>$\Longleftrightarrow$ | Multiplication<br>$\mathcal{B}(f) \times \mathcal{C}(f)$ |
| Differentiation<br>$\partial^p \mathbf{b}(t)/\partial t^p$ | $\Longleftrightarrow$<br>$\Longleftrightarrow$ | Differentiation<br>$[j2\pi f]^p \mathcal{B}(f)$ |
| Integration<br>$\underbrace{\int\int\cdots\int}_{p} \mathbf{b}(t)\, dt^p$ | $\Longleftrightarrow$<br>$\Longleftrightarrow$ | Integration<br>$[j2\pi f]^{-p} \mathcal{B}(f)$ |
| Symmetry<br>$\mathcal{B}(t)$ | $\Longleftrightarrow$<br>$\Longleftrightarrow$ | Symmetry<br>$\mathbf{b}(-f)$ |
| Data scaling<br>$\mathbf{b}(ct)$ | $\Longleftrightarrow$<br>$\Longleftrightarrow$ | Inverse scale change<br>$\frac{1}{\lvert c \rvert}\mathcal{B}(\frac{f}{c})$ |
| Inverse scale change<br>$\frac{1}{\lvert c \rvert}\mathbf{b}(\frac{t}{c})$ | $\Longleftrightarrow$<br>$\Longleftrightarrow$ | Frequency scaling<br>$\mathcal{B}(cf)$ |
| Data shifting<br>$\mathbf{b}(t \pm t_o)$ | $\Longleftrightarrow$<br>$\Longleftrightarrow$ | Phase shift<br>$\mathcal{B}(f)e^{\pm j2\pi f t_o}$ |
| Data Modulation<br>$\mathbf{b}(t)e^{\pm j2\pi t f_o}$ | $\Longleftrightarrow$<br>$\Longleftrightarrow$ | Frequency shifting<br>$\mathcal{B}(f \mp f_o)$ |
| Even function<br>$\mathbf{b_e}(t)$ | $\Longleftrightarrow$<br>$\Longleftrightarrow$ | Real ($R$)<br>$\mathcal{B}_\mathbf{e}(f) = R_e(f)$ |
| Odd function<br>$\mathbf{b_o}(t)$ | $\Longleftrightarrow$<br>$\Longleftrightarrow$ | Imaginary ($I$)<br>$\mathcal{B}_\mathbf{o}(f) = jI_o(f)$ |
| Real function<br>$\mathbf{b}(t) = \mathbf{b_r}(t)$ | $\Longleftrightarrow$<br>$\Longleftrightarrow$ | Real even, Imaginary odd<br>$\mathcal{B}(f) = R_e(f) + jI_o(f)$ |
| Imaginary function<br>$\mathbf{b}(t) = j\mathbf{b_i}(t)$ | $\Longleftrightarrow$<br>$\Longleftrightarrow$ | Real odd, Imaginary even<br>$\mathcal{B}(f) = R_o(f) + jI_e(f)$ |

**Table 10.1** summarizes additional analytical properties of the Fourier transform. The mathematical operations are obtained by taking the FT of the $t$-data domain operation or the IFT of the $f$-frequency domain operation. They are useful in considering the design of transforms and the related filtering operations. For example, any set of measured data describes a real function, which accordingly has the transform consisting of both real (i.e., cosine transform) and imaginary (i.e., sine transform) coefficients. The averaging operator for smoothing data, on the other hand, is an even function of real numbers where only the real coefficients of the transform are required in the design and application of this filter.

## 10.4 NUMERICAL FOURIER TRANSFORM

The bulk of the extensive literature on Fourier transforms deals with their applications to continuously differentiable analytical functions. Most data processing applications, however, involve inaccurately measured, discrete, gridded samples of unknown analytical functions that require the application of the numerical discrete Fourier transform (DFT). Developed mostly in the latter half of the $20^{th}$ century with the advent of digital electronic computing, the DFT can model any gridded dataset. This numerical modeling approach is also more straightforward to implement than analytical Fourier transforms and thus has wide application in modern data analysis.

The DFT is a streamlined adaptation of the general finite Fourier series (FS) used to represent non-analytical, discretely sampled, finite-length signals. In the general case of non-uniformly sampled data where the FS is applied, the full $\mathbf{A^tA}$ matrix must be computed. However, in electronic processing, the data are represented as an orthogonal data grid for which the $\mathbf{A^tA}$ matrix simply becomes the diagonal matrix because the off-diagonal elements involve cross-products of orthogonal wave functions that are zero [e.g., **[51, 20]**]. The language of these computations also becomes comparably abbreviated, invoking the term wavenumber to describe the gridded frequency or equivalent wavelength properties of the data array.

The simplicity of wavenumber analysis can be illustrated by considering the DFT of a finite-length data profile $\mathbf{b}(t_n)$ uniformly spanning the interval $0 \leq t_n \leq \Delta t(N-1)$. Here, the number of signal samples or coefficients, $N$, is the fundamental period of the signal, and $\Delta t$ is the uniform sampling or station interval. Thus, the signal consists of the sequence of $N$ uniformly spaced coefficients $b_n \;\forall\; n = 0, 1, 2, \cdots, (N-1)$ given by

$$\mathbf{b}(t_n) \equiv (b_0 = b[t_0], b_1 = b[t_1], b_2 = b[t_2], \cdots, b_{N-1} = b[t_{N-1}]). \tag{10.17}$$

In applying the DFT, it is assumed that the frequency interval of $N$ samples is the fundamental period over which the signal is uniformly defined and that outside this interval the signal repeats itself periodically. The example in **Figure 10.3** shows the relevant book keeping details for a uniformly sampled signal of $N = 21$ coefficients.

To Fourier analyze the signal $\mathbf{b}(t_n)$, its one-dimensional discrete Fourier transform $\mathcal{B}(\omega_k)$ is estimated using the discrete frequency variable $\omega_k = k\Delta\omega = k(2\pi/N) \;\forall\; k = 0, 1, 2, \cdots, (N-1)$. Here, $k \equiv$ wavenumber at which the coefficients $\mathsf{b}_k \in \mathcal{B}(\omega_k)$ are estimated by

$$\hat{\mathsf{b}}_k = \sum_{n=0}^{N-1} b_n e^{-j(n)\frac{2\pi}{N}(k)} \qquad \forall\; k = 0, 1, 2, \cdots, (N-1), \tag{10.18}$$

where $\hat{\mathsf{b}}_k \simeq \mathsf{b}_k$. This equation effectively samples the continuous Fourier transform $\mathcal{B}(\omega)$ at the uniform frequency interval of $(2\pi/N)$ so that the

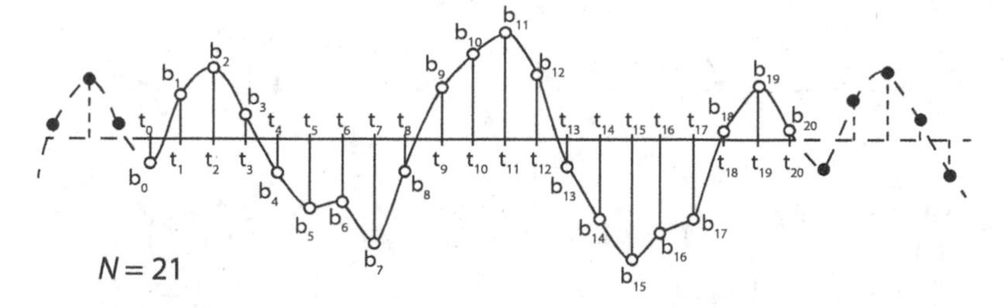

**Figure 10.3** The bookkeeping details in taking the Fourier transform of a 21-point signal. The transform assumes that the data (open circles) are infinitely repeated (black dots) on all sides of the signal. [Adapted from [**43**].]

wavenumber coefficients $\hat{\mathbf{b}}_k$ are actually the Fourier series coefficients to within the scale factor of $(1/N)$. **Figure 10.4** illustrates how the amplitude $CT_{22}$ of the 22 htz cosine transform component is taken by summing the products of the discrete $\mathbf{b}(t)$-signal amplitudes (solid dots) and the corresponding unit amplitude cosine values [solid dots]. Likewise, the amplitude $ST_{22}$ of the 22 htz sine transform component is obtained by the sum of the products of the signal amplitudes [solid dots] and the corresponding unit amplitude sine values [open dots]. The amplitude coefficients are commonly plotted as the *stick diagrams* or *peridograms* in the middle and bottom-left panels of **Figure 10.4**.

In theory these spectra involve both negative and positive frequencies, $f$, but only the amplitudes for $f \geq 0$ are generally computed and plotted in practice because they are symmetric about the origins of the cosine and amplitude spectra, and anti-symmetric about the origins of the sine and phase spectra [e.g., **Figure 10.2**]. Thus, the amplitude symmetries translate into considerable computational efficiency because the data modeling requires estimating only about half of the amplitudes of the complete spectrum.

The wavenumber $k$ relates the various uses of frequency and wavelength in different applications. For example, in geodata analyses, the focus is principally on linear frequency $f$ in the units of cycles per data interval $\Delta t$ and wavelength (or period) $\lambda$ $(= 1/f)$ in the units of the data interval $\Delta t$, whereas time series analysis commonly invokes the circular or angular frequency $\omega$ $(= 2\pi f)$ in the units of radians per data interval $\Delta t$. These various concepts are connected in terms of the wavenumber $k$ by

$$\omega_k = (\frac{2\pi}{N})k \longrightarrow f_k = (\frac{k}{N}) \ni \lambda_k = (\frac{N}{k})\Delta t = \frac{2\pi}{\omega_k}. \tag{10.19}$$

The harmonic frequencies defined by $k$ fix the range of frequencies or wavelengths over which the DFT can analyze the signal $\mathbf{b}(t_n)$. The $k = 0$ component, for example, corresponds to the infinite wavelength or zero frequency base level of the signal, which is also sometimes referred to in electrical terms

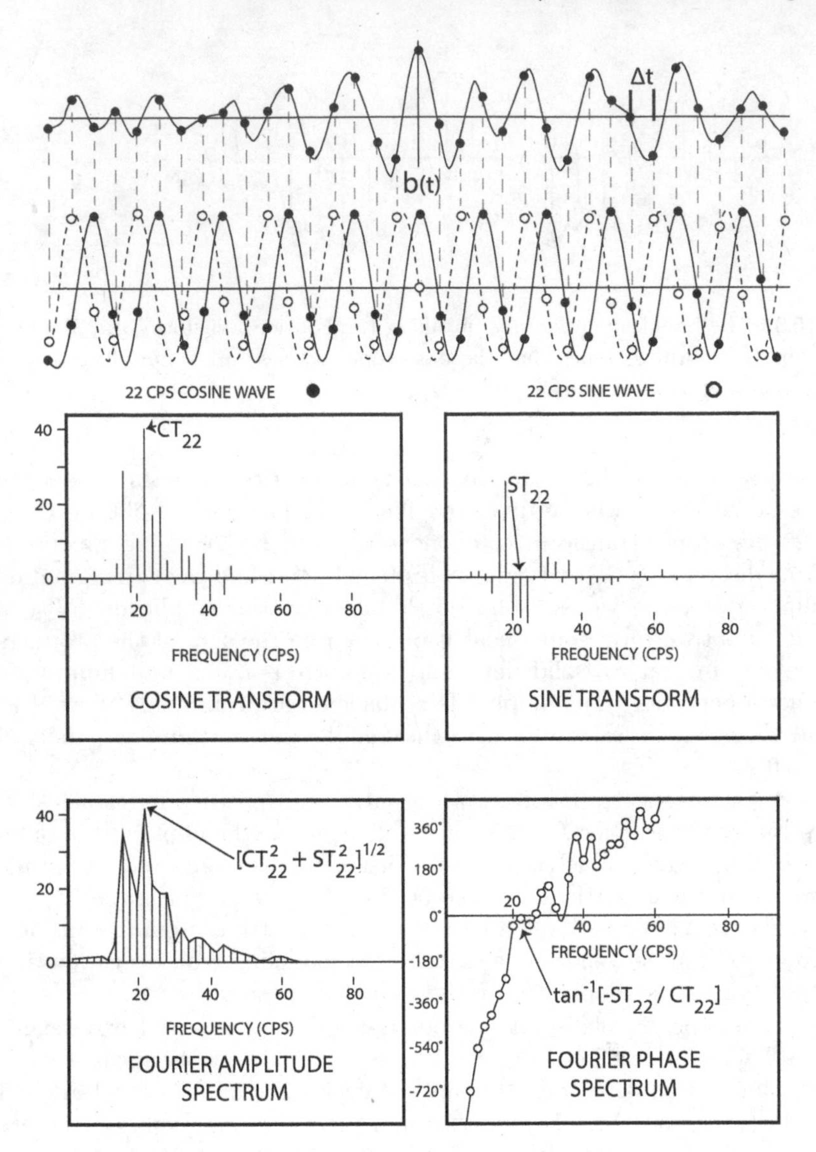


**Figure 10.4** Summing the cross-products of the $\mathbf{b}(t)$-signal values sampled at the $\Delta t$-interval with the corresponding unit amplitude cosine [solid dots] and sine [open dots] values [top panel] to take the discrete wavenumber amplitude components $CT_{22}$ and $ST_{22}$ for the respective 22 htz [or cps] cosine and sine transforms [middle panel], and amplitude and phase spectra [bottom panel]. [Adapted from **[76]**.]

**Table 10.2**
**Wavenumber structure of the discrete Fourier transform.**

| $k$ | NOMENCLATURE | $\omega_k$ | $f_k$ | $\lambda_k$ |
|---|---|---|---|---|
| 0 | Base level; DC-component | 0 | 0 | $\infty$ |
| 1 | Fundamental; First Harmonic | $2\pi/N$ | $1/N$ | $N\Delta t$ |
| 2 | Second Harmonic | $2\times\omega_1$ | $2\times f_1$ | $\lambda_1/2$ |
| $\vdots$ | $\vdots$ | $\vdots$ | $\vdots$ | $\vdots$ |
| $m$ | $m$-th Harmonic | $m\times\omega_1$ | $m\times f_1$ | $\lambda_1/m$ |
| $(N/2)-1$ | $[(N/2)-1]$-th Harmonic | $\pi-\omega_1$ | $\pi-f_1$ | $2\pi/(\pi-\omega_1)$ |
| $N/2$ | Nyquist | $\pi$ | $1/2$ | $2\times\Delta t$ |
| $\vdots$ | $\vdots$ | $\vdots$ | $\vdots$ | $\vdots$ |
| $N-1$ | $[N-1]$-th Harmonic | $-\omega_1$ | $-f_1$ | $-\lambda_1$ |

as the direct current or DC-component. The mean or a low-order trend surface is commonly removed from the data to make the DC-component zero so that the significant figures of the other coefficients are more sensitive to the remaining variations in the data.

The maximum finite wavelength component of the signal is the fundamental wavelength for $k=1$ given by $\lambda_1=N\Delta t$. This result seems counter-intuitive because the length of the signal with $N$ coefficients spaced uniformly at the interval $\Delta t$ should be $(N-1)\Delta t$. However, in applying the DFT, it is assumed that the signal repeats itself one data interval $\Delta t$ following the position of the last term or coefficient, $b_{(N-1)}$, so that the actual length of the signal is indeed $N\Delta t$.

The highest frequency (or minimum wavelength) signal component, on the other hand, is for $k=(N/2)$ where $\omega_{N/2}=\pi$. This limit of frequency (or wavelength) resolution is called the Nyquist frequency (or wavelength). At higher wavenumbers, the coefficients $\hat{\mathsf{b}}_{k>(\frac{N}{2})}$ can be found from the lower wavenumber coefficients $\hat{\mathsf{b}}_{k\leq(\frac{N}{2})}$ because $\hat{\mathsf{b}}_{(\frac{N}{2}+1)}=\hat{\mathsf{b}}^{\star}_{(\frac{N}{2}-1)}$, $\hat{\mathsf{b}}_{(\frac{N}{2}+2)}=\hat{\mathsf{b}}^{\star}_{(\frac{N}{2}-2)},\cdots,\hat{\mathsf{b}}_{(\frac{N}{2}+\frac{N}{2}-1)}=\hat{\mathsf{b}}^{\star}_{(1)}$. This folding or symmetry of coefficients makes for extremely efficient computation because only the relatively small number of coefficients up to and including the Nyquist wavenumber $k=N/2$ needs to be explicitly computed and stored to define the full DFT. **Table 10.2** summarizes the nomenclature, frequency, and wavelength attributes for each $k$-th wavenumber coordinate of the DFT(1D).

In summary, the wavenumber coefficient estimates $\hat{\mathsf{b}}_k$ (**Equation 10.18**) represent the discrete uniform sampling of the continuous transform $\mathcal{B}(f)$ at the interval $(1/N)$. The application of Euler's formula expands these coefficient estimates for their sine and cosine transforms, amplitude and phase spectra, and other components described in **Equation 10.4**.

The discrete signal coefficients $b_n \in \mathbf{b}(t_n)$, on the other hand, are synthesized or modeled from

$$\hat{b}_n = (\frac{1}{N}) \sum_{k=0}^{N-1} \hat{\mathsf{b}}_k e^{+j(n)\frac{2\pi}{N}(k)} \qquad \forall\, n = 0, 1, 2, \cdots, (N-1), \tag{10.20}$$

where $\hat{b}_n \simeq b_n$. To within the scale factor $(1/N)$, $\hat{\mathbf{b}}(t_n)$ represents the Fourier series estimate for the sequence $\mathbf{b}(t_n)$ from the Fourier coefficient estimates $\hat{\mathsf{b}}_k$. Hence, $\hat{\mathbf{b}}(t_n)$ is a periodic sequence which estimates $\mathbf{b}(t_n)$ only for $0 \leq n \leq (N-1)$ and $[\mathbf{b}(t_n) \simeq]\ \hat{\mathbf{b}}(t_n) \Longleftrightarrow \hat{\mathcal{B}}(f_k)\ [\simeq \mathcal{B}(f_k)]$.

The discussion of the DFT so far has focused on its application to single dimensioned profile data. Many studies, however, involve analyzing data in two dimensional arrays or maps. Fortunately, the one dimensional DFT can be readily extended to the representation and analysis of arrays in two or more dimensions.

Consider, for example, the $(N \times M)$ coefficients $b_{n,m}$ that sample the two dimensional signal $\mathbf{B}(t_{n,m})$ at uniform intervals $\Delta t_n$ and $\Delta t_m$, where $0 \leq n \leq (N-1)$, $0 \leq m \leq (M-1)$, and $N$, $M$, $\Delta t_n$, and $\Delta t_m$ take on arbitrary values. To Fourier analyze the signal $\mathbf{B}(t_{n,m})$, the coefficients $\mathsf{b}_{k,l} \in \mathcal{B}(f_{k,l})$ are estimated at wavenumber coordinates $(k, l)$ from

$$\hat{\mathsf{b}}_{k,l} = \sum_{n=0}^{(N-1)} \sum_{m=0}^{(M-1)} b_{n,m} e^{-j(m)\frac{2\pi}{M}(l)} e^{-j(n)\frac{2\pi}{N}(k)}, \tag{10.21}$$

for all $0 \leq k \leq (N-1)$ and $0 \leq l \leq (M-1)$ so that $\hat{\mathsf{b}}_{k,l} \simeq \mathsf{b}_{k,l}$. On the other hand, the discrete signal coefficients $b_{n,m} \in \mathbf{b}(t_{n,m})$ are synthesized from

$$\hat{b}_{n,m} = (\frac{1}{N \times M}) \sum_{k=0}^{(N-1)} \sum_{l=0}^{(M-1)} \hat{b}_{k,l} e^{+j(m)\frac{2\pi}{M}(l)} e^{+j(n)\frac{2\pi}{N}(k)}, \tag{10.22}$$

for all $0 \leq n \leq (N-1)$ and $0 \leq m \leq (M-1)$ so that $\hat{b}_{n,m} \simeq b_{n,m}$ and $\mathbf{B}(t_{n,m}) \simeq \hat{\mathbf{B}}(t_{n,m}) \Longleftrightarrow \hat{\mathcal{B}}(f_{k,l}) \simeq \mathcal{B}(f_{k,l})$.

**Figures 10.5** and **10.6** illustrate the basic organizational details of the wavenumber spectra for an array of $(N = 16) \times (M = 32)$ data values. For these Nyquist-centered spectra, the quadrants are arranged clockwise with the outside corners defining the baselevel or DC-component of each quadrant. Digital data are real numbers so that quadrants #3 and #4 are the complex conjugates of quadrants #1 and #2, respectively (**Figure 10.6a**). This symmetry of quadrants means that only one quadrant out of each of the two conjugate quadrant pairs needs to be computed to define the full spectrum. On the other hand, if the real numbers define an even function, then only one quadrant needs to be computed because the other three quadrants are symmetric to it (**Figure 10.6b**). Thus, for even, real functions like a radially symmetric averaging operator for data smoothing, for example, the filter's design is considered in only a single spectral quadrant.

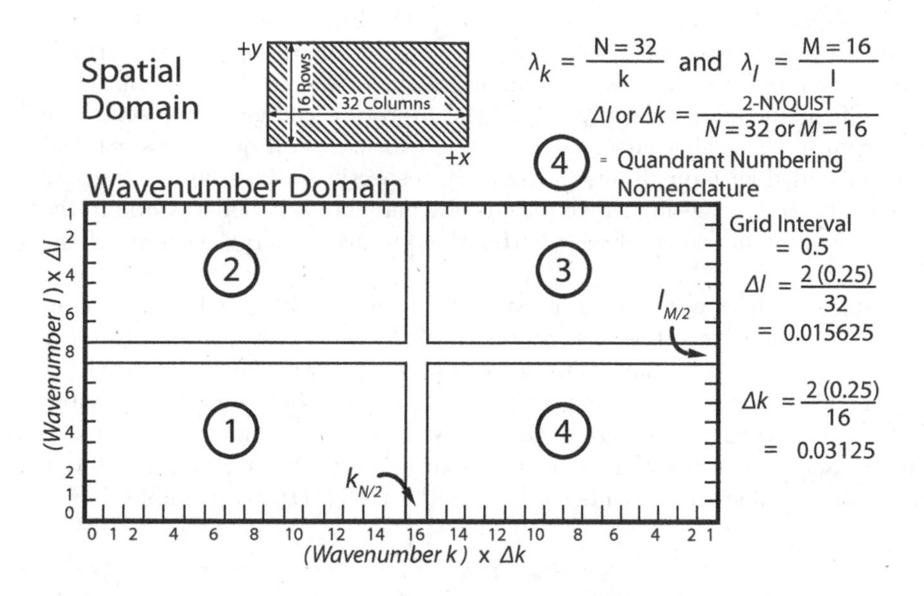


**Figure 10.5** The machine storage format for representing the discrete Nyquist-centered wavenumber transform of a 16 × 32 data array. [Adapted from [78].]

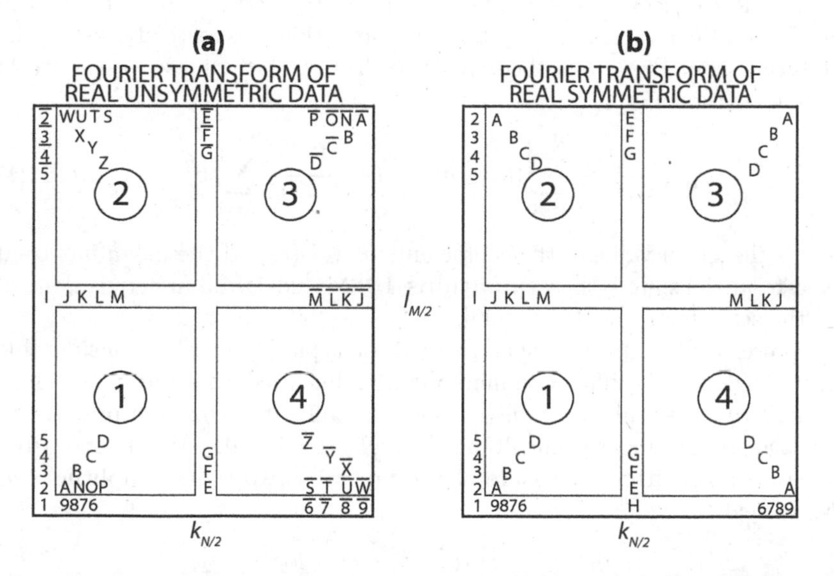


**Figure 10.6** Symmetry properties of the Nyquist-centered wavenumber spectrum for **(a)** non-symmetric and **(b)** symmetric real gridded data. [Adapted from [78].]

The Nyquist-centered arrangement of spectral quadrants facilitates computer processing objectives, but it is not ideal for displaying the spectrum's symmetry properties because the amplitudes near the Nyquist are marginal and difficult to follow visually. Thus, the literature commonly presents the spectrum in the higher energy, DC-centered format with quadrants #1 and #2 transposed with quadrants #3 and #4, respectively. This type of spectral plot concentrates the higher amplitude wavenumber components about the center of spectrum to emphasize better the dominant energy patterns of the data.

Using the linear addition property of transforms (**Table 10.1**), the dominant wavenumber components can be inversely transformed to estimate the corresponding data domain variations and the possible sources of this behavior. The amount of total signal power or energy that these wavenumber components account for is also quantified with the extension of this approach called Parseval's theorem. This theorem expresses the signal's power in terms of its wavenumber components so that for the DFT(1D), for example,

$$\sum_{n=0}^{N-1} b_n^2 = (1/N)\sum_{k=0}^{N-1} \mathsf{b}_k^2. \tag{10.23}$$

Thus, if a solution for a subset of wavenumber components is offered that accounts for say 49% of the signal's power, then the solution can also be said to explain this amount of the signal's power or 70% of its energy.

The physical concepts of signal power and energy can also be expressed in terms of statistical variance and standard deviation, respectively, using the $k \geq 1$ terms from Parseval's theorem. Specifically, the unbiased estimate of the signal's variance is

$$\sigma_{\mathsf{b}}^2 = \frac{1}{N-1}\sum_{n=0}^{N-1}(b_n - \bar{b})^2 = \frac{1}{N(N-1)}\sum_{k=1}^{N-1} \mathsf{b}_k^2, \tag{10.24}$$

where $\bar{b}$ is the mean value of the coefficients $b_n \in \mathbf{b}(t_n)$. If the mean has been removed from the signal, then **Equations 10.24** and **10.23** are equivalent to within the scale factor of $[1/(N-1)]$.

The power and amplitude spectra of data typically exhibit considerable dynamic range that is difficult to map out with linear contour intervals. Thus, to visualize the spectral variations most completely, it is common practice to contour the spectra logarithmically in decibels. Technically, the decibel (dB) is the base-10 logarithmic measure of the power of two signal amplitudes $b_1$ and $b_2$ defined by

$$(b_1^2/b_2^2)|_{dB} \equiv 10\log(b_1^2/b_2^2) = 20\log(b_1/b_2). \tag{10.25}$$

Hence, 6 dB is equivalent to the amplitude ratio of 2 or power ratio of 4, whereas 12 dB reflects the amplitude ratio of 4 and power ratio of 16. In application, each spectral coefficient is simply evaluated with **Equation 10.25** assuming that the denominator of the amplitude ratio is normalized to unity (i.e., $b_2 \equiv 1$). Plotting the power or amplitude spectral coefficients in dB-contours brings out the details of their variations much more comprehensively than conventional linear contouring can.

Up to this point, the basic numerical mechanics have been outlined for representing gridded data with the DFT. In practice, however, the numerical operations of the DFT are further streamlined into the fast Fourier transform (FFT). The FFT involves numerical engineering to evaluate the DFT using the computer operation of binary bit reversal to sort the data into the even numbered and odd numbered points.

For the FFT(1D), for example, the data profile is sorted into ordered even (i.e., $b_0, b_2, b_4, \cdots, b_{2n}$) and odd (i.e., $b_1, b_3, b_5, \cdots, b_{2n+1}$) point strings. The sorting is continued on the new data strings until strings are obtained containing only a single data point which is the transform of itself. Using weights of simple functions of $N$, the $(N)$ single-point transforms are combined into $(N/2)$ two-point transforms, which in turn are combined into $(N/4)$ four-point transforms, and so on recursively until the final N-point transform is obtained.

The FFT computation is extremely fast and efficient involving, for example, $N\ln(N)$ multiplications compared to the $N^2$ multiplications of the DFT [e.g., **[11]**]. The savings of computational effort that the FFT offers over the DFT in data analysis is especially significant for $N \geq 64$. In addition, the same efficient FFT computer code can compute the IDFT from the spectral coefficients $\hat{\mathsf{b}}_k$, but with the sign of the exponential reversed and the $(1/N)$ normalization applied. Thus, the FFT has greatly advanced modern data analysis and signal processing in general [e.g., **[51, 11]**].

### 10.4.1 NUMERICAL ERRORS

In spectral analysis, the effects of Gibbs' error $(G_e)$ must be limited in computing forward and inverse numerical transforms, as well as the inevitable errors of the data grid in representing the actual variations of the data. Data gridding errors can result in wavenumber leakage, aliasing, and resolution problems in estimating effective spectral properties.

#### 10.4.1.1 *Gibbs' Error*

**[32]** showed that even in the limit where $k \longrightarrow \infty$, the errors $(G_e)$ in estimating the forward and inverse numerical transforms are never zero (i.e., $G_e^2 \neq 0$). The signal in one domain always involves oscillating errors of approximation in the other domain upon transformation - i.e., Gibbs' error is ever present in **Equations 10.18** and **10.20** so that $\hat{\mathsf{b}}_k = \mathsf{b}_k + G_{e(DFT)}$ and $\hat{b}_n = b_n + G_{e(IDFT)}$, respectively. However, for nearly any application, Gibbs' error is or can be made sufficiently small that the approximations $\hat{\mathsf{b}}_k \simeq \mathsf{b}_k$ and $\hat{b}_n \simeq b_n$ are very accurate.

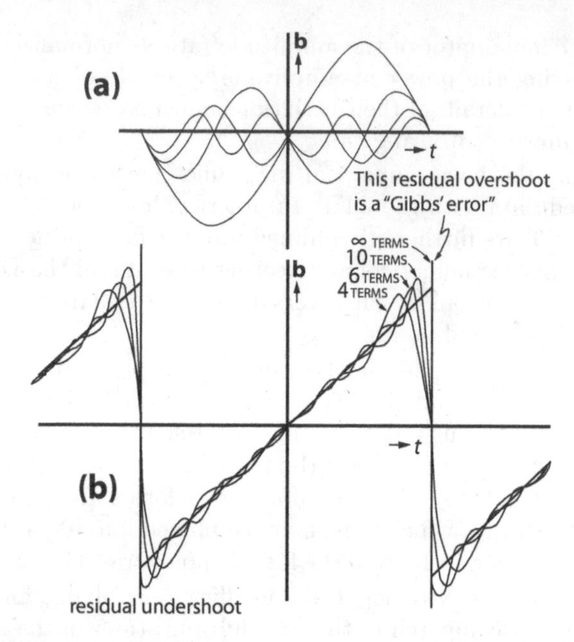


**Figure 10.7** Fourier synthesis of a saw tooth function. **(a)** The first four Fourier components of a sawtooth are $\sin(t), -(1/2)\sin(2t), (1/3)\sin(3t)$, and $-(1/4)\sin(4t)$. **(b)** Superposition of the Fourier components reconstitutes the saw tooth function to within over- and under-shooting Gibbs' errors. [Adapted from **[43]** after **[84]**.]

Gibbs' error commonly occurs as edge effects at the data margins and oscillating shadow effects or ringing around data spikes, discontinuities, and other sharp higher frequency features. **Figure 10.7** gives ***Example 10.4.1.1a*** of ringing in the synthesis of a saw tooth function for various numbers of terms in the IDFT series. The synthesis clearly becomes a better approximation of the saw tooth function as the number of terms in the series increases, but the approximation can never be perfect because even with an infinite series, some over- and under-shooting of the function's higher frequency features occurs.

Digital datasets typically exhibit smooth enough behavior that Gibbs' error is manifest mostly in edge effects involving a few data rows and columns along the data margins. The severity of Gibbs' error can be checked in practice by examining the residuals $[\mathbf{b}(t_n) - \hat{\mathbf{b}}(t_n)]$. Adding a rind or border of several data rows and columns to the dataset will help to migrate the edge effects out of the study area. Where no real data are available, a synthetic data rind constructed by folding out the rows and columns along the margins of the dataset can be tried. The most common approach, however, is to use window carpentry to smoothly taper the margins of the dataset to zero or some other base level. **[51]** summarize the statistical properties of the Bartlett, Hamming, Hanning, Parzen, Tukey, and other functions for window carpentry applications in spectral analysis.

#### 10.4.1.2 *Wavenumber Leakage*

Window carpentry to taper the edges of the dataset also reduces wavenumber leakage or errors in the estimated spectrum that occur when the length of the gridded dataset is not an integer multiple of the fundamental period of the data. This condition can result in sharp edge effects in the data domain that introduce or leak significant Gibbs' error into the wavenumber components.

As ***Example 10.4.1.2a***, consider **[11]**'s illustration of the wavenumber leakage problem with a single-frequency cosine wave in the $t$-domain. The signal was uniformly sampled at the rate of $T/8$ for $N = 32$ discrete data points over the length $T$ that was 8 times its fundamental period. The digitizing produced a dataset that would begin to repeat itself at the $[N + 1]$ coordinate of the its fundamental period so that the related DFT estimated the signal's correct frequency.

However, digitizing the cosine wave over the length 9.143 times its fundamental period produced a dataset that was sharply offset between the $N$ and $[N + 1]$ coordinates. The requirement for the digitized dataset to repeat itself at its fundamental period [i.e., at the $(N + 1)$ coordinate] introduced a sharp offset or edge with Gibbs' error that significantly corrupted the DFT.

To help mitigate the Gibbs' error from the artificial offset introduced by the sampling procedure and improve the dataset's DFT, window carpentry was applied that smoothly tapered the dataset's edges to zero amplitude. In particular, the tapered cosine wave coefficients were estimated from $\mathbf{b}'(t) = \mathbf{b}(t) \times \mathbf{h}(t)$, where $\mathbf{b}(t) = \cos(\frac{2\pi}{9.143}t)$ is the original signal and $\mathbf{h}(t)$ is the tapering function [e.g., the Hanning window **[11]**].

The tapering greatly muted the artificial edge effect and related Gibbs' error so that the tapered signal's DFT provided a much improved estimate of the waveform's frequency. In general, spectral analyses of data commonly employ window carpentry to limit wavenumber leakage, as well as edge effects in the synthesized data, and other Gibbs' errors generated by the margins of the datasets.

#### 10.4.1.3 *Wavenumber Aliasing*

Gridding at an interval that is too large to capture significant shorter wavelength features of the data can also corrupt the spectrum. Shannon's sampling theorem **[82]** points out that the shortest wavelength that can be resolved in a gridded dataset is given by the Nyquist wavelength $\lambda_{N/2} = 2 \times \Delta t$, where $\Delta t$ is the grid interval. In other words, proper gridding of a dataset requires that the Nyquist frequency $f_{N/2} = 1/\lambda_{N/2}$ of gridding be at least twice the highest frequency in the data.

Any data feature with frequency greater than $f_{N/2}$ by the amount $\Delta f$ will be represented in the data grid by an artificial longer wavelength feature with frequency $(f_{N/2} - \Delta f)$. This artificial feature is said to be the alias of the actual shorter wavelength data feature with frequency $(f_{N/2} + \Delta f)$. Thus, aliasing is purely an artifact of a sampling rate that is too low to represent the higher

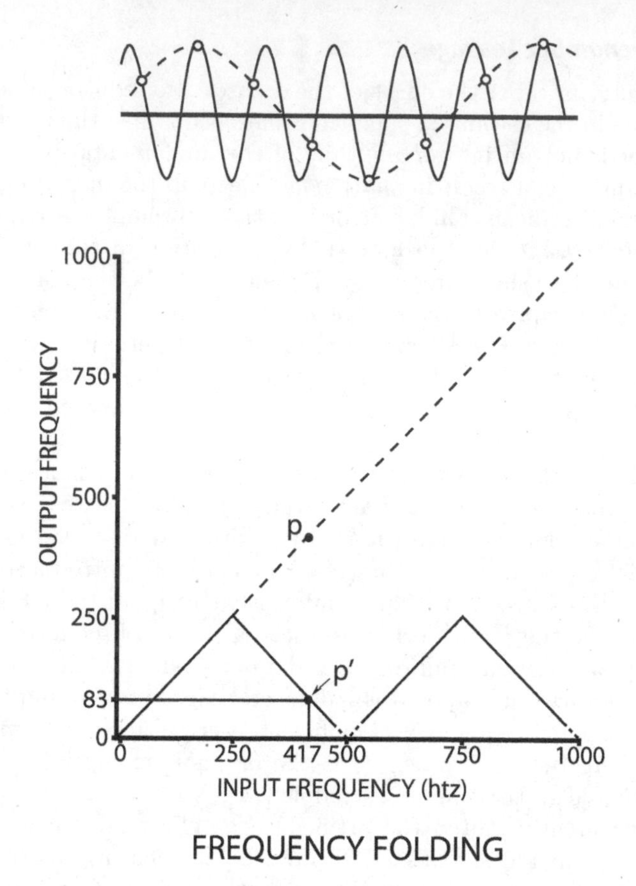


**Figure 10.8** Aliasing of a 417 htz signal sampled at 500 htz occurs at the folding frequency of 250 htz, which is the Nyquist frequency of sampling. [Adapted from [**43**] after [**76**].]

frequency data features. Inclusion of aliased features yields an incorrect data grid and a spectrum that result in data analysis errors.

As ***Example 10.4.1.3a***, consider the simple illustration of signal aliasing in **Figure 10.8**, where gridding of the solid 417 htz wave at the too-low rate of 500 htz (open dots) estimated the completely erroneous alias marked by the dashed 83 htz wave. The solid line folded in accordion fashion in the lower diagram gives the aliased frequencies for data frequencies above the sampling Nyquist rate $f_{N/2} = 250$ htz. Thus, to avoid aliasing in gridding the 417 htz signal, gridding rates of at least 834 htz and higher are necessary. Doubling the 500 htz sampling rate, for example, would sample at the open dots and midway between them for an effective grid estimate of the 417 htz signal.

Establishing an appropriate sampling rate that avoids aliasing for any set of observations is difficult in practice because the frequency components in the observations are usually poorly known. Excessive sampling of the dataset is always possible, but this can lead to unacceptable data processing requirements. Another, perhaps more effective, approach is to grid and plot the data at increasingly smaller grid intervals. An appropriate grid interval yields a data plot that does not change appreciably at smaller grid intervals.

#### 10.4.1.4 *Wavenumber Resolution*

The wavelength or equivalent frequency components between the harmonics are impossible to resolve with the DFT. Thus, the DFT reduces the degrees of freedom available to the problem by half because it transforms the $N$ data coefficients into $(N/2)$ unique wavenumber coefficients. Most digital datasets are large enough that the degrees of freedom lost in using the DFT are not a problem for achieving the objectives of the studies. However, the loss can seriously limit studies involving a small number ($N \leq 10$) of observations. Here, however, the statistical maximum entropy estimate of the power spectrum can be invoked, which can provide $N$ unique wavenumber coefficients at computational efficiencies that approach those of the FFT [e.g., [**13, 14, 43, 77, 80, 94**]].

## 10.5 KEY CONCEPTS

1. The principal format in modern digital data processing is the data grid or array that the fast Fourier transform (FFT) can model with minimal assumptions and maximum computational efficiency and accuracy. The FFT is a spectral model that accurately represents the data as the summation of cosine and sine series over a finite sequence of discrete uniformly spaced wavelengths. This sequence of wavelengths inversely corresponds to an equivalent discrete sequence of frequencies called wavenumbers which are completely fixed by the number and spacing of gridded data points. Because of its computational advantages, the FFT dominates modern data analysis.
2. Spectral analysis is based on the inverse processes of Fourier analysis and synthesis. Fourier analysis uses the Fourier integral of the data domain signal coefficients to find the frequency domain cosine and sine transforms. These transforms are simply the amplitudes of the respective cosine and sine waves defined over a finite number of discrete frequencies [i.e., wavenumbers] that inversely correspond to wavelengths ranging from the fundamental [$\equiv$ (number of data) $\times$ (station interval)] to the Nyquist [$\equiv$ (2) $\times$ (station interval)].
3. Alternate representations of the cosine and sine transforms include the power spectrum of summed squares of the cosine and sine transform coefficients, the amplitude spectrum of square roots of the power spectrum coefficients, and the phase spectrum of arctangents of the ratios of the negative sine transform coefficients to the positive cosine transform coefficients.

4. The inverse Fourier synthesis integrates the wavenumber amplitudes to model the data domain signal and its derivative and integral attributes. The synthesis can be derived from the combined coefficients of either the cosine and sine transforms or the phase and power [or amplitude] spectra.
5. All applications of forward and inverse Fourier transforms are accompanied by Gibbs' errors. These oscillating errors of approximation include the ringing effects along grid margins and around data spikes, discontinuities, and other high-gradient grid values.
6. In addition, wavenumber leakage errors result where the edges of the data do not repeat themselves. The mis-matched data margins, in turn, corrupt the wavenumber amplitude components of the spectrum with significant Gibbs' error.
7. Window carpentry involving numerical modifications of the wavenumber and data components can enhance the accuracy of spectral analysis. For example, significant reductions of Gibbs' and wavenumber leakage errors can result from tapering data smoothly to zero along the grid margins, or padding the edges with a rind of zeros or possibly grid values folded out across the margins.
8. Wavenumber aliasing is another gridding error that corrupts the longer wavelength components when the grid interval is too large to capture significant shorter wavelength data variations - i.e., when the grid's Nyquist wavelength [frequency] is greater [lower] than the data's Nyquist wavelength [frequency]. The data's effective Nyquist grid interval may be set by the largest grid interval at which smaller interval data plots no longer show appreciable change.
9. The wavenumber resolution offered by the Fourier transform can become problematic when dealing with small signals of $N \leq 10$ coefficients, where the spectral analysis proceeds with only $N/2$ degrees of freedom because of the symmetry of Fourier transform coefficients about the Nyquist wavenumber. However, where the spectral degrees of freedom need to be enhanced to $N$ unique wavenumber coefficients, a statistical maximum entropy-based spectrum can be constructed with the computational efficiences of the FFT.

# 11 Data Interrogation

## 11.1 OVERVIEW

*The purpose of the inversion process is to obtain a model that can be interrogated for the derivative and integral attributes of the data that satisfy the analysis objectives. These objectives tend to focus mainly on isolating and enhancing data attributes using either spatial domain convolution methods or frequency domain (i.e., spectral) filtering operations.*

*Any linear operation performed on data [e.g., integration, differentiation, interpolation, extrapolation, image processing, filtering, etc.] is a forward modeling or convolution operation. Although convolution can be defined and sometimes performed analytically, it is mostly performed numerically using an appropriately designed operator consisting of a finite set of discrete coefficients.*

*Convolution is a 4-stage process that involves folding or reversing the operator about the ordinate axis, displacing or shifting the operator by a lag amount, multiplying the paired operator and signal coefficients, and integrating or summing the products for the convolution signal's value at the lag position. Repeating these steps for all shifts or lags yields the complete convolution signal.*

*Deconvolution is the inverse operation of finding the set of coefficients [e.g., the 'black box' signal* **x**] *that when convolved with the input signal [e.g., the 'impulse response' signal* **A**] *yields the observed output signal [e.g., the 'convolution' signal* **b**].

*The convolution theorem states that convolution in one domain [i.e., data or frequency] is computationally equivalent to much simpler coefficient-by-coefficient multiplication in the other domain [i.e., frequency or data]. Thus, the data domain convolution of two signals is spectrally equivalent to cross-multiplying the coefficients of their amplitude spectra and summing the phase spectra coefficients.*

*Signal correlations in the data domain are essentially convolutions without signal folding so that in the frequency domain the phase spectra are differenced rather than summed. Thus, the signal's data domain autocorrelation, which measures how much a signal correlates with itself, is spectrally equivalent to its power spectrum with no [i.e., zero] phase information. And the data domain cross-correlation of two signals, which measures how much they correlate with each other, is spectrally equivalent to cross-multiplying their amplitude spectra and differencing the phase spectra.*

*Data measurements invariably include the 'residual' features of interest to the analysis superimposed on features beyond the focus of the analysis, which typically involve larger scale 'regional' variations and smaller scale 'noise' variations from observational and data reduction errors and other sources.*

*The residual separation problem is commonly solved by either isolating the residual signal through elimination or attenuation of the regional and noise components or enhancing the residual relative to the interfering features. This critical step in data interpretation is subjective, and thus a major potential limitation in data analysis.*

*Both data and frequency domain techniques are available to help solve the residual separation problem. Data domain methods include environmentally constrained quantitative modeling, graphical smoothing of profiles or contours, computing trend surfaces, and developing convolution grid operators for data smoothing, differentiation, integration, wavelength separation, interpolation, continuation to different altitudes, etc. In modern data analysis, however, the analytical grid methods have been largely superseded by more efficient spectral filtering techniques.*

*Spectral filtering involves multiplying the data's Fourier transform coefficients by filter or transfer function coefficients that achieve one or more processing objectives. Standard filters, for example, with pass and reject transfer function coefficients of 1's and 0's, respectively, can be applied to study the wavelength, directional, and correlation properties of any gridded dataset.*

*Specialized filters, on the other hand, have non-integer tranfer function coefficients that account for the data's horizontal derivatives and integrals and any theoretical extensions of the gridded data's geometrical attributes. Data following Laplace's equation, for example, have vertical derivatives that are related to the derivatives in the two orthogonal map dimensions. Thus, Laplace's equation can be used to design vertical derivative and integral filters to any order, as well as upward and downward continuation filters to assess the data respectively above and below the mapped observations.*

*The composite filter is the product of two or more tranfer function coefficients that achieve two or more filtering objectives. To enhance circular data features of specific diameters, for example, the transfer function product of an appropriately designed band-pass filter and the directional wedge filter that passes roughly equal wavelengths in the orthogonal directions of the amplitude spectrum may be used.*

## 11.2 INTRODUCTION

The objective of inversion is to obtain a model ($\mathbf{Ax}$) for the data ($\mathbf{b}$) that can be used to predict additional data attributes like their values at unsurveyed coordinates, gradients, integrals, and other properties. Of course, to realize these additional predictions, new design matrix coefficients are needed. For example, to take the partial derivative of $\mathbf{b}$ with respect to $t$, the original coefficients in $\mathbf{A}$ must be exchanged for another set of coefficients in $\mathbf{D}$ defined by $\partial \mathbf{A}/\partial t$ because

$$\frac{\partial \mathbf{b}}{\partial t} = \frac{\partial (\mathbf{Ax})}{\partial t} = (\frac{\partial \mathbf{A}}{\partial t})\mathbf{x} = [(\frac{\partial \mathbf{A}}{\partial t})(\mathbf{A}^t\mathbf{A})^{-1}\mathbf{A}^t]\mathbf{b} = [\mathbf{D}]\mathbf{b}. \tag{11.1}$$

Obviously, the solution **x** does not have to be taken explicitly to estimate the derivative and integral properties of **b**. All that is required here is an appropriate set of **D**-coefficients derived from the known coefficients of **A** to multiply against the observed data **b**.

An equivalent approach to finding **Db** was discussed in **Chapter 1** that involved taking a ruler to a plot of **b** and measuring the data's rise-to-run ratios. Another equivalent approach favored in many signal processing and other applied science and engineering applications considers any manipulation or processing of *input* coefficients [e.g., **b**] for an *output* [e.g., **Db**] as the *convolution* of the input with the coefficients of an appropriate *operator* [e.g., **h**]. Thus, assuming **h** is the $t$-derivative operator and using the symbol $\otimes$ for convolution, the matrix product becomes $\mathbf{Db} = \mathbf{h} \otimes \mathbf{b}$, where the coefficients of the convolution operator **h** are from the inverse operation of *deconvolution* of the observed input **b** and desired output **Db** coefficients.

## 11.3 CONVOLUTION

In general, any linear operation that can be performed on data [e.g., integration, differentiation, interpolation, extrapolation, image processing, filtering, etc.] can be expressed as a forward modeling or convolution operation. Convolution can be defined and sometimes performed analytically, but it is mostly performed numerically in practice using appropriately designed operators or sets of discrete coefficients.

### 11.3.1 ANALYTICAL CONVOLUTION

For continuous signals $\mathbf{x}(t)$ and $\mathbf{h}(t)$, the convolution or *superposition* integral is defined as

$$\mathbf{y}(t) = \int_{-\infty}^{\infty} \mathbf{x}(\tau)\mathbf{h}(t-\tau)d\tau = \mathbf{x}(t) \otimes \mathbf{h}(t), \tag{11.2}$$

where $\tau$ is the dummy variable of integration. As ***Example 11.3.1a*** after [11], consider the convolution of $\mathbf{h}(t) = e^{-t}$ and $\mathbf{x}(t) = 1 \ \forall \ t \geq 0$ given by

$$\mathbf{y}(t) = \int_{-\infty}^{\infty} \mathbf{x}(\tau)\mathbf{h}(t-\tau)d\tau = \int_{0}^{t} (1)e^{t-\tau}d\tau = e^{-t}[e^{\tau}|_0^t] = e^{-t}[e^t - 1] = 1 - e^{-t} \tag{11.3}$$

that is graphically illustrated in **Figure 11.1**

In general, any convolution involves the following four numerical operations: **1)** *folding* or reversing the the operator $\mathbf{h}(\tau)$ about the ordinate axis; **2)** *displacing* or shifting $\mathbf{h}(-\tau)$ by the *lag* or amount $t$; **3)** *multiplying* coefficient-by-coefficient the lagged operator $\mathbf{h}(t-\tau)$ by $\mathbf{x}(\tau)$; and **4)** *integrating* or summing the coefficient products for the area under $[\mathbf{h}(t-\tau) \times \mathbf{x}(\tau)] = \mathbf{y}(t)$. These steps are repeated for all shifts or lags $t$ to form the coefficients of $\mathbf{y}(t)$.

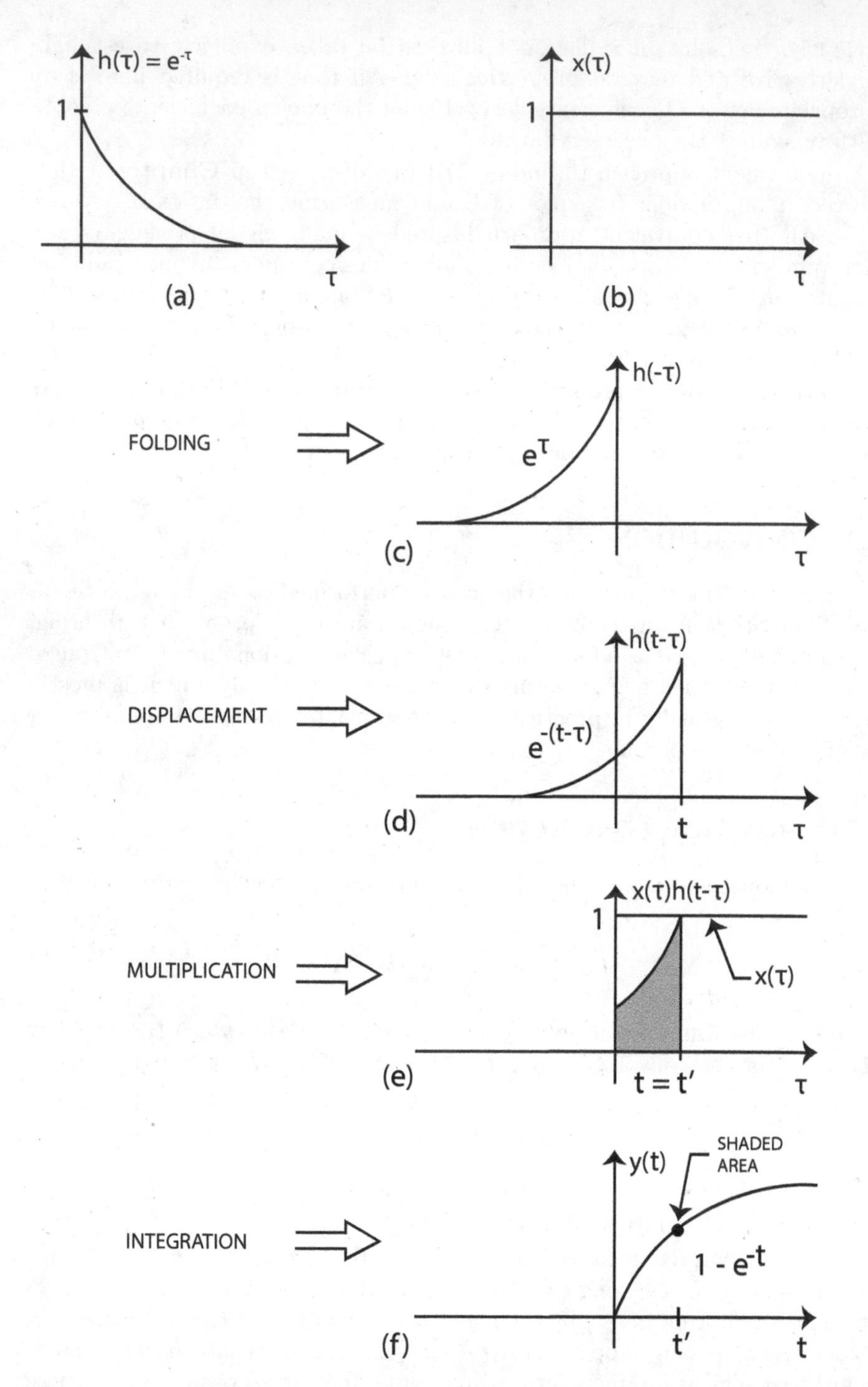


**Figure 11.1** Graphical view of the convolution procedure. [[**11**] by permission of Pearson Education, Inc., New York.]

Convolution also is an associative process where $\mathbf{x}(t) \otimes \mathbf{h}(t) = \mathbf{h}(t) \otimes \mathbf{x}(t)$. In addition, like the Fourier transform, it is most commonly taken numerically with gridded datasets and the analytical integration suppressed.

### 11.3.2 NUMERICAL CONVOLUTION AND DECONVOLUTION

For two finite-length, discrete signals $\mathbf{x_m}\ \forall\ m = 0, 1, 2, \ldots, m$ and $\mathbf{h_k}\ \forall\ k = 0, 1, 2, \ldots, k$, the numerical convolution contains $[m + k - 1]$ coefficients $\mathbf{y_n}$ $\forall\ n = 0, 1, 2, \ldots, (m + k)$ given by

$$\mathbf{y_n} = \mathbf{x_m} \otimes \mathbf{h_k} = \sum_{j=j_1}^{j_n} \mathbf{x_j h_{n-j}}, \tag{11.4}$$

where $j_1 = 0\,\forall\ n \leq k$ or $= n-k\,\forall\ n > k$; and $j_n = n\,\forall\ n \leq m$ or $= m\,\forall\ n > m$. For each lag $n$, the convolved sum $\mathbf{y_n}$ contains signal products for only those coefficients with subscripts that satisfy $n = m + k$. ***Example 11.3.2a*** in **Figure 11.2** illustrates the numerical convolution of a digital 3-point operator with a 4-point signal along with the simple executing code in FORTRAN.

*Deconvolution* is the inverse operation whereby a set of coefficients is determined so that the convolution of the coefficients with the coefficients of an input signal produces the coefficients of a desired output signal or result. In taking the $t$-derivative of **b** in **Equation 11.1**, for example, the deconvolution might involve estimating the unknown derivative operator coefficients from the known input **b**- and output **Db**-coefficients. ***Example 11.3.2b*** in **Figure 11.3** illustrates a numerical deconvolution.

Convolution in the **b**-data domain involves the computationally laborious operations of folding, shifting, multiplying, and integrating that is numerically executed much more efficiently and accurately by simple multiplication in the $\mathcal{B}$-frequency domain and vice versa [**Table 10.1**]. In data processing, the domain of multiplication is invariably preferred over that of convolution because the errors and number of numerical operations involved are minimized. These efficiencies result from the convolution theorem which is the basis of modern data processing. Furthermore, phase reversal of the convolution theorem leads to a theorem describing various signal correlations.

### 11.3.3 CONVOLUTION AND CORRELATION THEOREMS

The convolution of signals in the signal or data domain is equal to the frequency domain coefficient-by-coefficient multiplication and addition of the respective amplitude and phase spectra of the signals. More generally, by the *convolution theorem*, if $\mathbf{g}(t) \Longleftrightarrow \mathcal{G}(f)$ and $\mathbf{h}(t) \Longleftrightarrow \mathcal{H}(f)$, then

$$\mathbf{g}(t) \otimes \mathbf{h}(t) \Longleftrightarrow \mathcal{G}(f)\mathcal{H}(f) \text{ and } \mathbf{g}(t)\mathbf{h}(t) \Longleftrightarrow \mathcal{G}(f) \otimes \mathcal{H}(f). \tag{11.5}$$

```
      SUBROUTINE CONVLV (A, LA, B, LB, C, LC)
      DIMENSION A(LA), B(LB), C(LC)
C     SUBROUTINE CONVLV PERFORMS DISCRETE CONVOLUTION
C           OF SIGNAL A OF LENGTH LA AND SIGNAL B OF LENGTH LB.
C           THE RESULT IS SIGNAL C OF LENGTH LC = LA + LB - 1.

      LC=LA + LB - 1
      DO 10 I = 1, LC
10    C(I) = 0.0
      DO 20 I = 1, LA
      DO 20 J = 1, LB
      K = I + J - 1
20    C(K) = C(K) + A(I) * B(J)
      RETURN
      END
```

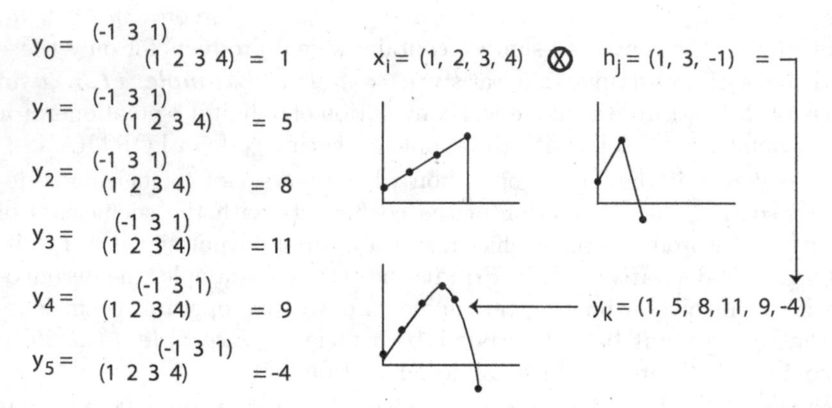


**Figure 11.2** FORTRAN code [box] used to carry out the digital convolution of a 3-point operator **h** with a 4-point signal **x** - i.e., $\mathbf{y_k} = \mathbf{x_i} \otimes \mathbf{h_j}$.

To establish the above-left Fourier transform pair, for example, the Fourier transform of the convolution

$$\mathbf{g}(t) \otimes \mathbf{h}(t) = \int_{-\infty}^{\infty} \mathbf{g}(\tau)\mathbf{h}(t-\tau)d\tau, \tag{11.6}$$

is taken, which yields

$$\int_{-\infty}^{\infty} [\int_{-\infty}^{\infty} \mathbf{g}(\tau)\mathbf{h}(t-\tau)d\tau]\mathbf{e}^{-j2\pi ft}dt. \tag{11.7}$$

Since the exponential term is not a function of $\tau$, the order of integration may be interchanged so that

$$\int_{-\infty}^{\infty} \mathbf{g}(\tau)[\int_{-\infty}^{\infty} \mathbf{h}(t-\tau)d\tau]\mathbf{e}^{-j2\pi ft}dt. \tag{11.8}$$

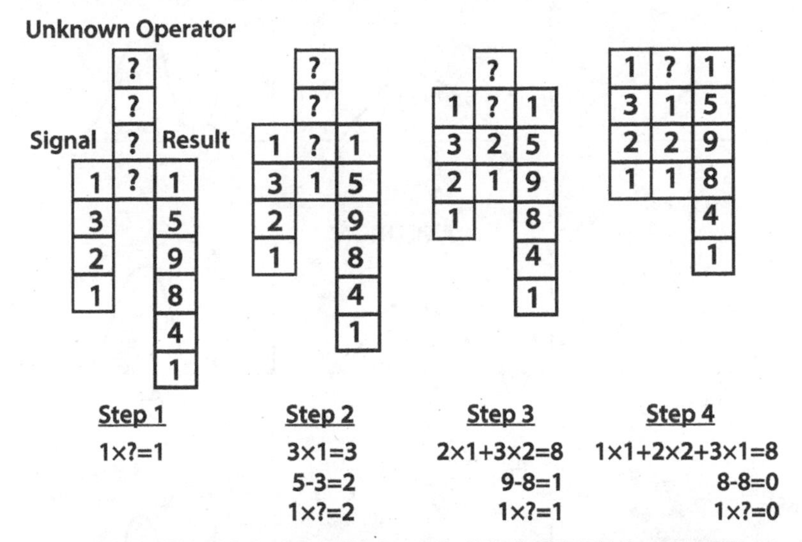


**Figure 11.3** Numerical deconvolution of input and output signal coefficients for a set of operator coefficients such that the convolution of operator and signal coefficients gives the result of output coefficients.

But by the *shift theorem* [**Table 10.1**], the term in brackets is just $\mathbf{e}^{-j2\pi f\tau}\mathcal{H}(f)$, and thus

$$\int_{-\infty}^{\infty} \mathbf{g}(\tau)\mathbf{e}^{-j2\pi f\tau}\mathcal{H}(f)d\tau = \mathcal{H}(f)\int_{-\infty}^{\infty} \mathbf{g}(\tau)\mathbf{e}^{-j2\pi f\tau}d\tau = \mathcal{H}(f)\mathcal{G}(f). \quad (11.9)$$

The converse $\mathbf{g}(t)\mathbf{h}(t) \Longleftrightarrow \mathcal{G}(f) \otimes \mathcal{H}(f)$ is similarly proven.

Likewise, it can be readily shown that

$$\int_{-\infty}^{\infty} \mathbf{g}(\tau)\mathbf{h}(t+\tau)d\tau = \mathcal{G}(f)\mathcal{H}^{\star}(f), \quad (11.10)$$

which is the *Wiener-Khintchine theorem* that describes signal correlation equivalences in the data and spectral domains. For example, where $\mathbf{g}(t) \neq \mathbf{h}(t)$, the *cross-correlation* of the signals in the data domain involves all of the mechanical operations of convolution except for the folding of $\mathbf{h}$. In the spectral domain, by contrast, cross-correlation involves the coefficient-by-coefficient multiplication and subtraction of the respective amplitude and phase spectra of the signals. On the other hand, where $\mathbf{g}(t) = \mathbf{h}(t)$, the theorem describes the signal's *auto-correlation*, which in the spectral domain is simply its power spectum with no [i.e., zero] phase information.

***Example 11.3.3a*** in **Figure 11.4** graphically illustrates the spectral properties of convolution, cross- and auto-correlation, and band-pass filtering. Accordingly, convolution in the spectral domain is done by cross-multiplying

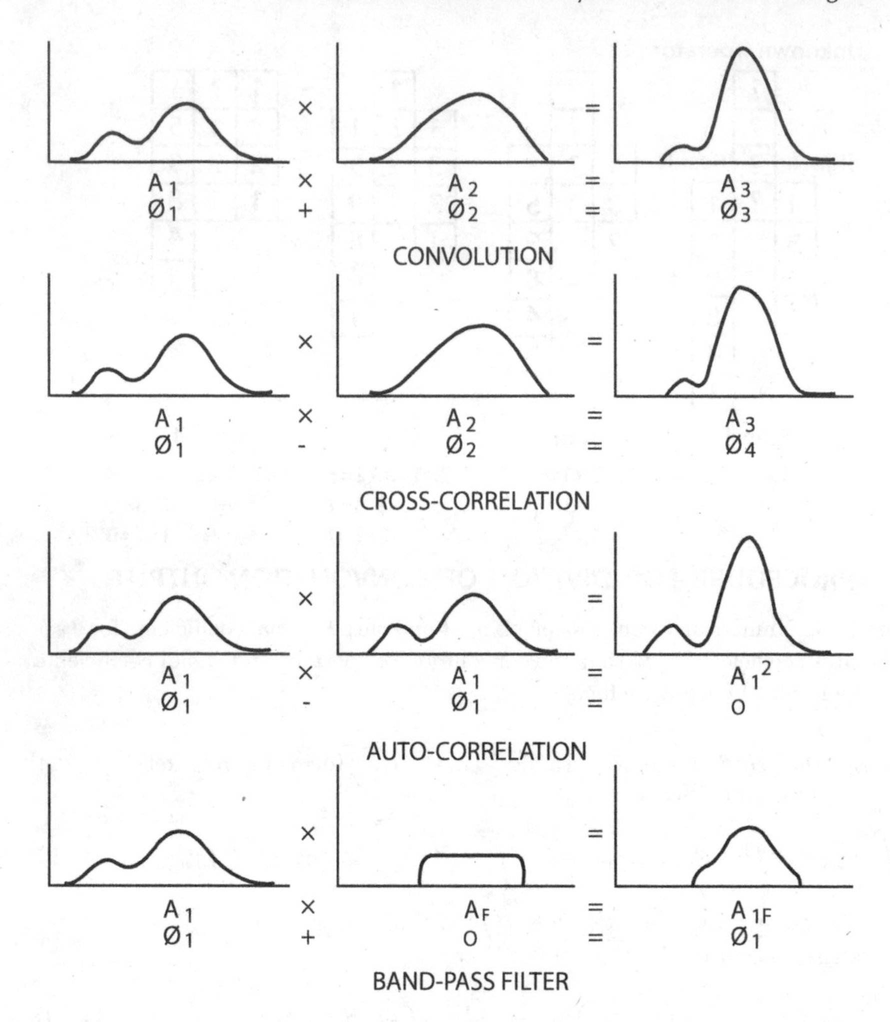


**Figure 11.4** Examples of taking the convolution [top row], cross-correlation [2*nd* row], and auto-correlation [3*rd* row] of signals #1 and #2 with respective $A_1$- and $A_2$-amplitude spectra and $\Phi_1$- and $\Phi_2$-phase spectra. The band-pass filtering example [4*th* row] shows the amplitude spectrum for the filter [$A_F$] with coefficients ranging from 0 to 1, as well as the filtered output's amplitude spectrum [$A_{1F}$].

coefficient-by-coefficient the spectral amplitude values of the signal [e.g., $A_1$] and operator [e.g., $A_2$] at each wavenumber to obtain the amplitude spectrum of the output. Simultaneously, the values of the phase spectrum of the signal are added to the corresponding phase values of the operator to obtain the output's phase spectrum. Inverse transforming or synthesizing the output amplitude and phase spectra yields the convolution in the data domain.

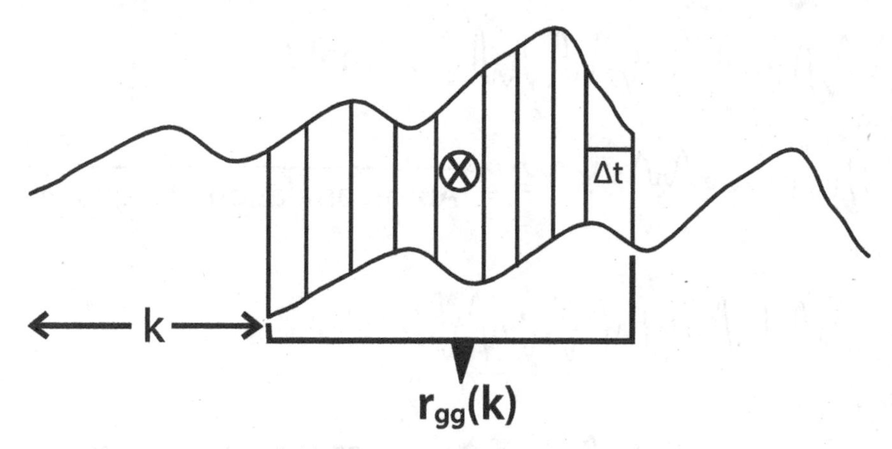


**Figure 11.5** The data domain auto-correlation of the $n$-point signal $\mathbf{g}(t)$ involves dragging it past itself at lags $k = [0, 1, \ldots, (n-1)] \times \Delta t$ and computing at each lag the correlation coefficient between the overlapping signal components.

The conceptual ease of manipulating data in the spectral domain carries over in actual application. For example, the data domain convolution of the $m$-point operator with an $n$-point signal requires $m$-multiplications and $m$-sums to be repeated $[n+m-1]$ times. Thus, a 50-point operator on a $1,000$-point signal requires over $50,000$ sums and products to be taken, whereas the same operation in the spectral domain requires no more than $1,000$ multiplications. The operation is even simpler for symmetric operators, like the zero-phase band-pass filter [e.g., *4th* row in **Figure 11.4**], which have no phase components and thus require no phase modifications.

Spectral correlations involve cross multiplying the amplitude spectra and differencing the phase spectra for outputs that are synthesized for data domain application. Auto-correlation, for instance, is a measure of how much a dataset correlates with itself. ***Example 11.3.3b*** in **Figure 11.5** illustrates data domain auto-correlation for a finite, discrete, uniformly sampled data sequence $\mathbf{g}_i \ \forall \ i = 1, 2, \ldots, n$ given by

$$r_{\mathbf{gg}}(k) = \frac{\sum_{i=1}^{n-k} \mathbf{g}_{i+k}\mathbf{g}_i}{\sum_{i=1}^{n} \mathbf{g}_i^2}, \tag{11.11}$$

where $k$ is the lag or offset by which the replicate of $\mathbf{g}$ is delayed at each computation of $r_{\mathbf{gg}}(k)$

Because $r_{\mathbf{gg}}(k)$ is normalized to the signal's power where $r_{\mathbf{gg}}(0) = 1$, the auto-correlation function varies between $\pm 1$. Values of $r_{\mathbf{gg}}(k)$ near $+1$ or $-1$ respectively indicate high positive [i.e., direct] or negative [i.e., inverse] correlations, and values near zero are commonly interpreted for low correlations of the dataset with itself at lag $k$. Non-normalized coefficients, on the other hand, describe the *auto-covariance* function.

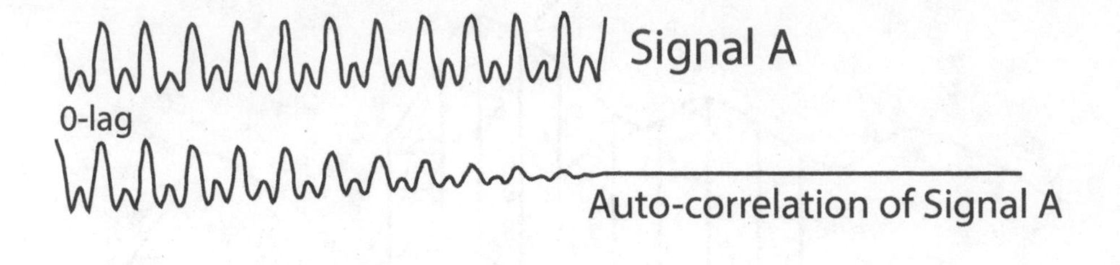


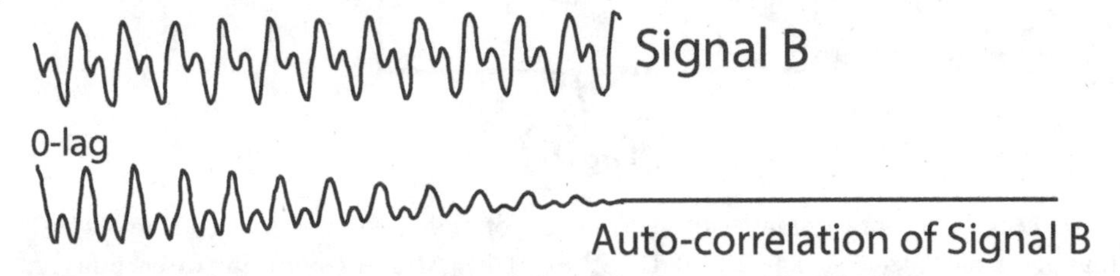


**Figure 11.6** The auto-correlograms for signals A and B degrade as $k \longrightarrow n$ and are considered reliable only for the first 10% to 50% of the lags. Signals A and B have the same frequency components but different phases and so their auto-correlations are the same.

***Example 11.3.3c*** in **Figure 11.6** shows that auto-correlation estimates become increasingly suspect as $k \longrightarrow n$, because fewer and fewer points are overlapping and contributing to $r_{\mathbf{gg}}(k)$. By the theorem of **Equation 11.10**, the Fourier transform of the auto-correlogram clearly is the power spectrum of the signal with no phase dependence. Thus, auto-correlograms of any signals with identical frequency content will also be identical irrespective of the phase differences in the signals [**Figure 11.6**].

The mean value of the dataset is usually removed before computing the auto-correlation, which is an even function in that values for negative lags $-k$ are the same as for positive lags $+k$ as shown by ***Example 11.3.3d*** in **Figure 11.7**. Auto-correlograms identify the frequency content of signals, where for example the single frequency auto-correlation corresponds to an input signal with the same frequency as shown in the top row of **Figure 11.7**. On the other hand, a single frequency auto-correlogram with maximum correlation at zero lag [second row] is indicative of the same frequency input signal contaminated with random noise. In addition, the decay or fall-off rates in auto-correlograms reveal the frequency bandwidths in the input signal, with narrow band signals [third row] decaying at slower rates than for broad band signals [bottom row].

Cross-correlation is a measure of how much two datasets correlate or are similar to each other as a function of lag [position]. ***Example 11.3.3e*** in **Figure 11.8** illustrates the data domain cross-correlation for finite, discrete, uniformly sampled data sequences $\mathbf{g}_i \ \forall \ i = 1, 2, \ldots, n$ and $\mathbf{h}_j \ \forall \ j = 1, 2, \ldots, m$

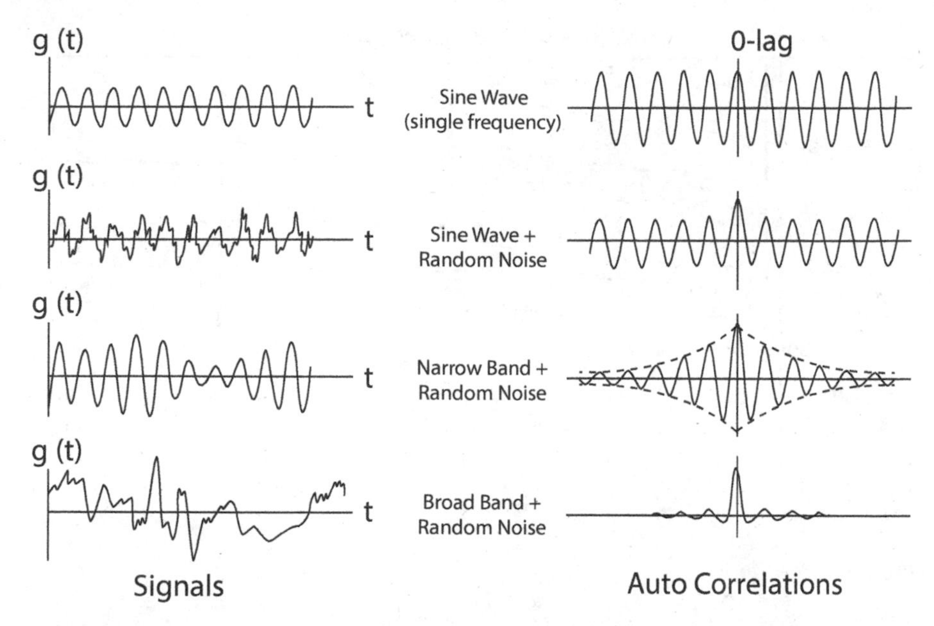


**Figure 11.7** Auto-correlograms are even functions and indicative of the frequency content of the signals $\mathbf{g}(t)$. Random noise components in the signals yield auto-correlograms with maximum values at zero lag. Also the decay rate of auto-correlograms from zero lag are sharp from broad band signals with many frequency components relative to the lower decay rates of narrower band signals.

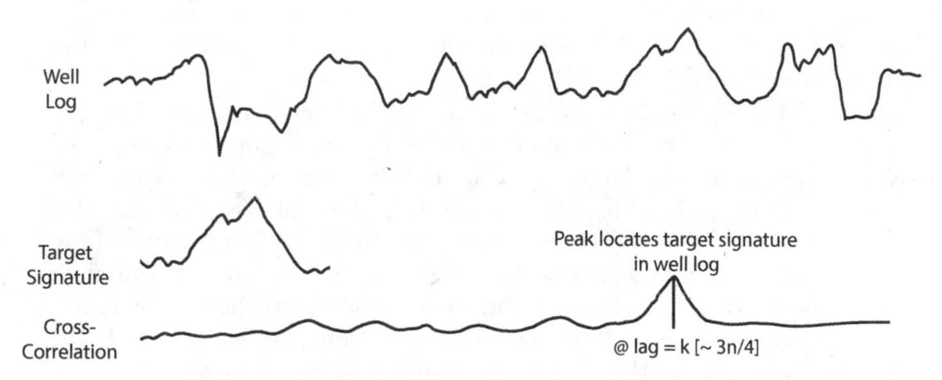


**Figure 11.8** The data domain cross-correlation of an $n$-point well log $\mathbf{g}(t)$ with the $m$-point target signature $\mathbf{h}(t)$ involves dragging them past themselves at lags $k = [0, 1, \ldots, (n-1)] \times \Delta t$ and computing at each lag the correlation coefficient between the overlapping signal components. The maximum in the cross-correlogram locates the target signature in the well log at lag $k \approx 0.75n$.

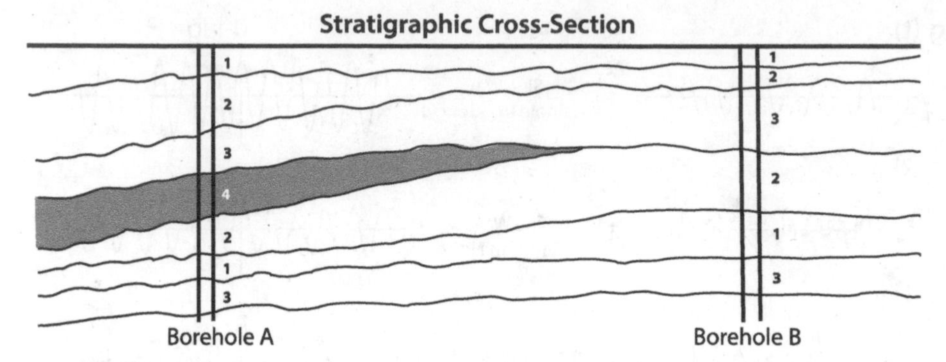


**Figure 11.9** Cross-associations between two well logs are a maximum $3/6 = 50\%$ at the top two lag positions and inconsequential at all other lags due to the pinch-out of unit #4 between the boreholes.

given by

$$r_{\mathbf{gh}}(k) = \frac{\sum_{i=1}^{n} \mathbf{g}_{i+k}\mathbf{h}_i}{\sqrt{r_{\mathbf{gg}}(0)}\sqrt{r_{\mathbf{hh}}(0)}}, \tag{11.12}$$

where $r_{\mathbf{gg}}(0) = \sum_{i=1}^{n} \mathbf{g}_i^2$, $r_{\mathbf{hh}}(0) = \sum_{j=1}^{m} \mathbf{h}_j^2$, and $\mathbf{g}_i$ and $\mathbf{h}_j$ are assumed to be zero outside the ranges of $i$ and $j$. Because $r_{\mathbf{gh}}(k)$ is normalized to the cross power $[\sqrt{r_{\mathbf{gg}}(0)}\sqrt{r_{\mathbf{hh}}(0)}]$ of the two signals, the cross-correlation function varies between $\pm 1$, where values near $|1|$ indicate either high positive or negative correlation at the $k$-th lag, and values near zero are indicative of low correlation. Non-normalized coefficients, on the other hand, describe the *cross-covariance* function.

In general, the maximum value for $r_{\mathbf{gh}}(k)$ will not necessarily be at $k = 0$. Furthermore, $r_{\mathbf{gh}}(k)$ will not normally be an even function, and thus the cross-correlogram also needs to be computed for negative lags, where the total number of lag positions considered can be as large as $\pm[n + m - 1]$. But again, for large values of $k$, $r_{\mathbf{gh}}(k)$ becomes increasingly suspect because of the decreasing overlap between signals. Cross-correlogram maxima and minima also can reflect strong perodicity in one signal when the other is lacking in correlative features, and so it is good interpretation practice to check the cross-correlations against the auto-correlograms of the signals.

*Cross-association* is a non-parametric correlation scheme whereby the statistical number of matches in events between signals as a function of lag is used to help resolve correlative features [**20**]. ***Example 11.3.3f*** in **Figure 11.9** illustrates the process for where the maximum number of possible matches between the well logs is six. Thus, the cross-associations in this case are maximum at the top two lag positions due to the pinch-out of unit #4 between the wells.

### 11.3.4 SUMMARY

In electronic data processing, $\mathbf{b}$ is invariably gridded at the uniform interval $\Delta t$ as noted in **Equation 10.17**. Thus, numerical differentiation and integration are possible using efficient electronic convolution of the signal with the appropriate coefficients of window or mask operators in the $t$-data domain. As ***Example 11.3.4a***, effective numerical estimates of $\partial\mathbf{b}/\partial t$ with increasing $t$ are obtained by the convolution of $\mathbf{b}(t_N)$ with the 3-point first derivative operator $\mathbf{d}_1(3_t) = (\frac{0.5}{\Delta t}\ 0\ \frac{-0.5}{\Delta t})$. Thus, the 3-point operator $\mathbf{d}_1(3_t)$ at $t_1$ gives $\partial\mathbf{b}(t_1)/\partial t = \mathbf{b}\otimes\mathbf{d}_1 = \frac{b_2 - b_0}{2\Delta t} = \partial b_1/\partial t$. However, the convolution $\mathbf{b}\otimes\mathbf{d}_1$ gives $(N-2)$ numerical estimates of $(\partial\mathbf{b}(t_n)/\partial t)\ \forall\ n = 1, 2, \cdots, (N-1)$, and the interpreter must invoke assumptions on the signal's extrapolated properties or other special boundary conditions at the end points $b_0$ and $b_N$ to obtain the complete convolution derivative estimates. In general, for the $M$-point mask operator, potentially corrupting edge effects must be considered for $[M-1]$ estimates of the convolution.

Another grid-based approach is to evaluate the derivative from the signal's spectral model **Equation 10.21**. For the gridded signal, $t$ changes with $n$ so that

$$\begin{aligned}\frac{\partial b_n}{\partial t} \simeq \frac{\partial \hat{b}_n}{\partial t}\frac{\partial t}{\partial n} = \frac{\partial \hat{b}_n}{\partial n} &= \frac{1}{N}\sum_{k=0}^{N-1} \hat{\mathsf{b}}_k \frac{\partial}{\partial n}(e^{j(n)\frac{2\pi}{N}(k)}) \\ &= \frac{1}{N}\sum_{k=0}^{N-1} \hat{\mathsf{b}}_k (j\frac{2\pi}{N}k)(e^{j(n)\frac{2\pi}{N}(k)}) \end{aligned} \tag{11.13}$$

with scalar transform coefficients $\hat{\mathsf{b}}_k \in \hat{\mathcal{B}}(\omega_k)$ and $j(\frac{2\pi}{N})k = \hat{\mathsf{d}}_k \in \hat{\mathcal{D}}_1(\omega_k)$. In other words, **Equation 11.13** is the IDFT of the product $\hat{\mathcal{B}}(\omega_k) \times \hat{\mathcal{D}}_1(\omega_k)$.

Thus, the IDFT of the products of the signal's transform coefficients $\hat{\mathsf{b}}_k$ with the coefficients $\hat{\mathsf{d}}_k$ of the transfer function or filter $\hat{\mathcal{D}}_1(\omega_k)$ estimate the first derivatives of the signal along the $n$-axis. However, by the Convolution Theorem of **Equation 11.5**, which is the basis of modern signal processing, multiplication in one domain is equivalent to convolution in the other domain. Thus, the derivative operations in the $t$-data and $f$-frequency domains form the Fourier transform pair $\mathbf{b}\otimes\mathbf{d}_1 \Longleftrightarrow \hat{\mathcal{B}}(\omega_k) \times \hat{\mathcal{D}}_1(\omega_k)$, which involves the coefficient-by-coefficient multiplication of the amplitude spectra and the addition of the phase spectra.

For signal processing, the domain of multiplication is generally preferred over that of convolution because it greatly minimizes the errors and number of mechanical operations involved in achieving the desired result. For example, to taper the edges of a frequency domain filter to reduce Gibbs' error, the ideal filter's response is typically Fourier synthesized into the data domain where window carpentry is applied to smooth the edges by coefficient-to-coefficient multiplication. The smoothed response is then Fourier analyzed back into the frequency domain for the desired filtering applications.

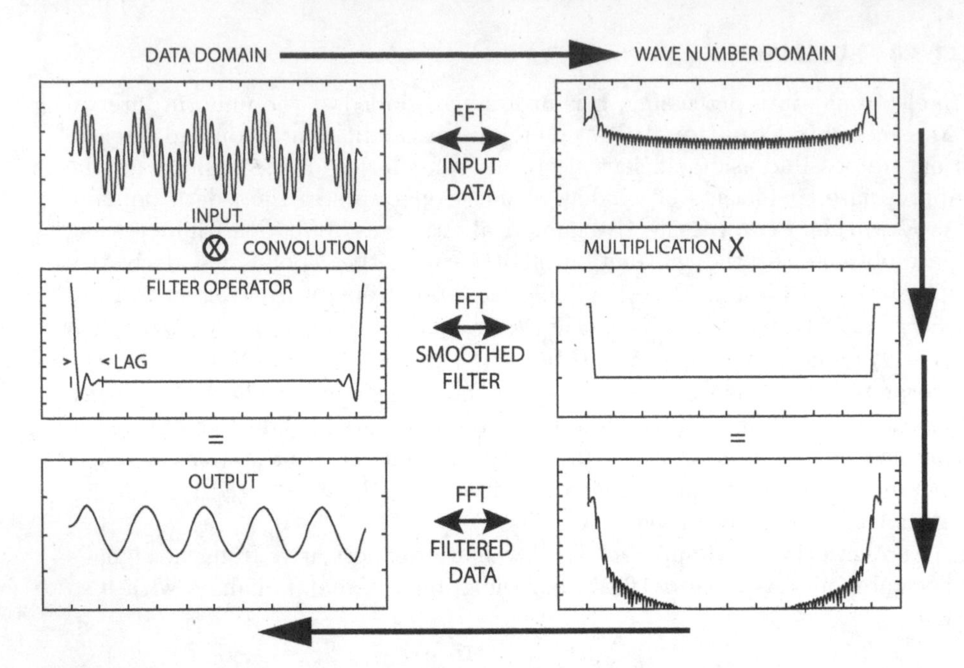


**Figure 11.10** Typical processing sequence where convolution operations such as the data domain application of a smoothing filter is done most efficiently and accurately in the equivalent wavenumber domain. [Adapted from [**78**].]

In summary, several approaches are available for interrogating observations for insight on the related processes. An explicit linear model $\mathbf{Ax}$ can be directly related to the observations, for example, using matrix inversion methods to solve for $\mathbf{x}$. Here, the integral and derivative properties of the data can be explored by exchanging the original design matrix coefficients for the coefficients of the integrated and differentiated model, respectively. However, because of the fundamentally graphical nature of integration and differentiation, these properties are also accessible numerically from convolution operations that spectral filters, in turn, can perform with utmost accuracy and efficiency.

***Example 11.3.4b*** in **Figure 11.10** illustrates the typical processing strategy in modern data analysis. In this example, the investigator seeks to extract the long wavelength signal shown in the 'output' panel from the mutli-frequecy signal in the 'input' panel. In the data domain, the output signal may be obtained by the numerically intensive convolution of the complex symmetric mathematical function shown in the 'operator panel' with the input data. However, in the wavenumber domain, the output can be realized with significantly greater numerical efficiency and accuracy by taking the amplitude spectrum of the input signal and multiplying it coefficient-by-coefficient against the amplitude spectrum of the symmetric zero-phase filter operator. The modified spectrum plus the phase spectrum of the input signal are then inverse transformed or synthesized back into the data domain to obtain the desired output. In ***Example 11.3.4b***, the full amplitude spectra are shown centered on their Nyqyist wavenumber, but in practice only about half of these coefficients are computed due to the symmetry of the spectra.

A variety of mathematical techniques, including those explained above, find extensive use in modern data processing and analysis. The next section considers common applications of these mathematical methods to help identify and analyze specific attributes of digital data.

## 11.4 ISOLATION AND ENHANCEMENT

The measured data variations consist of the superposition of a variety of effects. Specifically, the so-called *residual* or targeted features of the study exist in a background of longer-wavelength *regional* components and shorter-wavelength *noise* that includes observational and data reduction errors. The process of removal of the interfering regional and noise components in the data is the classical *residual-regional separation problem* which is a critical step in data interpretation. It is solved by either isolating the residual signal through elimination or attenuation of the regional and noise components or enhancing the residual relative to the interfering effects.

Isolation techniques attempt to eliminate from the observed data all components that do not have a certain specified range of attributes that are defined by the anticipated residual effects or anomalies from the environmental sources of interest. Depending on the effectiveness of this procedure the isolated feature is amenable to quantitative analysis. In contrast, enhancement techniques accentuate a particular characteristic or group of attributes which are definitive of the desired residual anomalies. The enhancement process increases the perceptibility of the residual feature in the data. As a result the features are distorted from those of the observed data limiting their usefulness for quantitative analysis, but improving their use in qualitative visual inspection analysis.

The isolation and enhancement of data features is a filtering process in which only the spatial wavelengths of interest are passed and the remainder are eliminated or at least highly attenuated. Unfortunately, the wavelengths of interfering anomalies may overlap considerably to complicate the effectiveness of the filtering process. Furthermore, the cut-offs in the characteristics of filters are usually not sharp, leading to possible amplitude and phase distortion in the results. Thus, the residual determination process is subjective and is a major potential limitation in data analysis.

Residual feature separation may be based on a variety of feature attributes such as magnitude, shape, sharpness or gradients, orientation, and correlation with other data. Both spatial and spectral analysis techniques have been used for this purpose. They range from simple graphical techniques employing manual procedures to a host of filtering methods based on the computationally-intensive spectral analysis procedures described above. Current usage is focused on spectral analysis of data, but graphical methods and mathematical procedures which emulate them still find use particularly in the analysis of limited-extent studies in which the residual effects contrast prominently against the regional and noise components.

### 11.4.1 SPATIAL FILTERING

A number of methods can be applied directly to data maps to isolate and enhance various spatial attributes of anomalies. In the hands of the experienced interpreter, these methods can be very effective in bringing out a wide range of data details for analysis.

#### 11.4.1.1 *Geological Methods*

The most direct and satisfying method of isolating local anomalous signals is to remove the regional effect of regional geological sources that obscure the local signal by modeling it using available environmental constraints. For example, the regional geophysical signal derived from seismically mapped depths to the base of the crust and bedrock depths constrained by borehole information can be calculated and removed from the observed signals to isolate the local feature. Alternatively, regional signals surrounding an observed local signal can be modeled to obtain the physical property distribution which best reproduces the observed field and the regional field derived from this distribution can be subtracted from the observed field to isolate the local anomalous signal. [**60**] have described such an approach for isolating residual magnetic anomalies that limits distortion of the local data feature with minimal effects from topography and spectral overlap of regional and residual anomalies. Variations on this so-called inversion-based signal separation for gravity data have also been suggested by [**38**], [**74**], [**73**], and others using least-squares linear Wiener filters designed to remove the spectra of the regional geology as defined by the measured power spectrum of the data pattern, modeling of known geology, or inversion of the regional anomalous field.

#### 11.4.1.2 *Graphical Methods*

Graphical methods are appropriately considered non-linear in the sense that it is impossible to exactly duplicate the results, but they are flexible, permitting the use of the interpreter's experience and auxiliary data in their application. The success of the methods is dependent on the experience of the interpreter, especially in the specific geological terrane, the simplicity of the regional signal, and the perceptibility of the residual signal. The methods were used extensively before computers were commonly available to rapidly and inexpensively perform analytical linear residual-regional separations.

Graphical methods are labor-intensive, and thus are not used except where the survey area is small, the regional is simple and noise subdued, and the residual anomalies are relatively easily discerned. In this case, the separation process can be carried out with a minimum distortion of the residual signal and quantitative calculations are readily performed on the isolated residual features. When graphical methods are used, it is advisable to ascertain the geological reasonability of the regional as a means of validating the viability of the separation process.

Profile methods are the simplest of the graphical methods. They visually establish the regional signal as a smooth curve through the values of an observed data profile that excludes the anomalous portion of the profile associated with the residual signal. The selection of the exclusion segment, of course, is a subjective decision depending greatly upon the analyst's experience and ability. Successful application of the method requires data well beyond the limits of the target area. The regional signal is subtracted from the observed data to determine the residual signal. However, in a situation where the regional is complex and the residual less identifiable, the separation of regional from residual is difficult and likely to be non-reproducible.

Where the data are in map form the graphical method can be used to isolate the residual signals by establishing the regional signal on the individual profiles which form a grid of intersecting profiles across the map area. Control on the regional is determined by adjusting the regionals so that they are the same at the intersection points. The resulting regional map is subtracted from the observed data to obtain the residual map. ***Example 11.4.1.2a*** in **Figure 11.11** shows another method of drawing smooth, consistent contours

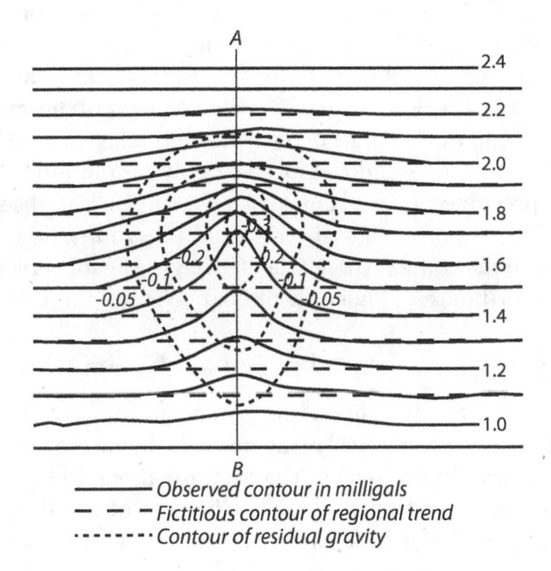


**Figure 11.11** Schematic gravity anomaly contour map of the residual anomaly superimposed on a regional anomaly gradient illustrating the method of isolating residual anomalies by subtracting fictitious smooth contours of the regional. [Adapted from [43] after [23].]

through the data map that represent the regional effects. These regional contour values are subtracted from the observed data values to obtain the residual signal's pattern. The regional anomaly obtained in this manner as is the case with all graphical methods should be checked to determine that it has a geologically reasonable origin - i.e., the derived regional may be geologically justified.

#### 11.4.1.3 *Trend Surface Analysis*

An analytical approach to residual-regional separation by graphical methods is to represent the residual-regional component anomalies as least squares polynomial approximations or trend surfaces [e.g., [**17, 87**]]. The method is used where the regional anomalies are complex and the differences between the regional and residual anomalies are subtle. It produces objective quantitative computations that facilitate the rapid exploration of observed anomalies for meaningful regional-residual components. However, the application and interpretation of the method is still subject to the concerns regarding subjectivity and non-uniqueness of the graphical methods. As in graphical methods, trend surface analysis can be applied to both profiles and maps.

The more common implementation of the method involves modeling the regional component as a low-order polynomial that fits the observed anomalies so as to minimize the sum of squared residuals. The residual data component then is estimated by subtracting the polynomial model of the regional from the observed anomalies. At the other extreme, possible noise components may be estimated by extending the method to higher degree polynomials, where the maximum degree possible is $(n-1)$ for $n$ observed data values.

By **Equation 9.11**, the least squares $p$-th degree polynomial model for the regional-residual component can be established using only the $(n \times 1)$ **b**-matrix of observed data values and the forward polynomial model of unknown coefficients that is presumed to account for the variations in observations. As ***Example 11.4.1.3a***, suppose that the regional ($\mathbf{b_R}$) may be approximated by the first degree equation of the plane through the observed anomalies mapped in $(x, y)$-coordinates. Thus, the linear system for establishing the regional is

$$\mathbf{b} = \alpha + \beta x + \gamma y \equiv \mathbf{b_R} \ni \mathbf{b_r} = \mathbf{b} - \mathbf{b_R}, \tag{11.14}$$

where $\mathbf{b_r}$ estimates the residual anomalies, and the $m = 3$ unknown coefficients, $\alpha$, $\beta$ and $\gamma$, are respectively the plane's mean value, and $x$- and $y$-directional slopes. Simply evaluating the forward model with the unknown coefficients set to unity gives the $(n \times m)$ coefficients of the design matrix $\mathbf{A}$ from which the least squares solution is

$$(\alpha \ \beta \ \gamma)^{\mathbf{t}} = (\mathbf{A^t A})^{-1}\mathbf{A^t b}. \tag{11.15}$$

Of course, the characterizations of the regional $\mathbf{b_R}$ and related residual $\mathbf{b_r}$ are never unique. In **Figure 11.12**, for example, the residual estimate at $O$

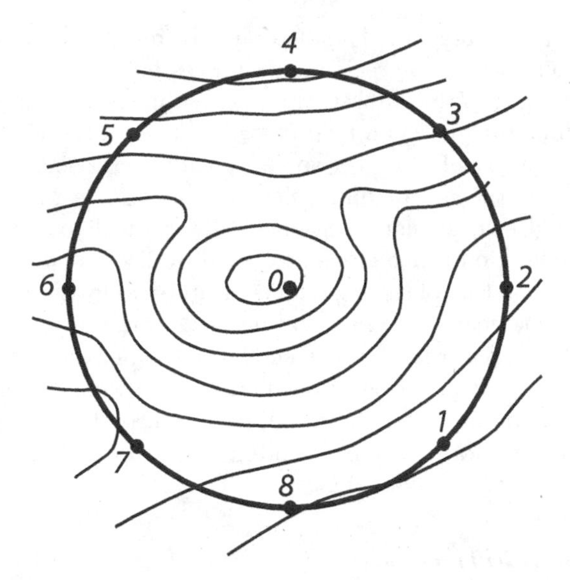


**Figure 11.12** Computation of the residual gravity anomaly at $O$ based on eight regional anomaly values selected by the interpreter. The non-unique regional anomaly at $O$ may be taken as an average of the selected values or some estimate from a polynomial or another function fit to them. [Adapted from **[43]** after **[34]**.]

may be based on subtracting the average or some estimate from a polynomial or another function fit to the eight values that the interpreter selected to characterize the regional for this application. Clearly, to make the analysis most effective, the interpreter must use available ancillary geological and other environmental data to justify the viabilities of the regional, noise, and residual anomalies as fully as possible.

In general, the trend surface analysis technique for regional-residual separation effectively operates like a low pass filter. However, the wavelength characteristics of the method are poorly understood and only can be controlled in a limited way. The prominent wavelength of the polynomial approximation can be crudely determined by noting that there are $(n-1)$ maxima and minima in a least squares approximation where $n$ is the maximum degree of the equation. Thus, for example, considering a profile, a regional feature with a prominent wavelength of the order or degree 2 can be anticipated that corresponds to the length of the profile divided by $(n-1)$. As a result the degree of appropriate polynomial equation must take into consideration the size of the profile or area as well as the character of the regional signal.

The problem, however, is much more complex than suggested by this approximation. For example, the application of trend surface analysis to maps also involves a directional bias depending on the shape of the area involved [e.g., **[93]**]. The result is incorrect orientation of the regional field and distortion of the residual anomalies. The distortion of residual anomalies is particularly troublesome where the regional anomalies have high gradients [e.g., [**87**]] and near the edges of the dataset. However, supplementing the data with rinds of real or simulated data can help mitigate some of these effects.

An alternative, but equally subjective approach to least squares trend surface computations is the use of finite element analysis [e.g., [**62**]]. In this method the regional field at map positions is determined to a first- through third-order approximation using a weighted sum of discrete values surrounding the position. The nature of the regional is established by a weighting factor determined by the element shape function based on finite-element analysis.

As a result of the hazards of the trend surface analysis technique, the method must be used with caution in residual-regional separation, especially where anomalies will be subjected to quantitative interpretational procedures. However, the method does serve well in preliminary separation of major regional effects by low order polynomials.

#### 11.4.1.4 *Analytical Grid Methods*

Analytical spatial methods include a variety of techniques for residual-regional separation based on determining the regional value at a location on an equidimensional grid from the surrounding data values. These methods are free of personal bias that limits the graphical procedures, but they are still subjective because the results are highly dependent on the method of determining the surrounding values and the weighting applied to these values. The simplest of procedures is to average the data on a circle of fixed radius around the point at which the regional value is being approximated and subtract the average from the central value to determine the residual signal at the location. This procedure is repeated at successive grid points until the entire map grid is evaluated. Other methods use values obtained on multiple rings surrounding the central location and the method can be used on profile data as well by using adjacent data to determine the regional value.

The method can be adapted for data differentiation, integration, continuations to different elevations, wavelength filtering, and other filtering applications. This is accomplished by convolving the input data with a filter designed to pass specified wavelengths [e.g., [**21, 30**]]. The filtering function, that is the convolution operator of coefficients, is designed by specifying the desired wavenumber response and taking its inverse Fourier transform. These transforms are modified from the continuous data case to the discrete dataset condition and made practical to implement by applying a smoothing function causing a gradual cutoff to where the amplitude of the function is negligible. The direct transform can be used to calculate the response of the filter in the wavenumber domain. Although this spatial method is readily applied to rapid analysis of gridded data by digital computers, the convolution theorem [**Equation 11.5**] has resulted in it being largely superseded by more efficient spectral filtering techniques that are described in the next section.

### 11.4.2 SPECTRAL FILTERING

Spectral filters efficiently process gridded data for their basic geometric properties and the theoretical extension of these properties into other data forms. The routine act of electronically plotting data involves the preparation of a data grid that provides access to these data properties directly via the FFT. Thus, spectral filtering has become widely used for data analysis.

Filtering in the wavenumber domain involves multiplying the Fourier transform of the data by the coefficients of a filter or transfer function that achieves one or more data processing objectives. *Standard filters* with transfer function coefficients of 1's and 0's can be applied to any gridded dataset to study the wavelength, directional, and correlation properties of the anomalies. Wavelength and directional filters suppress or enhance data features based on their spatial dimensions and attitudes, respectively. Spectral correlation filters, by contrast, extract positively, negatively, or null correlated features between two or more datasets.

*Specialized filters*, on the other hand, have non-integer coefficients that account for the horizontal derivatives and integrals of anomalies and their extensions via the underlying theory. As ***Example 11.4.2a***, suppose the signal $\mathbf{g}(x, y, z)$ follows *Laplace's Equation* given by

$$\frac{\partial^2 \mathbf{g}}{\partial x^2} + \frac{\partial^2 \mathbf{g}}{\partial y^2} + \frac{\partial^2 \mathbf{g}}{\partial z^2} = 0. \tag{11.16}$$

Thus, the horizontal differentials and integrals of the signal estimated by derivative and integral filters with respect to the independent horizontal variables [e.g., $x$ and/or $y$] can be extended by Laplace's equation to predict data variations along the vertical [e.g., $z$] axis. In general, Laplace's equation may be used to design vertical derivative and integral filters, as well as upward and downward continuation filters to assess the data respectively above and below the mapped observations.

#### 11.4.2.1 *Wavelength Filters*

Ideally, these standard filters consist of simple binary coeficient patterns of 1's and 0's that respectively pass and reject the signal's transform components $\hat{\mathfrak{b}}_k$ upon multiplication. The equivalent data domain convolution operator is a real symmetric or even function so that only a single quadrant of the real transform is needed for designing the pass/reject pattern that by symmetry extends to the other real quadrants. The imaginary transform is zero and thus these filters are called zero phase or minimum phase filters.

***Example 11.4.2.1a*** in **Figure 11.13** illustrates the low-pass/hi-cut filter designed to pass frequencies in a data profile up to the 40th harmonic and reject the higher harmonics. The top right panel is a linear, DC-centered plot of both real quadrants of the ideal filter. The solid dots on the filter edges locate the 50% amplitude levels. The inverse transform of this box car filter

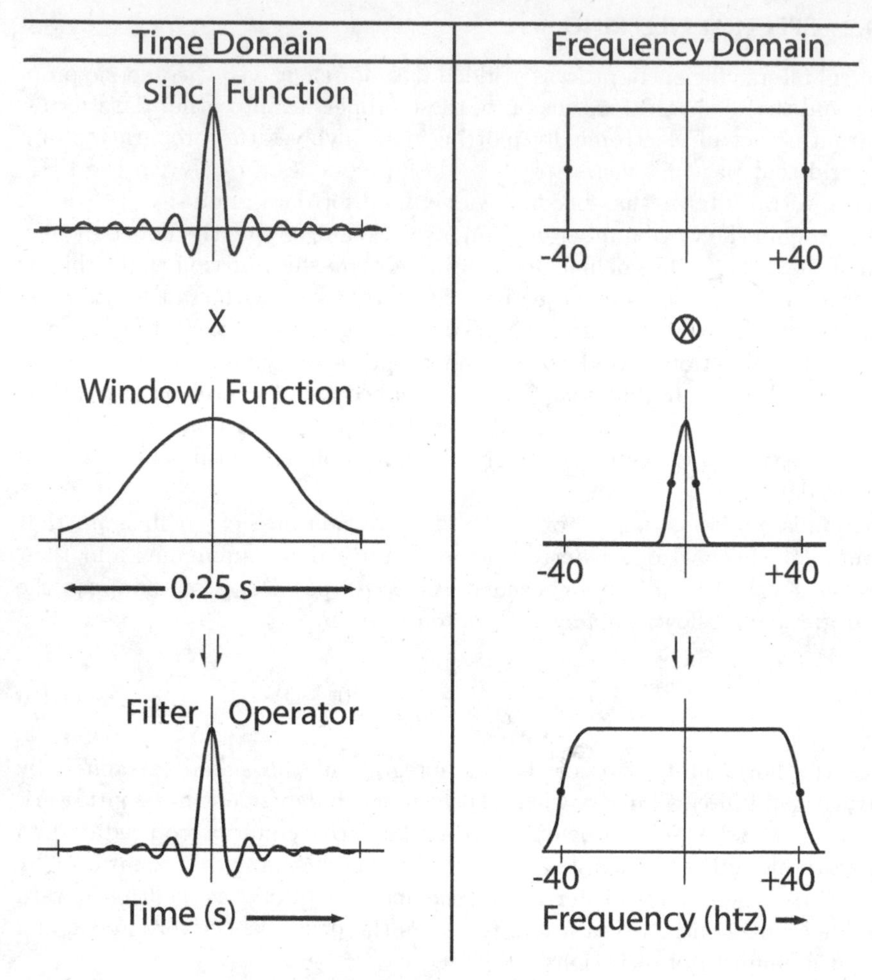


**Figure 11.13** Design of the digital low-pass filter operator with 50% amplitude cut-off at 40 htz. Gibbs' error is more efficiently minimized in the data domain by coefficient-to-coefficient multiplication of the window function against the ideal filter [i.e., the sinc function] rather than in the frequency domain where the effort requires the more computationally laborious convolution of the two equivalent frequency responses. [Adapted from **[43]** after **[76]**.]

is the equivalent data domain convolution operator given by the sinc function ($\equiv \sin(t)/t$) in the top left panel. This convolution operator is defined over all data points and clearly is not a simple averaging function.

To minimize Gibbs' error, window carpentry is used to taper the sharp edges of the box car function in the data domain (left panels) rather than in the frequency (wavenumber) domain (right panels) where the frequency responses of the window and ideal filter functions must be convolved. The left middle panel shows the Hamming window function that was applied. It consists of a single cycle cosine wave raised on a slight pedestal over the fundamental period of the data profile and has the frequency response shown in the right central panel. The Hamming window was multiplied against the ideal filter operator (left, top panel) to obtain the tapered filter operator (bottom, left panel) with the tapered frequency response given in the lower right panel. The tapered filter has nearly the same performance characteristics as the ideal filter, but with greatly reduced Gibbs' error. Typically, the performance specifications of the tapered filter are reported in terms of its 50% cut-off, which for the low-pass filter in the lower right panel is at the 40th harmonic of the data profile.

***Example 11.4.2.1b*** in **Figure 11.14** shows the design of digital high-pass/low-cut and band-pass filters for a data profile. Because the wavenumber components in the Fourier transform follow the superposition principle, the 1's and 0's of the low-pass filter are interchanged to obtain the complementary transfer function of the high-pass filter. Suppose, for example, that it is desired to pass frequencies above the 20th harmonic of the data profile and sharply attenuate the lower frequencies. The tapered low-pass filter operator is first designed with the 20th harmonic cut-off as shown in the top panels. Then the all-pass filter is designed that consists of the data spike in the data domain (second left panel) with the uniform amplitude components in wavenumber domain (second central panel). Subtracting the low-pass filter from the all-pass filter yields the high-pass filter in the third panels with the 50% cut-off at the 20th harmonic of the data profile. Furthermore, subtracting the high-pass filter from the low-pass filter in the bottom panel of **Figure 11.13** gives the band-pass filter in the bottom panels of **Figure 11.14** that passes the wavenumber components essentially between the 50% cut-offs at the 20th and 40th harmonics.

Extending wavelength filtering to the two-dimensional transform with wavenumbers $k$ and $l$ in the respective $n$- and $m$-directions is also straightforward. Specifically, the transfer function for the ideal wavelength filter is

$$\begin{aligned} WLF(k,l) &= 1 \quad \forall \;\; k_L^2 + l_L^2 < k^2 + l^2 < k_H^2 + l_H^2, \\ &= 0, \quad \text{otherwise}, \end{aligned} \tag{11.17}$$

where the lowest and the highest wavenumbers to be passed are $(k_L, l_L)$ and $(k_H, l_H)$, respectively. To taper the filter's response and reduce its Gibbs' error, window carpentry is applied in the data domain to the inversely transformed ideal filter. ***Example 11.4.2.1c*** in **Figure 11.15** illustrates some low-pass/hi-cut (left panel), hi-pass/low-cut (middle panel), and band-pass (right panel) filters designed for the 16 × 32 data array of **Figure 10.5**.

Only a single real quadrant is necessary for designing the essential elements of these wavelength filters, which consist predominantly of 1's and 0's and

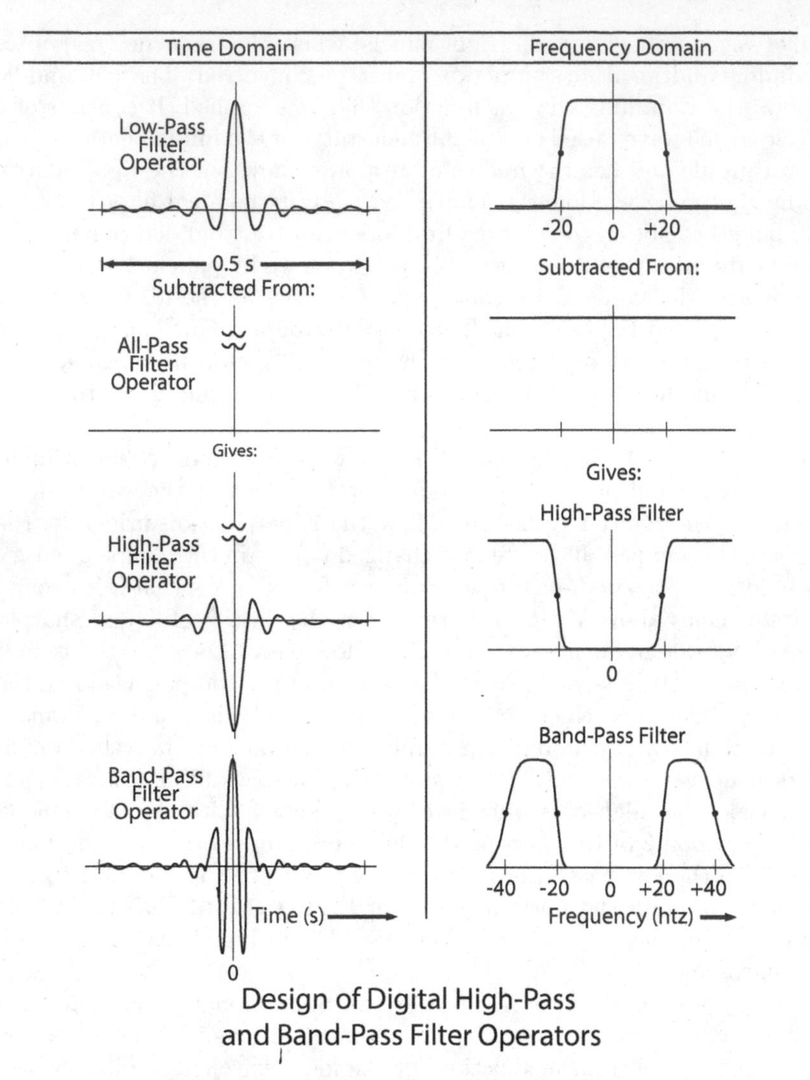


**Figure 11.14** Data domain [left column] and spectral domain [right column] constructions of digital high-pass and band-pass filter operators. [Adapted from **[43]** after **[76]**.]

some in-between values that reflect the smoothing of their edges to minimize Gibbs' error. ***Example 11.4.2.1d*** in **Figure 11.16(a)** gives an isometric view of a complete tapered, DC-centered band-pass filter designed to reject data features with wavenumber components corresponding to the 0's of the filter. In the next section, a further adaptation of 1's and 0's is described that brings out the filtered features with specific directional or strike orientations.

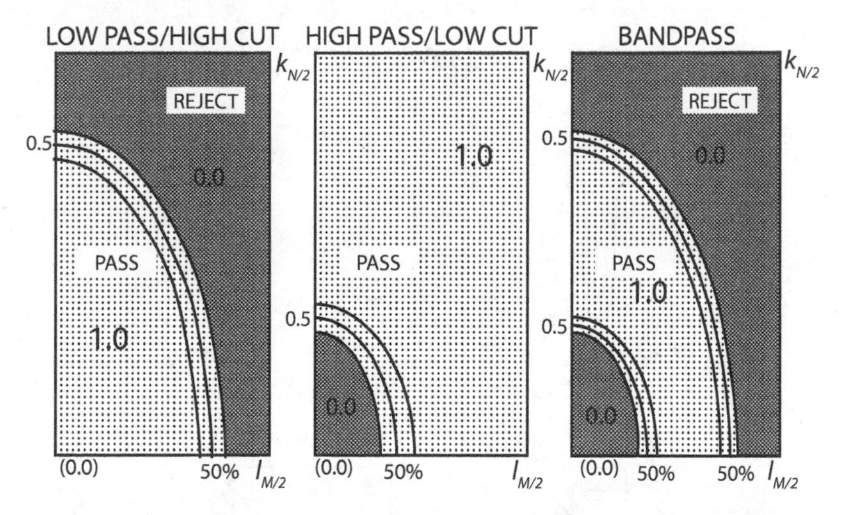


**Figure 11.15** Three idealized wavelength filter operators with edges tapered for enhanced suppression of Gibbs' error. The middle contour is the 50% value of the taper. The design of these filters requires only a single quadrant. [Adapted from [78].]

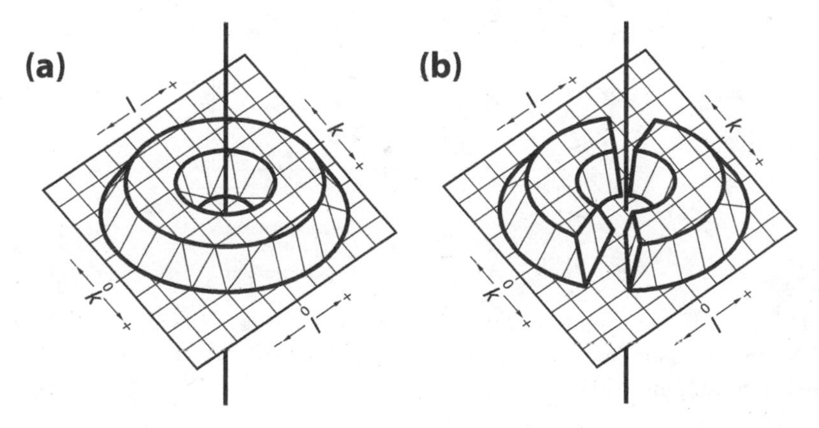


**Figure 11.16** (a) The smoothed or tapered doughnut-shaped response function of a band-pass filter for extracting gridded data features with maximum and minimum wavelengths defined by the wavenumbers at the respective outside edges of the doughnut hole and the doughnut. (b) Response function for the composite band-and-strike-pass filter that passes the maximum and minimum wavelength features which lack components with wavenumbers in the wedge-shaped breaks of the doughnut. [Adapted from [43] after [61].]

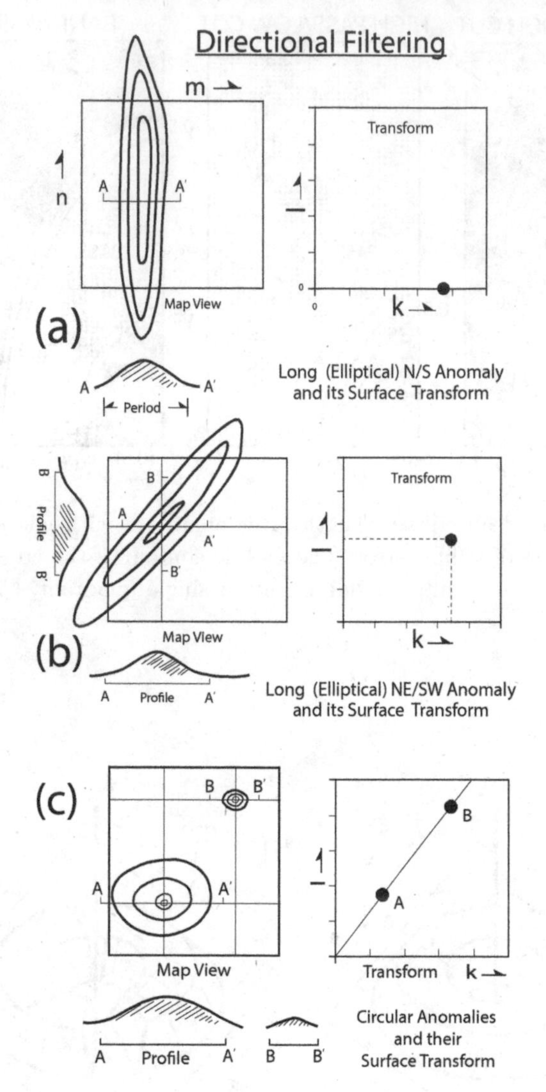


**Figure 11.17** Map and frequency responses for **(a)** elliptical N/S-striking, **(b)** elliptical NE/SW-striking, and **(c)** circular longer and shorter wavelength anomalies. [Adapted from [**43**] after [**61**].]

#### 11.4.2.2 *Directional Filters*

For gridded two-dimensional data, the ratios of the orthogonal wavenumbers reflect the azimuthal or strike orientations of the data variations. As ***Example 11.4.2.2a***, the N/S-striking elliptical data feature in **Figure 11.17(a)** has an effectively infinite wavelength component in the N/S-direction with a significantly shorter wavelength component in the E/W direction. Thus, the spectral energy for this feature plots towards the higher wavenumber end of the $k$-axis as shown in the top right panel.

The data can be tested for this feature using a standard wedge-pass filter centered on the $k$-axis, which also is sometimes called a fan or butterfly or pie-slice filter. To pass N/S-striking regional features at all E/W wavelengths, the edges of the wedge filter may be defined for example at 80° and 100° clockwise from the $l$-axis of the transform. Then 1's and 0's are assigned in-between and outside these azimuths, respectively, to define the ideal strike-sensitive filter, which is next inversely transformed to the data domain for window carpentry to slightly smooth the filter's edges to reduce Gibbs' errors. Transforming the tapered E/W-striking elliptical convolution operator into the wavenumber domain and multiplying it against the data's amplitude spectrum coefficient-by-coefficient and inversely transforming the filtered output into the data domain leads to an estimate of the N/S-trending anomalies in the data.

To pass E/W-striking data features, on the other hand, the wedge filter is simply rotated 90° to center it on the $l$-axis. At orientations between these two extremes, the wavenumber energies migrate predictably into other parts of the spectrum. As ***Example 11.4.2.2b***, the elliptical feature in **Figure 11.17(b)** defines wavelength components that are longer in the N/S-direction than in the E/W-direction. Thus, this orientation contributes energy to the portion of the spectrum involving the lower frequency $l$ and higher frequency $k$ wavenumbers. Of course, as the feature is rotated to strike E/W, the spectral energy migrates towards the Nyquist end of the $l$-axis.

***Example 11.4.2.2c*** in **Figure 11.17(c)** shows that the spectral energies of circular features are concentrated along the 45°-line in the transform where $l \simeq k$. Elliptical data features with azimuths at 45° to the map axes also contribute energies in this part of the spectrum. However, these features die out as the wedge is rotated away from the 45°-line in the transform, whereas circular features contribute energies at all orientations of the wedge filter.

To find elliptical features with specific across-strike wavelengths, the wedge filter is multiplied against the appropriate band-pass filter to construct a *composite filter* that further restricts the range of 1's such as those shown in **Figure 11.15(b)**. This approach also can help discriminate circular features of different wavelengths as shown in **Figure 11.17(c)**.

***Example 11.4.2.2d*** in **Figures 11.18(a)** and **(b)** shows the azimuthal orientations in the spatial and wavenumber domains involve a 90° rotation, whereas ***Example 11.4.2.2e*** in **Figures 11.18(c)** and **(d)** gives the typical computer storage schemes for designing and applying the directional filtering. In these schemes, passing or rejecting feature orientations is considered in the data domain between azimuths $A1$ and $A2$ in the range $0° < A1 < A2 < 180°$ measured clockwise positive from the left axis of the dataset [e.g., **Figure 11.18(c)**]. The minimum phase transfer functions are commonly designed with 1's and 0's in the first and second spectral quadrants that extend into

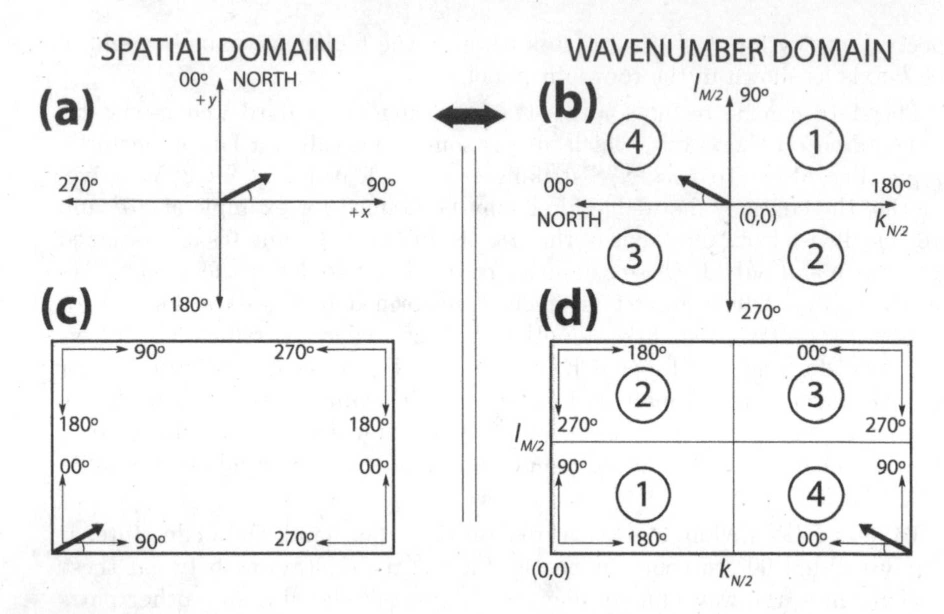


**Figure 11.18** Contrasting representations for the azimuthal orientations in the data domain [(**a**) and (**c**)] and the wavenumber domains [(**b**) and (**d**)]. Panels (**c**) and (**d**) also show the machine storage formats for the data and wavenumbers, respectively. [Adapted from [**78**].]

the two conjugate quadrants by symmetry. Thus, the transfer function for the ideal strike-pass filter, for example, is

$$\begin{aligned} SPF(k,l) &= 1 \quad \forall \quad -\tan^{-1}(A2) \leq l/k \leq -\tan^{-1}(A1), \\ &= 0, \quad \text{otherwise.} \end{aligned} \tag{11.18}$$

Simply interchanging the 0's and 1's in **Equation 11.18** gives the complementary transfer function of the ideal strike-reject filter.

In practice, azimuthal differences $(A2 - A1) < 20°$ typically affect relatively marginal levels of the signal energy and thus wedges of about 20° and wider are usually required to produce useful filtered outputs. ***Example 11.4.2.2f*** in **Figure 11.19** shows ideal directional filters designed for the 16 × 32 data array in **Figure 10.5**. In the left and middle panels, the filters reject azimuths in the data from 30° to 180° and 70° to 132°, respectively, whereas in the right panel the directional filter passes data azimuths from 160° to 180°.

The complete filters consist of the coefficients designed in quadrants #1 and #2 of the computer-formatted, Nyquist-centered spectrum shown in **Figure 11.18(d)** and mapped by symmetry into the conjugate quadrants #3 and #4, respectively. However, in this machine format, the filter's response is less than intuitive and thus for presentation purposes, the quadrants are typically transposed into the DC-centered spectrum like the one in **Figure 11.16(b)**.

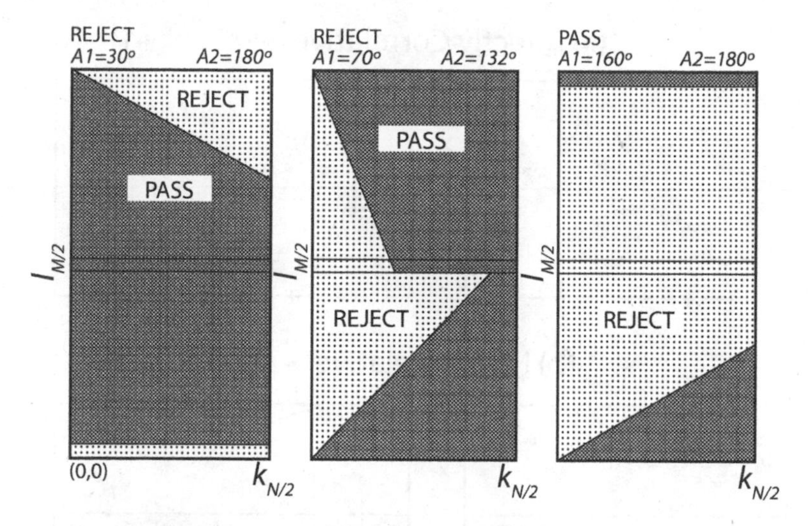


**Figure 11.19** Three idealized directional filters. The design of these filters requires two quadrants. [Adapted from [**78**].]

In general, a good practical test of the performance of a wedge or, for that matter, any filter is to sum the data passed and rejected by the filter for comparison with original input data. Well performing filters yield sums that closely match the input data with minimal residuals.

### 11.4.2.3 *Correlation Filters*

Standard correlation filters of 1's and 0's can also be developed to extract correlative features between two or more datasets [**96**]. A specialized form of this filter is the least-squares Wiener filter [**102**] designed to extract random signals. ***Example 11.4.2.3a*** in **Figure 11.20** summarizes the feature associations that the *correlation coefficient* ($r$) quantifies. In panel a, for example, peak-to-peak and trough-to-trough feature associations reflect direct or positive correlations where $r \longrightarrow +1$. Panel c, on the other hand, illustrates peak-to-trough and trough-to-peak feature associations that reflect inverse or negative correlations where $r \longrightarrow -1$.

In general, the above interpretations are unique only if $|r| = 1$, whereas interpretational uniqueness does not hold for other values where $-1 < r < +1$. Panel b, for example, illustrates two interpretations that quantitatively account for the null correlation where $r \longrightarrow 0$. The left profiles reflect the classical explanation assumed for the lack of correlation, namely that variations in one dataset do not match the variations in the other dataset. However, the null correlation also results when 50% of the signals directly correlate and the remaining 50% of the signals inversely correlate as in the right set of profiles. This example illustrates the folly of using some arbitrary threshold value ($\neq \pm 1$) of $r$ to indicate how signals actually correlate.

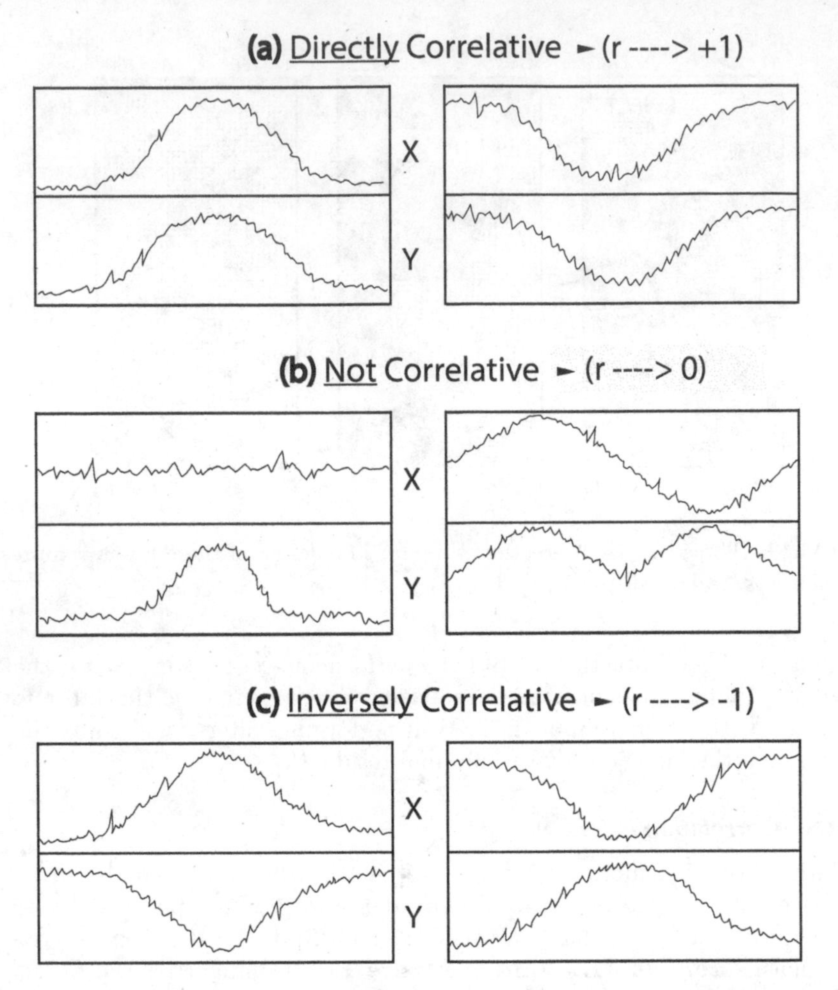


**Figure 11.20** Interpretations of the correlation coefficient $r$ for **(a)** positive, **(b)** null, and **(c)** negative feature associations. [Adapted from [**43**] after [**96**].]

Initial efforts to characterize the spectral properties of correlation focused on the squared correlation or *coherency coefficient* $r^2$ (see **Equations 9.21** and **9.22**). Following the pioneering studies of [**101**], the coherency coefficient between two signals **x** and **y** with Fourier transforms $\mathcal{X}$ and $\mathcal{Y}$ is defined by

$$\begin{aligned} r^2 &\equiv |\mathcal{X}\mathcal{Y}^\star|^2/[(\mathcal{X}\mathcal{X}^\star)(\mathcal{Y}\mathcal{Y}^\star)] \quad \forall \quad [(\mathcal{X}\mathcal{X}^\star)(\mathcal{Y}\mathcal{Y}^\star)] > 0 \\ &\equiv 0 \quad \forall \quad [(\mathcal{X}\mathcal{X}^\star)(\mathcal{Y}\mathcal{Y}^\star)] = 0, \end{aligned} \tag{11.19}$$

where $(\mathcal{X}\mathcal{X}^\star)$ and $(\mathcal{Y}\mathcal{Y}^\star)$ are the power spectra and $(\mathcal{X}\mathcal{Y}^\star)$ is the cross-power spectrum.

The coherency coefficient essentially tests signals that differ only in terms of their random noise components [e.g., [**29**]]. Thus, coherency provides very useful estimates of the noise ($N$)-to-($S$) signal ratio given by

$$(N/S) \simeq \sqrt{(1/\sqrt{r^2}) - 1} = \sqrt{(1/|r|) - 1}. \tag{11.20}$$

However, coherency is not sensitive to the sign of the correlation coefficient, or the individual wavenumber components for which it is always unity because

$$r^2(k) = \frac{|\mathcal{X}(\mathbf{k})\mathcal{Y}^{\star}(\mathbf{k})|^2}{[\mathcal{X}(\mathbf{k})\mathcal{X}(\mathbf{k})^{\star}][\mathcal{Y}(\mathbf{k})\mathcal{Y}(\mathbf{k})^{\star}]} = \frac{([\mathcal{X}(\mathbf{k})\mathcal{Y}(\mathbf{k})^{\star}][\mathcal{X}(\mathbf{k})\mathcal{Y}(\mathbf{k})^{\star}])^{\star}}{[\mathcal{X}(\mathbf{k})\mathcal{X}(\mathbf{k})^{\star}][\mathcal{Y}(\mathbf{k})\mathcal{Y}(\mathbf{k})^{\star}]} = 1. \tag{11.21}$$

Thus, coherency is implemented predominantly over band-limited averages of the spectra where it resolves only the relatively gross, positively correlated relationships between the signals.

The full spectrum of correlations between digital datasets becomes evident, however, when the transforms are considered as vectors in the complex plane at any given wavenumber, $k$. Analysis of the polar coordinate expressions for the wavevectors $\mathcal{X}(\mathbf{k})$ and $\mathcal{Y}(\mathbf{k})$ in **Figure 11.21**, for example, shows that their correlation is

$$r(k) = \cos(\Delta\theta_k) = \frac{\mathcal{X}(\mathbf{k}) \cdot \mathcal{Y}(\mathbf{k})}{|\mathcal{X}(\mathbf{k})||\mathcal{Y}(\mathbf{k})|}, \tag{11.22}$$

where the numerator is the dot product between the wavevectors with the phase difference $\Delta\theta_k = [\theta_{\mathcal{Y}(\mathbf{k})} - \theta_{\mathcal{X}(\mathbf{k})}]$ **[96]**. In other words, the correlation coefficient between two wavevectors is their normalized dot product, which in turn is a simple cosinusoidal function of their phase difference. The meaning

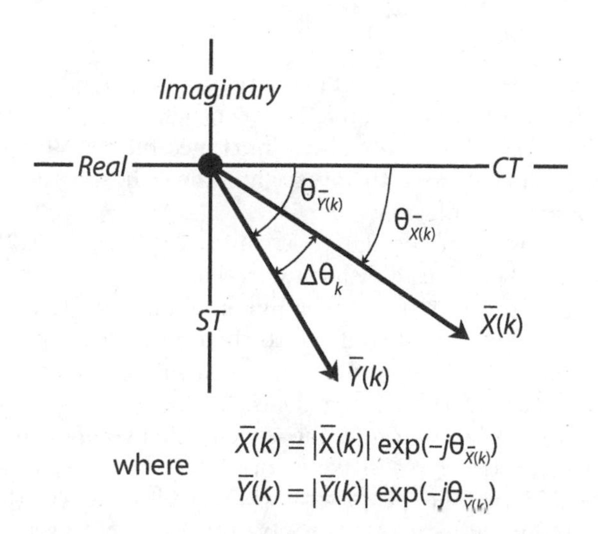


**Figure 11.21** The $k$-th wavevectors for maps $\mathbf{X}$ and $\mathbf{Y}$ represented in complex polar coordinates. [Adapted from **[43]** after **[96]**.]

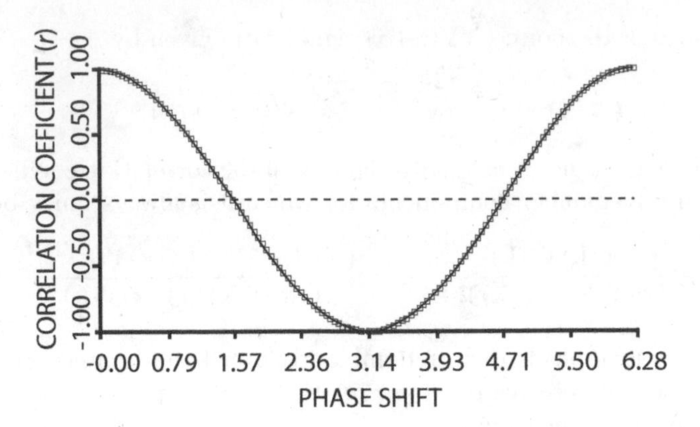


**Figure 11.22** The meaning of the correlation coefficient described by **Equation 11.22**. [Adapted from [**43**] after [**96**].]

of this result is evident from ***Example 11.4.2.3b*** in **Figure 11.22**, which illustrates the variations of the correlation coefficient computed as a function of $\Delta\theta_k$ as any sine or cosine of period $k$ is lagged past itself.

This fundamental result relates several important spectral concepts in the literature, including

$$\begin{aligned} \cos(\Delta\theta_k) &= \frac{\sigma^2_{\mathcal{X}(\mathbf{k}),\mathcal{Y}(\mathbf{k})}}{\sqrt{\sigma^2_{\mathcal{X}(\mathbf{k})}\sigma^2_{\mathcal{Y}(\mathbf{k})}}} && \text{e.g., [3]} \\ &= \frac{Re[\mathcal{X}(\mathbf{k})\mathcal{Y}(\mathbf{k})^\star]}{\sqrt{[\mathcal{X}(\mathbf{k})\mathcal{X}(\mathbf{k})^\star][\mathcal{Y}(\mathbf{k})\mathcal{Y}(\mathbf{k})^\star]}} && [54] \qquad (11.23) \\ &= Re[\frac{\mathcal{X}(\mathbf{k})}{\mathcal{Y}(\mathbf{k})}][\frac{|\mathcal{Y}(\mathbf{k})|}{|\mathcal{X}(\mathbf{k})|}], && [26] \end{aligned}$$

where the raw, unnormalized ratio $[\mathcal{X}(k)/\mathcal{Y}(k)]$ is commonly called the *in-phase response* or *admittance function* between signals $\mathbf{x}$ and $\mathbf{y}$. [**26**] and subsequently others [e.g., [**100**]] use the admittance function between elevation or bathymetry and gravity to investigate the lithospheric response to varying topographic and geologic loads.

To investigate $\mathbf{x}$ and $\mathbf{y}$ for correlative features, **Equation 11.22** is evaluated for all $k$-wavenumbers to establish the correlation spectrum of the signals. From the correlation spectrum, the wavenumber components are chosen with correlation coefficient values that conform to the desired feature associations. Inversely transforming the selected wavenumber components in both signals maps out the correlative features for analysis.

For example, inversely transforming the wavenumber components for which $r(k) \geq 0.70$ favors the stronger positively correlated data variations between the signals, whereas the components with $r(k) \leq -0.70$ reflect the stronger negatively correlated data features. Inversely transforming wavenumber components for which $-0.30 \leq r(k) \leq +0.30$, on the other hand, enhances the presence of null correlated data features.

Modifying the spectral properties of the data by window carpentry or padding the data margins with zeroes greatly affects the correlation spectrum [**52**]. Thus, these modifications are avoided in correlation filtering that are very beneficial for other spectral applications. When edge effects are unacceptably severe in the correlation filtered output, every effort is made to expand the data margins with real observations. Where this is not possible, folding out the data rows and columns across the data margins always results in more meaningful correlation filtered output than adding rinds of zeroes to the datasets [**52**].

In general, meaningful filter outputs account for significant levels of the input energy. Thus, the standard deviations of the input and output data are typically assessed to monitor the energy levels. These efforts also help establish effective $r$-threshold levels of the correlation filter. For example, filtering for wavenumber components satisfying $r \geq 0.95$ may significantly improve the signal-to-noise ratio by **Equation 11.20**, but the result will be meaningless if the output energy is in the noise level of the data.

As with any standard filter, spectral correlation filtering cannot produce results that are not already evident in the original datasets. Indeed, an important test of the veracity of the correlation filtered anomalies is to verify their existence in the original datasets. Spectral correlation filters can be effective in isolating feature correlations in large and complex datasets. However, the occurrence of correlated features is not necessarily indicative of a common source due to the lack of solution uniqueness in inversion.

#### 11.4.2.4 *Derivative and Integral Filters*

To this point, only standard filters constructed from simple patterns of 1's and 0's have been considered. For derivative and integral filters, however, specialized transfer functions involving non-integer coefficients are required that are simple extensions of the signal's wavenumbers. For example, taking the first order partial derivative of the signal $\mathbf{B}$ along its $n$-axis in **Equation 11.13** showed that the transfer function coefficients are $j\frac{2\pi}{N}k = j\omega_k = j2\pi f_k$ so that $\partial\mathbf{B}/\partial n \iff [j\frac{2\pi}{N}k]\mathcal{B}$ and $\partial/\partial n \iff [j\frac{2\pi}{N}k]$ form transform pairs. Furthermore, it can be shown that $\partial^p/\partial n^p \iff [j(\frac{2\pi}{N})k]^p$ by taking the $p$-th order derivative of **Equation 11.13**.

These results also hold for derivatives taken along the $m$-axis of the two-dimensionally gridded dataset modeled by the transform **Equation 10.22** for which $\partial^p/\partial m^p \iff [j(\frac{2\pi}{M})l]^p$. Now, data that follow Laplace's equation [e.g., potential fields]

$$\nabla^2\mathbf{B}_{M,N} = (\frac{\partial^2}{\partial m^2} + \frac{\partial^2}{\partial n^2} + \frac{\partial^2}{\partial z^2})\mathbf{B}_{M,N} = 0, \tag{11.24}$$

have data variations measured in the orthogonal map dimensions $m$ and $n$ that define the variations of the data in the map's perpendicular vertical $z$ direction. Thus, in theory, the simple transfer function coefficients for the measured horizontal derivatives of the data predict the data's vertical derivatives.

To see this, take the Fourier transform of **Equation 11.24** to obtain the linear second order differential equation

$$\frac{\partial^2 \mathcal{B}_{l,k,z}}{\partial z^2} = (2\pi)^2[f_l^2 + f_k^2]\mathcal{B}_{l,k,z} = a\mathcal{B}_{l,k,z} \tag{11.25}$$

for the data array at altitude $z$. This equation has the general solution

$$\mathcal{B}_{l,k,z} = Ce^{+z\sqrt{a}} + De^{-z\sqrt{a}} \tag{11.26}$$

with constant coefficients $C$ and $D$. However, the physically realistic form of the solution requires that $(\partial\mathcal{B}_{l,k,z}/\partial z) \longrightarrow 0$ as $z \longrightarrow \infty$. Invoking these boundary conditions gives

$$\lim_{z\to\infty} \mathcal{B}_{l,k,z} = Ce^{+\infty\sqrt{a}} + De^{-\infty\sqrt{a}} = 0, \tag{11.27}$$

which implies that $C \equiv 0$ for **Equation 11.27** to hold. Thus, the realistic form of the solution is

$$\mathcal{B}_{l,k,z} = De^{-z\sqrt{a}} \tag{11.28}$$

so that

$$\frac{\partial\mathcal{B}_{l,k,z}}{\partial z} = [-2\pi\sqrt{f_l^2 + f_k^2}]\mathcal{B}_{l,k,z}. \tag{11.29}$$

In other words, these results show that the vertical derivative of the signal forms the Fourier transform pair $(\partial\mathbf{B}_{M,N}/\partial z) \Longleftrightarrow [-2\pi\sqrt{f_l^2 + f_k^2}]\mathcal{B}_{l,k}$, or more generally that $(\partial^p\mathbf{B}_{M,N}/\partial z^p) \Longleftrightarrow [-2\pi\sqrt{f_l^2 + f_k^2}]^p\mathcal{B}_{l,k}$ holds for any order $p$ of the vertical derivative. **Table 11.1** summarizes the transfer function coefficients for horizontal and vertical differentiation and the corresponding inverse operations of integration.

***Example 11.4.2.4a*** in **Figure 11.23** shows ideal first derivative filters designed for the $16 \times 32$ data array in **Figure 10.5**. The ***Example 11.4.2.4b*** in **Figure 11.24** gives the ideal vertical second and fourth derivative filters in the left and right panels, respectively. Derivative filters clearly amplify the higher frequency over the lower frequency components and thus work

**Table 11.1**

**Spectral filter coefficients for estimating $p$-order derivatives and integrals in the horizontal (H) and vertical (V) directions.**

| $p$-th Order | Differentiation | Integration | Filter Coefficients |
|---|---|---|---|
| $m$-th H-axis | $p > 0$ | $p < 0$ | $[j2\pi(l/M)]^p$ |
| $n$-th H-axis | $p > 0$ | $p < 0$ | $[j2\pi(k/N)]^p$ |
| $z$-th V-axis | $p > 0$ | $p < 0$ | $[-2\pi\sqrt{(l/M)^2 + (k/N)^2}]^p$ |

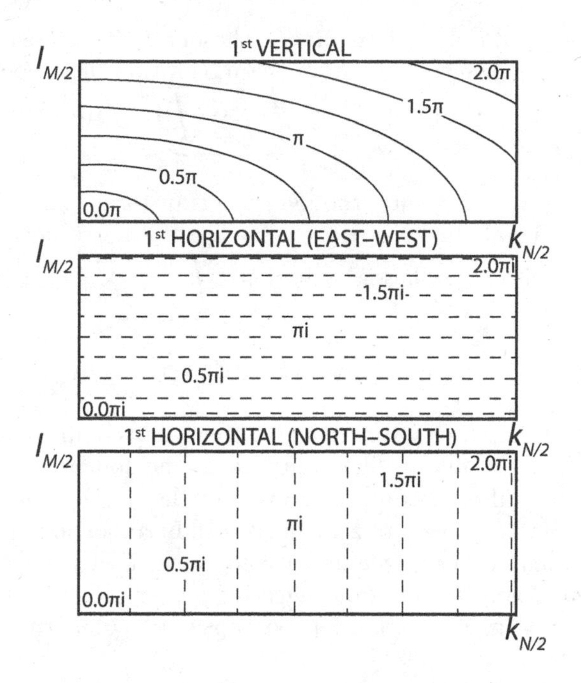


**Figure 11.23** Ideal filter operators for the vertical and horizontal derivatives to order $p = 1$. The design of these filters requires only a single quadrant. [Adapted from [**78**].]

like hi-pass/low-cut filters. However, the noise amplification in the higher frequency components due to measurement, sampling and other errors can limit substantially the veracity of derivative estimates, especially those with orders $p > 2$.

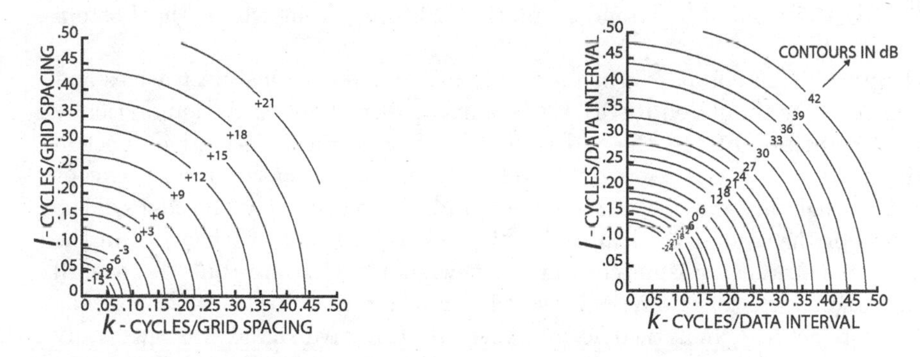


**Figure 11.24** Ideal filter operators for the vertical derivatives to orders $p = 2$ (left) and $p = 4$ (right). The design of these filters requires only a single quadrant. [Adapted from [**43**] after [**30**].]

For the inverse operation of integration, the derivative transfer function coefficients are simply inverted to obtain the integration filter coefficients as shown in **Table 11.1** [e.g., [**51**]]. Thus, $\underbrace{\int\int\cdots\int}_{p}\mathbf{B}_{M,N}(dn^p \text{ or } dm^p) \Longleftrightarrow$ $([j(\frac{2\pi}{N})k]^{-p} \text{ or } [j(\frac{2\pi}{M})l]^{-p})\mathcal{B}_{l,k}$ are the Fourier transform pairs for the $p$-integrals of the signal in the $n$- or $m$-directions, respectively. Furthermore, by virtue of Laplace's equation, $\underbrace{\int\int\cdots\int}_{p}\mathbf{B}_{M,N}(dz^p) \Longleftrightarrow$ $[-2\pi\sqrt{f_l^2+f_k^2}]^{-p}\mathcal{B}_{l,k}$ for the $p$-integrals of the signal in the vertical $z$-direction.

Integration filters amplify lower over higher frequency components and thus are like low-pass/high-cut filters. Their outputs are much more stable and less noisy than derivative filter outputs. However, in the application of integral filters, care must be taken to set to zero the zeroth harmonic component (i.e., $\mathfrak{b}_0 \equiv 0$), as well as any other wavenumber component that is effectively zero within working precision. In addition, effective numerical performance may require extensive padding of the dataset, perhaps out to one or two times the original dimensions of the dataset.

As ***Example 11.4.2.4c***, consider the profiled gravity effects of the five horizontal cylinders in the top panel of **Figure 11.25** as modeled by **Equation 9.3** from the subsurface parameters of the cylinders in the second panel. In practice, these gravity effects would be typically observed as the total field (i.e., TOTAL) consisting of their summed or integrated effects (i.e., RESIDUAL) superimposed on a regional field (i.e., REGIONAL). To evaluate the mapped TOTAL data for the subsurface parameters of the cyclinders requires extracting the RESIDUAL data for their gradients [e.g., [**43**]] that, in turn, may be spectrally estimated. To help optimize the performance of the derivative [and integral] filters, the value at each end of the simulated RESIDUAL signal was extended another 100 km as shown in the bottom panel.

**Figure 11.26** compares the spectrally determined vertical derivatives and integrals of the simulated RESIDUAL signal against those derived analytically from **Equation 9.3**. In the top panel, the first vertical derivative spectral estimate for $p = 1$ compares quite well with the analytical derivative except for some slight mismatches at the higher gradient peaks and at the edges of the signal. For the second vertical derivative spectral estimate with $p = 2$ in the middle panel, these mismatches are somewhat more severe, but the overall comparison with the analytical second derivative is still very good. In the bottom panel, the analytical second vertical derivative signal was spectrally integrated with $p = -2$ to obtain the original simulated RESIDUAL data with virtually no error.

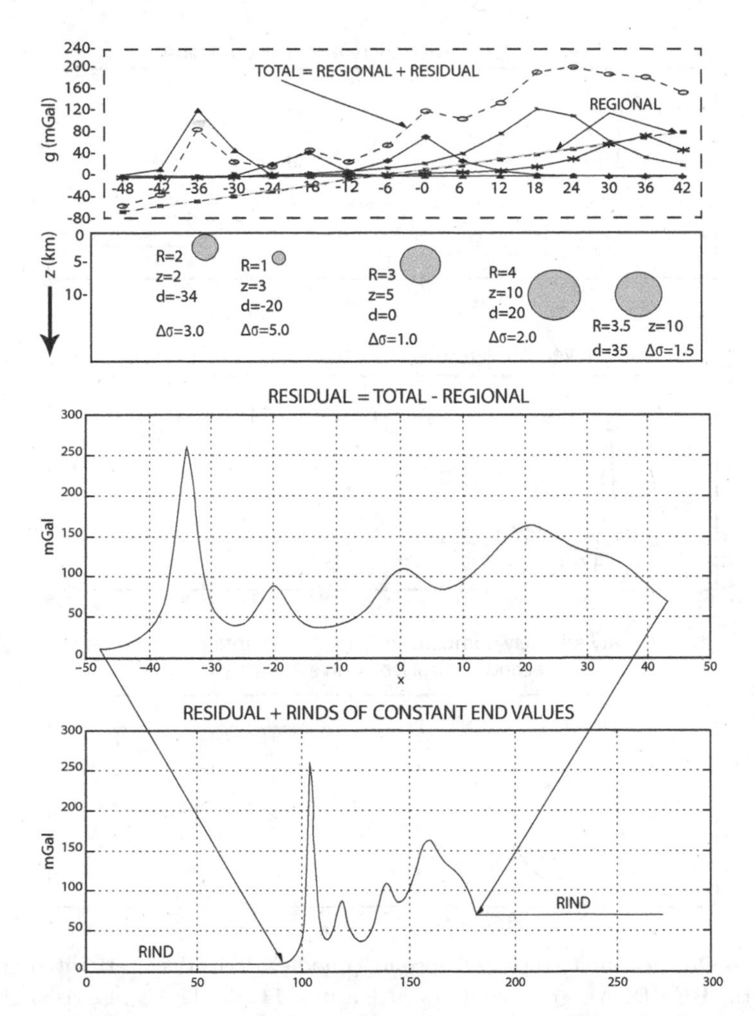


**Figure 11.25** Top panel shows the profiles of the gravity effects for five horizontal cylinders plus a regional field [REGIONAL]. The second panel gives the parameters of the cylinders, whereas the third panel shows their integrated gravity effects [i.e., RESIDUAL] over a 90-km survey at 1-km intervals, which are extended as rinds of constant values in the bottom panel to enhance the accuracy of the derivative and integral estimates given in **Figure 11.26**. [Adapted from [**43**].]

### 11.4.2.5 *Interpolation and Continuation Filters*

Specialized filters can also be developed to estimate data values at unsurveyed locations. For example, to densify the data grid for additional data estimates at one-half the original data interval, a zero can be inserted at the Nyquist end of the spectrum, or if additional estimates at one-quarter the data interval

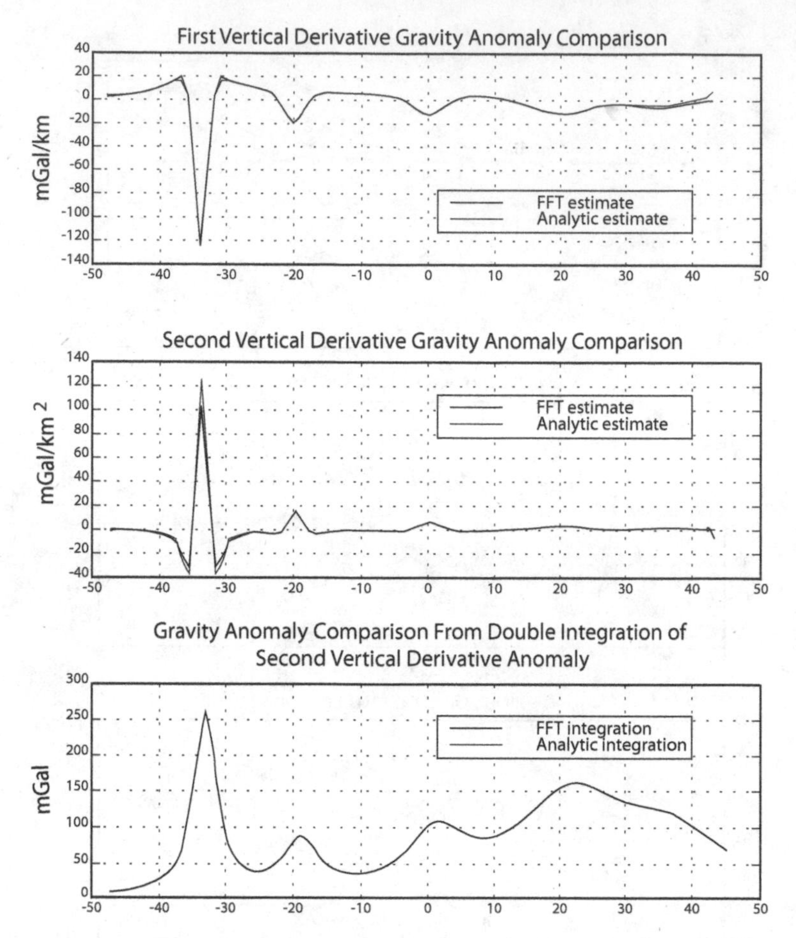


**Figure 11.26** Vertical first (top) and second (middle) derivatives estimated spectrally for the RESIDUAL gravity effects in **Figure 11.25**. The comparison (bottom) of the double spectral integration of the derivative in the middle panel to the original RESIDUAL gravity effects reveals negligible differences. [Adapted from [43].]

are desired, the transform can be padded with two zeroes, etc. This approach does not modify the spectral properties of the dataset and is purely an interpolation scheme. Sometimes slight smoothing by low-pass filtering helps to make the interpolated estimates appear more meaningful.

Where data satisfy Laplace's equation, upward and downward continuation filters can be developed to estimate data variations in the vertical $z$-direction above and below the data grid, respectively. To obtain the transfer function coefficients for upward continuation, for example, **Equation 11.29** is rearranged into the differential equation $(\partial\mathcal{B}_{l,k}/\mathcal{B}_{l,k}) = [-2\pi\sqrt{f_l^2+f_k^2}]\partial z$ and integrated to obtain

$$\ln(\mathcal{B}_{l,k})|_{\mathcal{B}_{l,k,z=0}}^{\mathcal{B}_{l,k,z}} = [-2\pi\sqrt{f_l^2+f_k^2}]z|_{z=0}^{z=z} \ni \ln(\mathcal{B}_{l,k,z}/\mathcal{B}_{l,k,z=0}) = [-2\pi\sqrt{f_l^2+f_k^2}]z. \quad (11.30)$$

Taking the exponential of both sides of the above equation gives

$$\mathcal{B}_{l,k,z}/\mathcal{B}_{l,k,z=0} = e^{[-2\pi\sqrt{f_l^2+f_k^2}]z} \ni \mathcal{B}_{l,k,z} = (\mathcal{B}_{l,k,0})e^{[-2\pi\sqrt{f_l^2+f_k^2}]z} \quad (11.31)$$

so that $\mathbf{B}_{M,N,z>0} \Longleftrightarrow (\mathcal{B}_{l,k,0})e^{[-2\pi\sqrt{f_l^2+f_k^2}]z}$ is the Fourier transform pair for upward continuation.

For downward continuation, the procedure is the same except that the limits of integration are interchanged so that

$$\ln(\mathcal{B}_{l,k})|_{\mathcal{B}_{l,k,z=0}}^{\mathcal{B}_{l,k,-z}} = [-2\pi\sqrt{f_l^2+f_k^2}]z|_0^{-z} \ni \ln(\mathcal{B}_{l,k,-z}/\mathcal{B}_{l,k,z=0}) = [2\pi\sqrt{f_l^2+f_k^2}]z. \quad (11.32)$$

Taking the exponential again of both sides gives

$$\mathcal{B}_{l,k,-z}/\mathcal{B}_{l,k,z=0} = e^{[2\pi\sqrt{f_l^2+f_k^2}]z} \ni \mathcal{B}_{l,k,-z} = (\mathcal{B}_{l,k,0})e^{[2\pi\sqrt{f_l^2+f_k^2}]z} \quad (11.33)$$

so that $\mathbf{B}_{M,N,z<0} \Longleftrightarrow (\mathcal{B}_{l,k,0})e^{[2\pi\sqrt{f_l^2+f_k^2}]z}$ is the Fourier transform pair for downward continuation.

Upward and downward continuations are obviously inverse operations with transfer function coefficients given by

$$\exp([-2\pi\sqrt{f_l^2+f_k^2}]z), \quad (11.34)$$

where the elevation of continuation $z > 0$ for upward continuation and $z < 0$ for downward continuation. ***Example 11.4.2.5a*** in **Figure 11.27** shows ideal upward (upper panel) and downward (lower panel) continuation filters to continue the $16 \times 32$ data array in **Figure 10.5** through elevation differences of three times the 1-km grid interval. ***Example 11.4.2.5b*** in **Figure 11.28** gives the ideal frequency responses for upward continuation filters that continue the data one (left panel) and two (right panel) grid intervals.

Clearly, upward continuation filters smooth data and emphasize their regional features much like low-pass/high-cut filters do. Downward continuations, on the other hand, tend to promote the higher frequency features in the data like high-pass/low-cut filters. Continuation estimates are most reliable over elevations within a few grid intervals of the observations due to measurement errors and the nonuniqueness of continuation [e.g., **[97]**]. To be certain of data behavior at more distant elevations, there is no recourse but to survey. For example, multiple altitude grids of potential field observations are becoming increasingly available from surface, airborne, and satellite surveys where both the $C$ and $D$ coefficients in **Equation 11.26** can be applied to evaluate data values at the intermediate altitudes beween the grids.

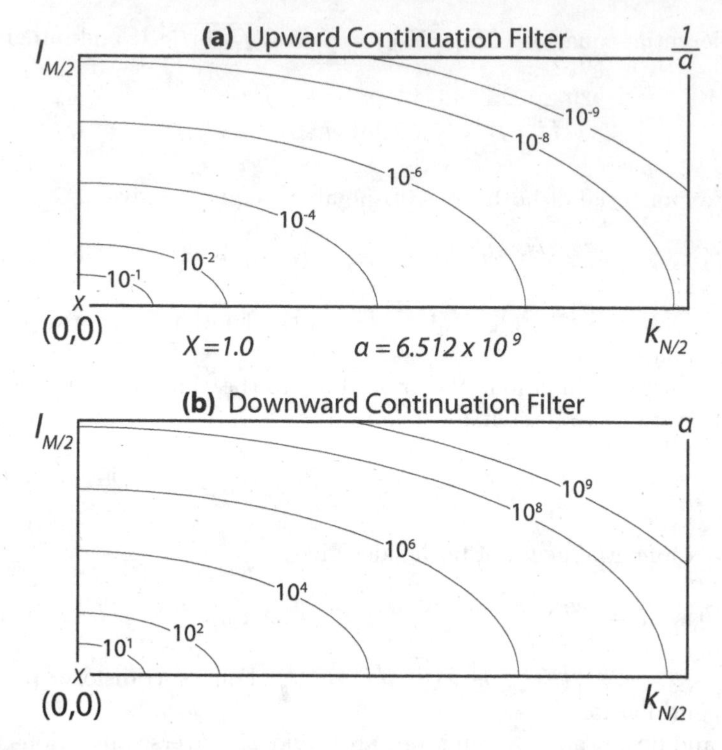


**Figure 11.27** Ideal upward (**a**) and downward (**b**) continuation filter coefficients for continuations of the data to 3 station intervals above and below the original altitude of the $16 \times 32$ data array. The design of these filters requires only a single quadrant with coefficients ranging from the $x$ to $\alpha$ values shown. [Adapted from [**78**].]

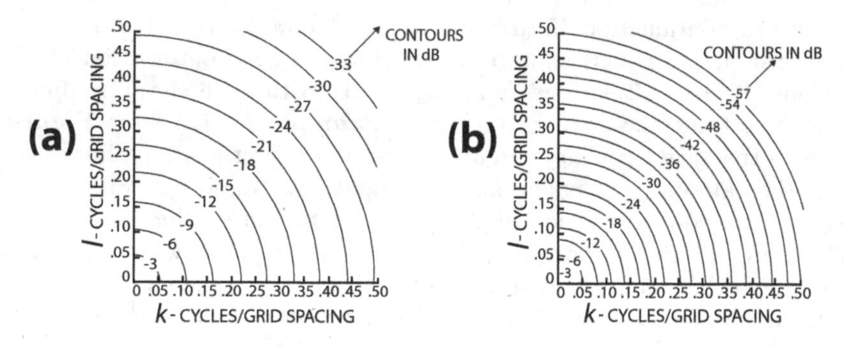


**Figure 11.28** Ideal upward continuation filters to continue data (**a**) 1- and (**b**) 2-station intervals. Changing the sign of the transfer function coefficients continues the data downward 1- and 2-station intervals, respectively. [Adapted from [**43**] after [**30**].]

In particular, let the lower altitude data grid be referenced to relative zero and the higher altitude grid referenced to altitude $H$ where the Fourier transforms of $\mathbf{B}(0)$ $(\equiv \mathbf{B}_{M,N,0})$ and $\mathbf{B}(H)$ $(\equiv \mathbf{B}_{M,N,H})$ are $\mathcal{B}(0)$ $(\equiv \mathcal{B}_{l,k,0})$ and $\mathcal{B}(H)$ $(\equiv \mathcal{B}_{l,k,H})$, respectively. Both transforms may be related through **Equation 11.26** to common $C$ and $D$ coefficients given by

$$C = \frac{\mathcal{B}(0)e^{-z\sqrt{a}} - \mathcal{B}(H)}{e^{-z\sqrt{a}}} \text{ and } D = \frac{\mathcal{B}(H) - \mathcal{B}(0)e^{+H\sqrt{a}}}{e^{-z\sqrt{a}}}. \tag{11.35}$$

Thus, the interval continued field at an intermediate $z$-altitude [i.e., $0 \le z \le H$] can be obtained by inversely transforming

$$\begin{aligned}\mathcal{B}(z) &= \frac{1}{q}\left[\mathcal{B}(0)e^{-H\sqrt{a}}e^{+z\sqrt{a}} - \mathcal{B}(H)e^{+z\sqrt{a}} + \mathcal{B}(H)e^{-z\sqrt{a}} - \mathcal{B}(0)e^{+H\sqrt{a}}e^{-z\sqrt{a}}\right]\\ &= \frac{1}{q}\left[\mathcal{B}(0)(e^{+[z-H]\sqrt{a}} - e^{-[z-H]\sqrt{a}}) + \mathcal{B}(H)(e^{-z\sqrt{a}} - e^{+z\sqrt{a}})\right]\end{aligned} \tag{11.36}$$

which in compact matrix notation is given by

$$\mathcal{B}(z) = \frac{1}{q}\begin{pmatrix} [e^{+(z-H)\sqrt{a}} - e^{-(z-H)\sqrt{a}}] \\ [e^{-z\sqrt{a}} - e^{+z\sqrt{a}}] \end{pmatrix}^t \begin{pmatrix} \mathcal{B}(0) \\ \mathcal{B}(H) \end{pmatrix} \tag{11.37}$$

with

$$q = e^{-H\sqrt{a}} - e^{+H\sqrt{a}}. \tag{11.38}$$

The interval continuation operator $\mathcal{B}(z)$ fully honors the boundary conditions imposed by multiple altitude slices of the data, and thus its predictions yield insights on the altitude behavior of anomalies that are not available from standard single-surface upward and downward continuations.

In summary, the spectral model is very powerful in data analysis because it has tremendous numerical efficiency and accuracy and invokes minimal assumptions about the source of the data variations. For hardly more than the effort of gridding, the spectral model provides access to the basic wavelength, directional, correlative, differential, integral, interpolation and continuation properties of the data. The grid is also the fundamental basis for electonically visualizing data in effective data analysis as the next section highlights.

## 11.5 KEY CONCEPTS

1. Data analysis fundamentally interrogates the forward model for the derivative and integral components of the data. These attributes in data applications are used mainly for feature isolation, enhancement, modeling, and graphical display purposes. They may also be obtained numerically by convolution of the data with an appropriately designed operator consisting of a finite set of discrete coefficients.

2. The convolution is carried out by reversing or folding the operator about the ordinate axis, shifting or displacing the operator by a fixed amount or lag, cross-multiplying definable products of the operator and signal coefficients, and integrating or summing all defined products for the convolution's value at the lag position. Repeating these steps for all possible lags yields the complete convolution.
3. Deconvolution is the inverse process of finding the set of operator coefficients that when convolved with the system's impulse response yields the desired output [i.e., the data's derivative and integral components].
4. The convolution theorem shows that data domain convolution is much more efficiently computed in the frequency domain and vice versa. In particular, the convolution of two data domain signals is taken most efficiently in the spectral domain by inversely transforming the amplitude spectrum consisting of the products of their amplitude spectra and the phase spectrum of the sums of their phase spectra.
5. Correlations of data domain signals are their convolutions without signal folding so that in the frequency domain the phase spectra are differenced rather than summed. Thus, a signal's autocorrelation in the data domain is obtained by inversely transforming its zero-phase power spectrum. The cross-correlation of two data domain signals, by contrast, is given by inversely transforming the cross-power spectrum of the products of their amplitude spectra and the phase spectrum of the differences of their phase spectra.
6. Isolation and enhancement methods respectively seek to isolate and enhance the data variations of interest (i.e., residuals) from the interfering effects of longer wavelength (i.e., regional) and higher frequency (i.e., noise) variations. The residual separation problem is subjective, and thus a major limitation in data analysis. It is commonly approached by the use of filters in the data and spectral domains.
7. Data domain filtering includes the use of environmentally constrained modeling, and smooth contouring and other graphical methods to draw and digitize the regional through a dataset. Trend surfaces also are used to identify and extract the regional. In addition, analytical grid or convolution methods are available for isolating and enhancing data derivative and integral properties, but these methods have been largely superseded by the more efficient spectral filtering techniques.
8. Spectral filtering involves the use of transfer function coefficients that are multiplied against the wavenumber coefficients from the FFT of the data. Inversely transforming the modified or filtered spectral model efficiently and accurately estimates the spatial attributes of the data.
9. Standard spectral filters with transfer function coefficients of 1's and 0's to respectively pass or reject wavenumber amplitudes are used to

study data wavelength and directional properties, and the correlative features between two or more datasets.

10. Specialized spectral filters with non-integer transfer function coefficients account for the data's mapped derivatives and integrals and their theoretical extensions. For example, data conforming to Laplace's equation have second vertical derivatives related to the mapped second horizontal derivatives in the grid's two dimensions. Thus, Laplace's equation can be used to design spectral filters to estimate the data's vertical derivatives and integrals of any order, and the respective upward and downward continuations of data above and below the mapped grid.
11. Composite spectral filters achieve multiple data analysis objectives using the product of multiple transfer functions. For example, cross-multiplying the zero phase transfer functions of standard wedge and wavelength filters yields a composite filter to enhance or suppress data variations of given orientation and wavelength characteristics, respectively.

# 12 Data Graphics

## 12.1 OVERVIEW

*Essential elements of modern data analysis include effective presentation graphics to display the data and data processing results. Graphical displays range from simple 1-D line plots and cartoons to elaborate colored 2-D map and 3-D isometric plots and animations of the data. They greatly enhance data use and understanding, but also must be applied with care to minimize bias and distortion in the presented data and results.*

*Data mapping projections include the simple Cartesian [x,y,z]-projection that is effective for smaller spatial scale applications with principal dimensions* $\leq 0.5^o$*, the transverse Mercator projection for larger scale applications with dimensions up to roughly* $6^o$*, and the universal transverse Mercator projection over more regional scale areas with dimensions typically* $\geq 6^o$*.*

*The fundamental format for electronic data analysis and graphics is the data grid. Thus, the investigator must be proactive in understanding how the data are gridded and presented for electronic analysis. For example, by the Nyquist theorem, the chosen grid interval must be at least half of the shortest data wavelength to avoid corrupting the lower frequency spectrum with erroneous aliased components.*

*The need for taking numerical derivatives and integrals has spawned numerous gridding schemes including the basic linear interpolation method, the more complex cubic spline and popular minimum curvature methods, the least squares equivalent source and polynomical interpolation methods, and the data covariance-based statistical kriging and collocation methods.*

*An effective legend for a data graphic lists key data attributes that include the amplitude range [AR = (min; max)] of minimum and maximum values, amplitude mean [AM], amplitude standard deviation [ASD], and amplitude units [AU] together with the contour interval [CI] and grid interval [GI]. The CI and color and shading scales are typically set to represent amplitude variations within* $\pm 1$*, or* $\pm 2$*, or* $\pm 3$ *ASD, whereas the GI defines the Nyquist wavelength of the gridded dataset.*

*Datasets with different measurement units can be standardized into dimensionless values with zero means and unit standard deviations to facilitate graphical [and numerical] analyses. Here, each standardized value corresponds to a datum minus the data's mean value that is divided through by data's standard deviation.*

*Normalized data are standardized data adjusted to a non-zero mean and arbitrary standard deviation. The correlation coefficient between a dataset and its standardized and normalized transformations is always positive maximum. In addition, dimensionless datasets with common means and standard devi-*

*ations can be compared with common graphical parameters [e.g., projection axes, contour intervals, color bars, etc.].*

*Directly correlated features between two datasets can be studied by adding their standardized [or normalized] transformations as summed local favorability indices [SLFI]. Thus, the map of positive coefficients [SLFI $> 0$] emphasizes the positive feature correlations in the two datasets [i.e., peak-to-peak associations], whereas the map of the negative coefficients [SLFI $< 0$] highlights the negative feature correlations [i.e., valley-to-valley associations].*

*Inversely correlated features between, say, map A and map B, on the other hand, can be studied by subtracting the standardized [or normalized] transformation of map B from the related transformation of map A as differenced local favorability indices [DLFI]. Thus, the map of positive coefficients [DLFI $> 0$] emphasizes positive features in map A that correlate with negative features of map B [i.e., map A peak-to-map B valley associations], whereas the negative coefficients [DLFI $< 0$] highlight negative features of map A that correlate with positive features in map B [i.e., map A valley-to-map B peak associations].*

*Common modes of presenting gridded data include 1-D profiles either individually or as a stacked set to provide perspective on data variations across the profiles. Line and color-filled contour maps provide widely accepted 2-D images of data, whereas 3-D perspective views include line and color-filled isometric plots, and shaded relief maps that emphasize lineaments and other data features perpendicular to an investigator-specified illumination angle. In general, the importance of effective graphics for achieving and promoting successful outcomes in earth and environmental data analysis is difficult to overstate.*

## 12.2 INTRODUCTION

Modern reporting of data analyses requires the use of effective graphics to represent the variations and their inversions. Graphics may range from simple line plots and cartoons to elaborate 3-D color plots and animations of the data and analytical results that are made possible by a variety of efficient computer processing and graphics software and hardware. Simple listing of data is often available, but seldom used. Rather the data are presented in graphical form in a spatial context or less frequently tied to a temporal base for ease of viewing and analysis. These presentations greatly enhance the use and understanding of the data, but care is required to use graphical formats and parameters that present the data and results with minimal bias and distortion.

In this section, rudimentary principles are outlined for developing useful presentations of the data and their interpretations with modern digital graphics. An important graphical consideration in any application, for example, is the choice of the coordinate reference frame for projecting data onto a map. Presenting data in various forms such as profiles or contour maps requires significant application-dependent choices. This section provides guidelines for the most effective imaging of data for analysis and interpretation.

Transforming data into data grids greatly facilitates electronic data processing and graphics efforts. However, gridding presents critical data issues because modern algorithms operate at a wide variety of performance levels that can make even unacceptable data look quite suitable for analysis. Thus, the interpreter must be proactive in understanding how the data are gridded for analysis and take suitable precautions when dealing with grids generated by inappropriate or unknown algorithms.

The statistical properties of the data facilitate choosing effective graphical parameters for revealing significant data variations. These statistical attributes also allow the interpreter to standardize and normalize datasets into common graphical parameters where feature associations and other interpretational insights are more apparent.

## 12.3 MAP PROJECTIONS AND TRANSFORMATIONS

Gridded data of areal studies are presented in map form for analysis and interpretation. This requires the projection of data observed on the earth's spherical surface, or more appropriately an oblate spheroid which represents the true earth shape, onto a flat surface, that is a map. Literally hundreds of projection methodologies have been developed [e.g., [**86**]]. All of these produce distortions from the actual spherical surface; thus, the choice of projection is based on minimizing a particular undesirable attribute of the map, e.g., area, shape, or scale, and its intended use. However, the map projection may be chosen to duplicate the projection of existing maps which are used in conjunction with the data map.

The geographic position of observations is generally given in degrees of longitude and latitude to 7 decimal places except in local surveys in which the earth is assumed to be flat and observations are measured in Cartesian coordinates. Currently, to achieve consistency in locations the International Terrestrial Reference Frame (ITRF) in conjunction with the Geodetic Reference System (GRS80) ellipsoid is used for the horizontal datum. The ITRF differs from the World Geodetic System-1984 (WGS84) which is used in specifying horizontal location in the global positioning system (GPS) by less than 10 cm.

A widely used projection for data maps of limited regions is the transverse Mercator projection which projects from the earth's surface onto a cylinder whose axis is in the equatorial plane (i.e., transverse to the normal Mercator projection which projects the earth's surface onto a cylinder whose axis is coincident with the earth's axis of rotation) and is tangential to the earth's surface along a meridian of longitude which is termed the central meridian of the map. This projection is conformable whereby relative local angles are shown correctly. As a result the scale is constant at the equator and at the central meridian and the parallels of latitude and meridians of longitude intersect at right angles. The distortion of the map and scale increases with the east/west $x$-distance and thus the projection is not suitable for world maps.

To achieve high-scale accuracy in the transverse Mercator projection the central meridian is changed to coincide with the map area. For world-wide consistency in the universal transverse Mercator projection, the earth between 84°N and 80°S is divided into 60 zones each generally 6° wide in longitude. The central meridian is placed at the center of each zone. True scale is achieved approximately 180 km on either side of the central meridian. Beyond this distance the scale is too great. Locations in a zone are referenced to the intersection of the central meridian of the zone and the equator. This position in the northern hemisphere is given a coordinate of $500,000$ m in the $x$-direction and zero in the $y$-direction with coordinates increasing in the East and North directions. In the southern hemisphere $x$ remains $500,000$ m and $y$ is $10,000,000$ m.

Whatever the projection, it is desirable to place the origin of the coordinates at a specified position that is held constant in producing maps of a variety of gridded datasets. This so-called registering of the coordinates facilitates joint computational analysis of the gridded datasets. Grids generally are made equidimensional for simplicity and convenience and to avoid anisotropic effects due to the different spacing in the orthogonal directions. Commonly grids are oriented in the cardinal directions, i.e., N-S, E-W, in dealing with regions measured in distances measured in hundreds of meters. However, with grids measured in kilometers or greater, the spherical shape of the earth necessitates modification of the grid from the cardinal directions to maintain the orthogonal, equispaced grid. Contour maps from equidimensional grids can be displayed on non-square grids (e.g., in latitude and longitude coordinates) by transforming the grid to the map projection, but consideration should be given to the bias resulting from the non-square grid values. In local surveys where data are acquired along transects, the grid may be oriented so that one coordinate of the grid coincides with the orientation of the transects to capture the detail available in the higher sampling interval along the transects.

## 12.4 GRIDDING

Uniformly sampled or gridded data are preferred for electronic data analysis because the data input and storage requirements are minimal. All that is necessary are the signal amplitudes in their proper spatial order, and a reference point and grid interval from which the independent variable coordinates of the data amplitudes can be computed internally. Converting a set of irregularly spaced, finite, discrete data values into gridded format involves critical choices of the grid interval and the interpolation procedure [**65**].

As discussed in **Chapters 5.3.1** and **10.4.1.3** above, the grid interval $\Delta t$ cannot be larger than one-half the shortest wavelength that is present in the data. In other words, the sampling or grid frequency must be at least twice the highest frequency feature in the dataset. Otherwise, the grid will include erroneous aliased data components that corrupt the data analysis. Gridding at an excessively short interval can of course limit sampling errors, but results in large and possibly unacceptable levels of computational effort. A more computationally efficient grid interval can be ascertained by gridding a representative segment of the data at successively smaller intervals and noting the interval giving data variations that do not change appreciably at smaller sampling intervals.

Generally, the grid spacing is chosen so that each grid square will contain at least a single point value or it is set at the approximate average spacing between point values. Where point values are closely spaced along widely spaced transects, it may be advantageous to use an anisotropic gridding scheme, or to smooth the data along the transects before gridding to reflect the across-track Nyquist wavelength defined by the track spacing. Further, it may be useful to insert dummy point values based on manual interpolation in regions of the data field with sparse or no data, but this should be done as a last resort in the gridding process with the dummy point values interpolated conservatively and the number of points held to a minimum.

Gridding schemes commonly are classified according to the prediction strategy employed. The *local gridding strategy*, for example, predicts a grid value from samples restricted to a window surrounding the grid point, whereas the *global gridding strategy* utilizes all the data for estimating the grid value. The choice of a gridding method in practice is controlled by such factors as the characteristics of the data, the desired application, and the interpreter's experiences. However, all gridding schemes are beset by problems of avoiding spurious features in regions of sparse data while at the same time representing short wavelength data features accurately. They also invariably suffer from edge effects or prediction errors due to the diminished data coverage at the dataset margins. It is advisable to screen the gridding for errors by visually checking contour maps prepared from the grid for closed contours that do not contain observations, series of elliptical contours centered on a succession of observations, and abnormally high gradients.

A broad variety of interpolation procedures is available that can be invoked for data gridding. Interpolation involves finding the dependent variable value (i.e., prediction) at a specific value of the independent variable (i.e., prediction point) from a local or global set of surrounding data. ***Example 12.4.0a*** in **Figure 12.1(a)** illustrates predicting $b_p = b(t_p)$ along a profile using linear (dashed line) and cubic spline (solid line) interpolation functions that pass exactly through the data observations, and a least squares polynomial function (dotted line) that approximately fits the observations.

### 12.4.1 LINEAR INTERPOLATION

For profile data, this local method predicts a grid value based on the two surrounding observations. The prediction is obtain by

$$b_p \equiv b(t_p) = b_{p-1} + \frac{(b_{p+1} - b_{p-1})(t_p - t_{p-1})}{(t_{p+1} - t_{p-1})}, \tag{12.1}$$

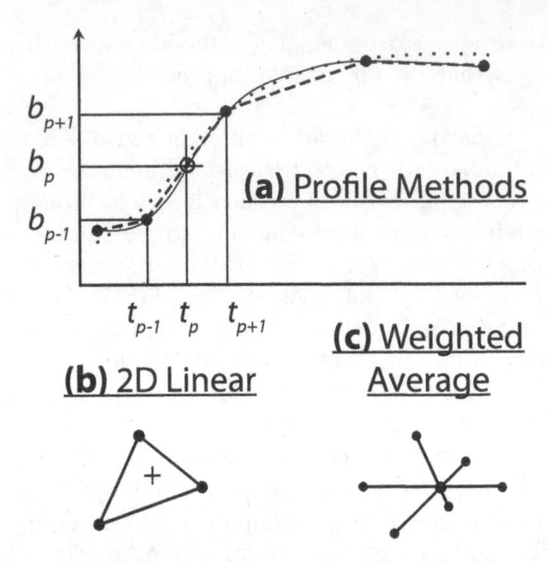


**Figure 12.1** Schematic illustration of predicting the value $b(t_p)$ for **(a)** a profile between known data points (dots) at the location $t_p$ by linear (dashed line), cubic spline (solid line), and least squares (dotted line) methods, and for maps using **(b)** linear interpolation by surrounding triangles and **(c)** statistical interpolation by inverse-distance weighted averages. [Adapted from [**43**] after [**8**].]

where $b_{p+1}$ and $b_{p-1}$ are the two observations on either side of the prediction $b_p$ as shown in **Figure 12.1(a)**.

For data mapped in $2D$ spatial coordinates, the linear interpolation can be implemented by fitting the first degree plane to the three nearest observations surrounding the prediction point as shown in **Figure 12.1(b)**. Thus, each grid estimate is evaluated via **Equation 11.14** from the $\alpha, \beta$ and $\gamma$ coefficients determined in **Equation 11.15** for the plane through the nearest three surrounding data values. The surrounding triangle method is effective for well distributed observations as shown by ***Example 12.4.1a*** in **Figure 12.2(b)**.

The linear interpolation function exactly matches the observations, but invokes no other assumptions concerning the analytical properties of the data and predictions. Thus, the grid predictions provide a relatively honest and pristine view of the dataset. However, the method does not smooth the data which may lead to unsuitable interpolations for analyzing higher wavenumber components, for example, in high-pass and derivative filtering applications.

Linear interpolation is a simple, very cost-effective gridding method that imposes no constraints on the data errors. Thus, its utility for rendering the derivative properties of the data is relatively limited. However, cubic spline interpolation smoothly honors the observations exactly while also maintaining the continuity of first and second horizontal derivatives at the observations.

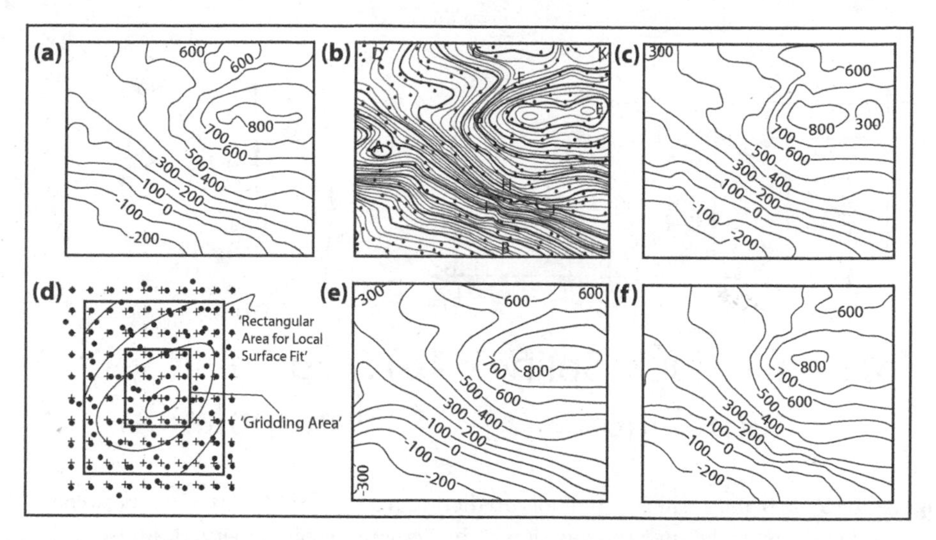


**Figure 12.2** Contour map comparisons of data gridded from 200 randomly located point samples (dots) of the aeromagnetic anomalies in Map **(b)**. Map **(a)** is based on the surrounding triangle method in **Figure 12.1(b)**, Map **(c)** is from the intersecting piecewise cubic spline method in **Figure 12.3(b)**, and Map **(f)** from the weighted statistical averaging method in **Figure 12.1(c)**. Map **(e)** is from the local polynomial surface fitting method illustrated in Map **(d)**. All maps are based on grids of 20 × 20 values at the interval of 2 km. Contour values are in nT. [Adapted from **[43]** after **[8]**.]

### 12.4.2 CUBIC SPLINE INTERPOLATION

Cubic splines are piecewise third degree polynomials that are constrained to maintain continuous first and second horizontal derivatives at each observation point [e.g., **[20, 31, 64]**]. For profile data, they are commonly used to predict a data value based on the two observations on either side of the prediction point as shown by ***Example 12.4.2a*** in **Figure 12.3(a)**. Over each $i$th interval or span between two successive observations $b_i$ and $b_{i+1}$, a cubic spline function can be determined for the interpolations given by

$$b_p = c_0 + c_1(t_p - t_i) + c_2(t_p - t_i)^2 + c_3(t_p - t_i)^3 \;\; \forall \;\; t_i \leq t_p \leq t_{i+1}. \quad (12.2)$$

The four coefficients defining the $i$th spline are

$$c_0 = b_i; \;\; c_1 = F_i - \frac{h_i(2S_i + S_{i+1})}{6}; \;\; c_2 = \frac{S_i}{2}; \text{ and } c_3 = \frac{S_{i+1} - S_i}{6h_i}, \quad (12.3)$$

where $h_i = t_{i+1} - t_i$ is the data or span interval, $F_i$ and $S_i$ are the respective first and second horizontal derivatives at the observed data value $b_i$, and $S_{i+1}$ is the second horizontal derivative at the successive data value $b_{i+1}$. The

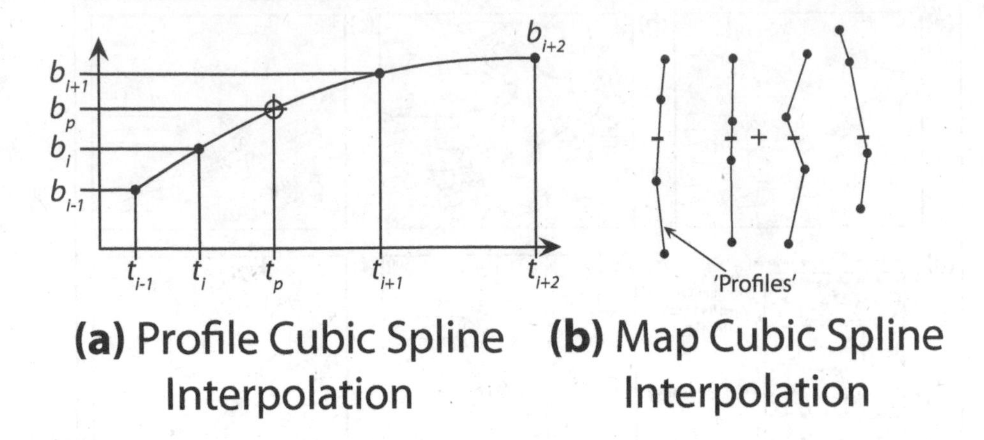


**Figure 12.3** Schematic illustration of predicting **(a)** a profile value $b_p$ between known data points (dots) at the location $t_p$ by the cubic spline method, and **(b)** a map value (+) by intersecting piecewise cubic splines. [Adapted from [**43**] after [**8**].]

horizontal derivatives can be taken numerically so that at $b_i$, for example,

$$F_i = \frac{\partial b_i}{\partial t} = \frac{b_{i+1} - b_{i-1}}{t_{i+1} - t_{i-1}} = \frac{b_{i+1} - b_{i-1}}{h_i + h_{i-1}} \quad \text{and} \quad S_i = \frac{\partial^2 b_i}{\partial t^2} = \frac{F_{i+1} - F_{i-1}}{h_i + h_{i-1}}. \tag{12.4}$$

To interpolate the $N$ data values to a grid, **Equation 12.2** must be fitted to each of the $(N-1)$ spans. However, only $(N-2)$ derivatives are defined so that the processor must specify the derivatives at the beginning and ending data points of the profile. A common default is to take the natural spline where these derivatives are set to zero. However, the end conditions are arbitrary, and thus are usually taken to minimize spurious grid estimates with wavelengths smaller than the span lengths in the beginning and ending spans. In addition, making the derivatives equal at the first and last points of the profile can force the grid to cover a full period [e.g., [**31**]] to help minimize spectral leakage. Simply adding artificial data values at each end of the profile is also effective where the processor can infer the extrapolated properties of the data profile.

***Example 12.4.2b*** in **Figure 12.4** shows the magnetic field behavior measured at a base station for a ground-based magnetic survey of a prehistoric archaeological site on the north bank of the Ohio River in Jefferson County, Indiana. The natural cubic spline was used to interpolate the drift curve (solid line) from 33 base station observations (dots). The drift corrections were applied relative to the first base station observation of the day as given by the lengths of the vertically upward and downward pointing arrows that extend from the successive base station readings. The corrections were respectively positive and negative for base station readings below and above the first value.

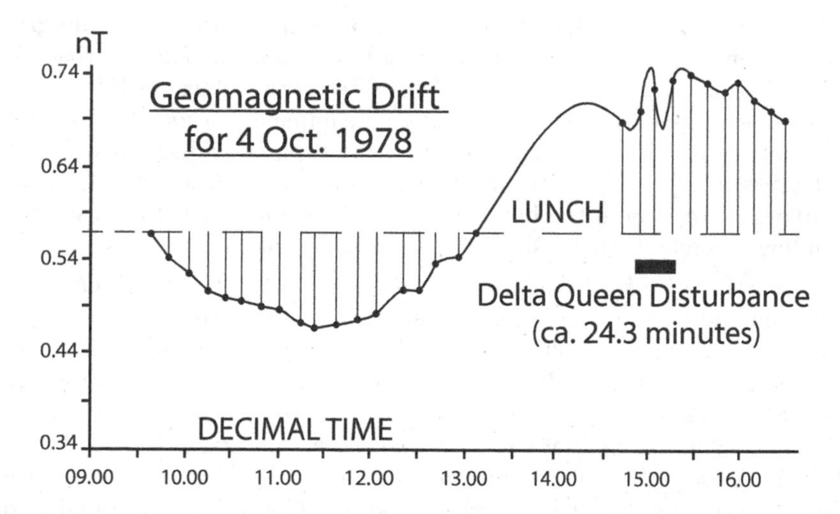


**Figure 12.4** Geomagnetic field drift for 4 October 1978 at a prehistoric archaeological site in Jefferson County, Indiana, on the north bank of the Ohio River across from Louisville, Kentucky. The diurnal field variation curve (solid line) was interpolated from the base station observations (dots) by the natural cubic spline with the mean value of 56,566 nT removed. The horizontal axis is in decimal time where for example 15.50 = 1530 hours = 3:30 p.m. [Adapted from [**43**].]

For the first half of the day, the cubic spline corrections appear to be well constrained by the observations and thus would be closely matched by simpler linear interpolations. However, over the lunch break, spurious cubic spline corrections were obtained due to the lack of constraining observations. This overshoot was exacerbated by additional spurious geomagnetic field behavior apparently related to electromagnetic emissions from the Delta Queen steamboat that had pulled up onto shore about 100 m from the base station. With the steamboat's departure some 24 minutes later, the cubic spline corrections again became consistent with the observations.

This application illustrates the relative advantages of using gridding schemes with advanced analytical properties over simpler schemes. For example, to maintain continuity of derivatives at the observation points, the cubic spline produced excessive over- and undershoots that flagged inadequately sampled regions of the drift curve. Simple linear interpolation, on the other hand, is indifferent to poor sampling and thus must be used with due caution in regions of sparse data coverage.

For gridding data mapped in $2D$ spatial coordinates, the cubic spline can be implemented by arranging the data into profiles and fitting intersecting splines to the profiles as shown by ***Example 12.4.2c*** in **Figure 12.3(b)**. At each grid node, the nearest two profiles on both sides are identified and four cubic spline estimates made at the profile coordinates marking the intersections of the crossing line from the grid node. The cubic spline estimate at the grid node is then made from the four intersecting line estimates. This approach is simple to implement and is particularly well suited for gridding track-line data from airborne, marine, and satellite surveys. The implementation also does not require that the profiles be straight and uniformly spaced and sampled. Thus, it is effective on randomly distributed data as shown by ***Example 12.4.2d*** in **Figure 12.2(c)** where the implementation was much faster than for the surrounding triangle method [**8**].

The piecewise $2D$ or biharmonic cubic spline can also be implemented to estimate the grid value from the nearest surrounding four (or more) observations. Indeed, this bicubic spline is applied by the minimum curvature technique as perhaps the most widely used current method for gridding data [e.g., [**10, 89, 85**]]. The technique is analogous to predicting thin-plate deflections from a biharmonic differential equation describing curvature. This equation is solved iteratively using finite differences modeled by bicubic splines with continuous derivatives at the observation points. Convergence is rapid and grid estimates tend toward the values of the observations as the observations approach the grid node.

The minimization of the curvature eliminates short-wavelength irregularities, providing a smooth result retaining the major features of the field. Generally, this gridding procedure produces acceptable results as long as the arbitrarily located data are well distributed over the field. Prior to gridding, noisy data commonly must be filtered to remove the short-wavelength components because the gridding requires the surface to pass through the observations.

In addition, undesirable spurious oscillations may develop in areas of the field where the data are poorly distributed. However, the results can be greatly improved by adding tension to the elastic sheet flexure equation [**85**], which is equivalent to using error variance (section **4.2.B**) in determining the spline coefficients. The advantage is illustrated by ***Example 12.4.2e*** in **Figure 12.5** where two contour maps of marine gravity data are compared. Map **(a)**, which was produced using the normal untensioned minimum curvature method, shows an unwarranted and unconstrained closed minimum south of the central positive data feature. Map **(b)** produced with tensioned splines, on the other hand, eliminates the unconstrained minimum with little effect on the other regions of the mapped field.

Minimum curvature gridding also is problematic when applied to anisotropic data with strong directional biases. Closed contours centered on maxima and minima can be produced along transects that sample field components having much larger length scales oblique to the transects than along them as is the case in most airborne surveying. These undesirable mapping artifacts are common for transects striking across elliptical anomalies. In addition, elevation deviations in flight lines from nominal specifications can produce undesirable line-centered anomalies as shown by ***Example 12.4.2f*** of the aeromagnetic data gridded in **Figure 12.6(a)**. Problems such as this

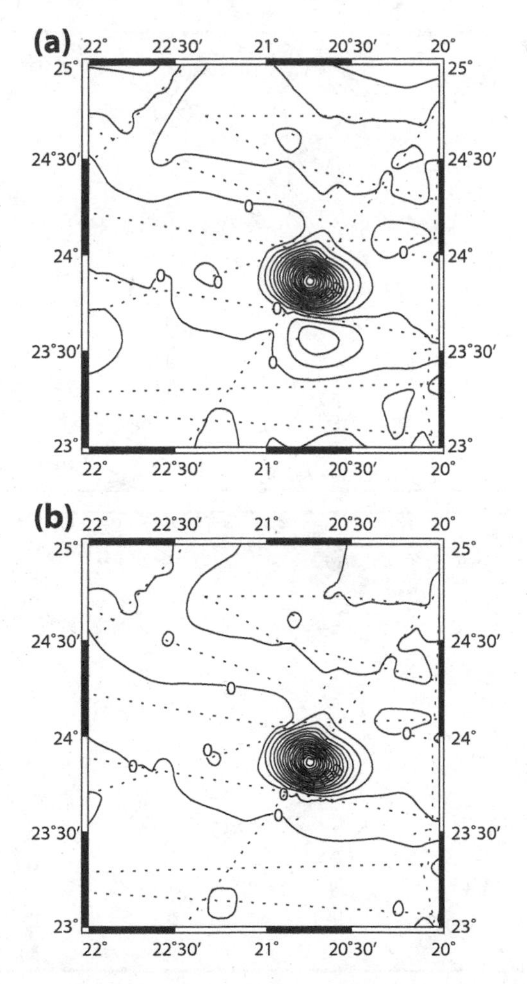


**Figure 12.5** Contour maps of marine gravity anomalies prepared from grids interpolated by the minimum curvature method. Map **(a)** was prepared from grid values obtained by the normal untensioned minimum curvature method. Map **(b)** was prepared from grid values estimated by tensioned minimum curvature. [Adapted from [**43**] after [**85**].]

may be mitigated by low-pass filtering the transect data, or using tensioned splines or another interpolation technique like the equivalent source gridding that produced **Figure 12.6(b)**.

### 12.4.3 EQUIVALENT SOURCE INTERPOLATION

Force field observations [e.g., gravity, magnetic, electrical, thermal, seismic, etc.] vary inversely with distance about the relevant physical property

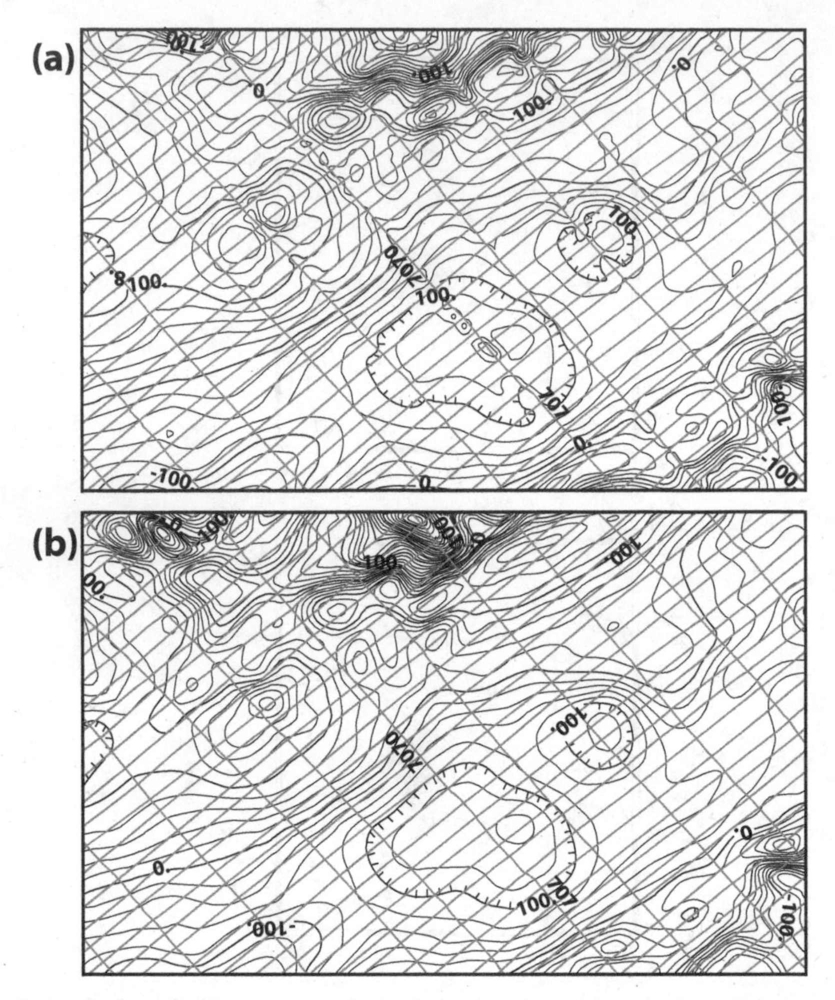


**Figure 12.6** Aeromagnetic survey data gridded **(a)** using a minimum curvature algorithm and **(b)** equivalent source interpolation. Both maps were produced at a contour interval of 20 nT. [Adapted from **[43]** courtsey of EDCON-PRJ, Inc.]

perturbation [e.g., the contrast in density, magnetization, electrical resistivity, thermal conductivity, elastic velocity, etc.], and thus can be modeled by an indefinite number of equivalent sources. Equivalent source solutions readily accommodate analysis of data that are variably distributed horizontally and in altitude. The solutions of course do not possess a physical reality, but can be used to recalculate the data and their derivative and integral components at any desired spatial coordinates including grids at constant or variable altitudes.

With gravity and magnetic potential field data, for example, this gridding method typically invokes elementary analytical sources such as the gravity point mass or magnetic point dipole or their simple closed-form extensions as spheres, cylinders, or prisms [e.g., [**43**]]. A common implementation relates the data by least squares matrix inversion to the physical property variations of an equivalent layer of uniformly thick blocks [e.g., [**67, 18**]]. As ***Example 12.4.3a*** in **Figure 12.7**, the gravity profile **(a)** of the two prismatic 2*D* vertical sources is reproduced almost exactly by the variable density effects of the equivalent layer in profile **(b)**. Assuming negligible conditioning errors for the inversion, the equivalent layer model also produces the data feature

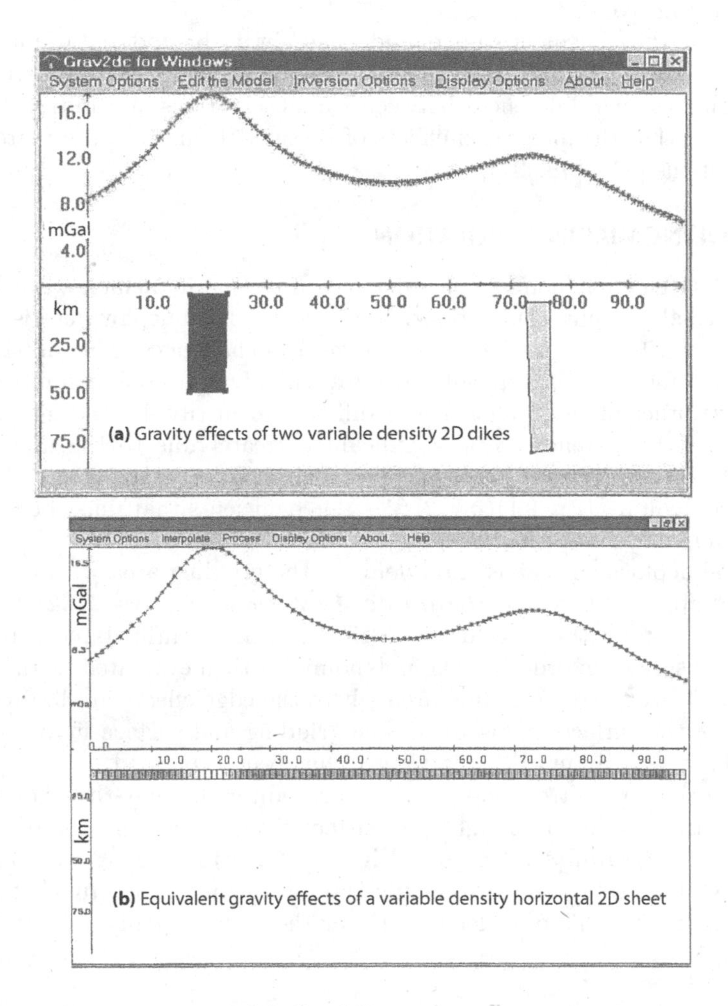


**Figure 12.7** Screen shots of **(a)** the 2*D* gravity effects of two prismatic bodies with different uniform densities equivalently modeled by **(b)** the effects of numerous density contrasts within a uniformly thick layer. [Adapted from [**43**] after [**18**].]

and its derivative and integral components with comparable accuracy at all spatial coordinates on or above the earth's surface.

In contrast to minimum curvature, equivalent source gridding is typically implemented using all the data, but with more computational labor. However, it accommodates the data gradients more comprehensively because of its basis on globally applicable force field equations. The artificial line anomalies in **Figure 12.6(a)**, for example, result from the ineffectiveness of the minimum curvature gridding to account for the altitude variations in the aeromagnetic data. However, the equivalent magnetic layer gridding readily incorporates these variable altitude effects to virtually eliminate the line anomalies as shown in **Figure 12.6(b)**.

In addition, the equivalent source model can always be updated to improve gridding accuracy and resolution. The updating involves fitting a subsequent model to the residual differences between the observations and initial model predictions so that the integrated effects of the models improve the match to the observations [e.g., **[98, 37]**].

### 12.4.4 POLYNOMIAL INTERPOLATION

Profile and surface polynomials of order $n > 3$ can be implemented in both local and global gridding applications. However, the local deployment is commonly preferred because it offers fewer numerical challenges. The global implementation, for example, typically requires a higher order polynomial where the effective order often is unclear and difficult to justify. Furthermore, the computational labor tends to be significantly greater due to the increased number of polynomial coefficients involved. The $2D$ polynomial of order $n$, for example, requires $([(n+1)(n+2)/2]-1)$ coefficients that must be solved for and manipulated to obtain the grid estimates.

The local deployment is based on defining the gridding area within a surface fitting area as shown by ***Example 12.4.4a*** in **Figure 12.2(d)**. All observations (dots) within the surface fitting area are identified and fitted to a polynomial surface of order $n$. The polynomial is then evaluated at the grid nodes (pluses) within the gridding area where the edge effects are limited by the choice of the surface fitting area. The gridding and surface fitting areas are moved successively until the entire grid has been processed.

The investigator controls the gridding procedure through the choices of the sizes of the gridding rectangle and surface fitting area, and the order of the polynomial. ***Example 12.4.4b*** in **Figure 12.2(e)** illustrates the contour map of the 2 km grid obtained from the 200 data points in **Figure 12.2(b)** using the $2D$ polynomial of order $n = 4$ over the surface fitting area of 6 km $\times$ 6 km **[8]**.

### 12.4.5 STATISTICAL INTERPOLATION

Statistical procedures have been extensively used for gridding earth science data [e.g., **[20]**]. Indeed, the weighted average procedure of ***Example 12.4.5a*** in **Figure 12.1(c)** often is included as a default gridding algorithm in commercial computer graphics packages. The inverse distance averaging procedure in the two $(x, y)$-spatial dimensions, for example, involves locating a selected number $k$ of data points nearest the prediction point. The distance $D_{ip}$ from the observation point $(x_i, y_i)$ to the prediction point $(x_p, y_p)$ is

$$D_{ip} = \sqrt{(x_p - x_i)^2 + (y_p - y_i)^2} \tag{12.5}$$

so that the grid estimate can be obtained from

$$b_p = \frac{\sum_{i=1}^{k}(b_i/D_{ip}^n)}{\sum_{i=1}^{k}(1/D_{ip}^n)}, \tag{12.6}$$

where $n$ is the order of the distance function that the investigator selects for the inverse distance weighting. This simple statistical method is effective as shown by ***Example 12.4.5b*** in **Figure 12.2(f)**, where the contour map was prepared from the gridded averages of the nearest $k = 9$ neighbors weighted by the first order (i.e., $n = 1$) inverse distance function [**8**].

Another widely used statistical approach in the earth sciences exploits theoretical or observed structure in the covariance function of the observations which reflects how the products of the observations behave in terms of their separation distances. This covariance modeling is known as kriging in mining and geological applications [e.g., [**20**]], and least squares collocation in gravity and geodetic applications [e.g., [**70**]].

***Example 12.4.5c*** in **Figure 12.8(a)** illustrates the covariance modeling of $k$ detrended observations where each observation $b_i$ is taken as the true signal $b'_i$ plus noise $\mathcal{N}$ or

$$b_i = b'_i + \mathcal{N}. \tag{12.7}$$

The objective is to estimate the true signal at the grid point given by

$$b'_p = \mathbf{c}^t(b'_p, b_i)\mathbf{C}(b_i, b_i)^{-1}\mathbf{b}, \tag{12.8}$$

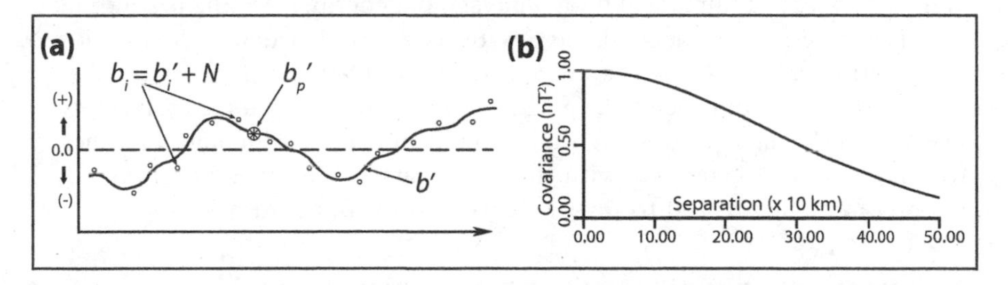


**Figure 12.8** (a) Gridding by least squares collocation. **(b)** A typical covariance function adapted to grid satellite altitude magnetic anomalies in nanotesla (nT). [Adapted from [**43**] after [**33**].]

where $\mathbf{c}(b'_p, b_i)$ is the $(k \times 1)$ cross-covariance vector between the signal and observations, $\mathbf{C}(b_i, b_i)$ is the $(k \times k)$ covariance matrix of observations, and $\mathbf{b}$ is the $(k \times 1)$ observation vector.

The symmetric covariance matrix contains elements that are the products of all pairs of observations normalized by the total variance of the observations or

$$\mathbf{C}(b_i, b_i) = \mathbf{b}\mathbf{b}^t / \sigma^2_{\mathbf{b}}. \tag{12.9}$$

The elements of the cross-covariance matrix, on the other hand, cannot be determined in this way because the value of $b'_p$ is not known. However, in many applications, the products show a dependence on offset (i.e., separation distance) between observations as in the theoretical gravity covariance function of **Figure 12.8(b)** from [**70**]. Thus, when available, the covariance function can be evaluated for the cross-covariance matrix elements in terms of the offsets defined by the observations relative to the prediction $b'_p$. Indeed, numerically efficient implementations of **Equation 12.8** commonly invoke the use of the covariance function to establish the elements of both the covariance and cross-covariance matrices.

In ***Example 12.4.5d*** from [**33**], $3D$ collocation was used to process multi-altitude Magsat crustal magnetic anomalies over South America into a $2^o$-grid at the uniform altitude of 405 km. The Magsat crustal anomalies were derived from dawn-side observations with minimal external field perturbations that were collected over unevenly distributed orbital tracks at altitudes varying from about 345 to 465 km. The procedure involved sorting the Magsat data into $2^o \times 2^o \times 120$ km bins and processing each bin for the least squares grid estimate using **Equation 12.8**. However, effective gridding required the use of a trade-off diagram as described in **Figure 9.2** to test the conditioning of the covariance matrix inverse. More generally, this practice is recommended whenever a matrix inverse must be taken to protect against using unstable and poorly performing solutions in analysis.

The foregoing descriptions illustrate the range of statistical and other gridding methods available for data analysis. However, they cover only a small fraction of the great number of gridding methods that have been developed through the history of numerical data analysis. In general, selecting a gridding method is subjective and depends largely on the experience and capabilities of the investigator, as well as the presumed nature of the surface, characteristics of the dataset, and use of the gridded values. To assess its effectiveness, however, the grid must ultimately be rendered into a contour map or other appropriate graphic to compare with the input data. Thus, effective graphical parameters must be invoked to reveal the significant data variations.

## 12.5 GRAPHICAL PARAMETERS AND PRACTICE

Every graphical presentation of the dataset should include statistical attributes that summarize its key variational properties like the amplitude range

(AR = [min; max]) of the minimum and maximum values, amplitude mean (AM) or average value, amplitude standard deviation (ASD) or energy, and amplitude unit (AU) together with the contour interval (CI) and grid interval (GI). The amplitude range, of course, indicates the total amplitude variation, whereas the grid interval defines the Nyquist wavelength of the dataset.

However, the most critical map attributes are the amplitude mean, $\mu_{\mathbf{b}}$, and standard deviation, $\sigma_{\mathbf{b}}$, because they set the appropriate contour (or color) range and interval for representing the data. For example, where the data have been effectively detrended, and thus are statistically stationary, roughly 95% of the variability of the dataset can be expected to occur within $\mu_{\mathbf{b}} \pm 2\sigma_{\mathbf{b}}$. Thus, choosing a contour interval that fits roughly 20 contour levels into the 95% variability range should reveal the key features and behavior of the dataset.

A simple non-parametric test for stationarity is to check if roughly 50% of the data amplitudes are in the range $\mu_{\mathbf{b}} + 2\sigma_{\mathbf{b}}$. If not, then the data should be explored for fitting and removing a low-order trend surface, as is done to suppress the DC-component in Fourier transform analysis. This approach is especially warranted where the trend surface may be rationalized for unwanted regional effects that are not the focus of the analysis.

### 12.5.1 STANDARDIZED AND NORMALIZED DATA

Another important application of the mean and standard deviation is to standardize data into dimensionless coefficients

$$s_i(\mathbf{b}) = \frac{b_i - \mu_{\mathbf{b}}}{\sigma_{\mathbf{b}}} \ \forall \ s_i \in \mathbf{s}, \tag{12.10}$$

which have the mean $\mu_{\mathbf{s}} = 0$ and standard deviation $\sigma_{\mathbf{s}} = 1$. However, the standardized coefficients can also be transformed into normalized coefficients

$$n_i(\mathbf{b}) = s_i \times \sigma_{\mathbf{n}} + \mu_{\mathbf{n}} \ \forall \ n_i \in \mathbf{n}, \tag{12.11}$$

where the dimensions of the normalized coefficients can be set to any desired units. Furthermore, the normalized mean $\mu_{\mathbf{n}}$ and standard deviation $\sigma_{\mathbf{n}}$ also may be set to any desired values.

In practice, however, non-zero means are commonly removed from the coefficients to enhance scaling the residual variations to the working precision of the calculations. Suppressing the non-zero means also simplifies the coefficients because **Equation 12.11** becomes $n_i(\mathbf{b}) = (\sigma_{\mathbf{n}}/\sigma_{\mathbf{b}})b_i$. In presenting the normalized coefficients, the normalization factor

$$NF = (\frac{\sigma_{\mathbf{b}}}{\sigma_{\mathbf{n}}}) \tag{12.12}$$

is typically included with the other statistical attributes so that the corresponding coefficients of the original signal $b_i$ may be readily estimated as the normalized data are considered for meaningful patterns. The normalization modifies the data gradients only by the multiplicative constant $NF$ so that the correlation coefficient is unity between the dataset **b** and its normalization **n** by **Equation 12.11**.

Obviously, standardized or normalized coefficients can be used to compare disparate datasets in common graphical parameters. As ***Example 12.5.1a***, note that gravity measurements collected on variable elevation terrain are commonly reduced to Bouguer gravity anomalies, where the terrain's differential mass effects from the elevation changes are suppressed to better image the effects of deeper mass variations [e.g., [**43**]]. The Bouguer gravity anomaly is obtained by subtracting from each measurement the approximating effect of a simple infinite horizontal slab with thickness and density respectively equal to the measurement site's elevation change and terrain density. To test the veracity of the terrain densities assumed in making the Bouguer slab reductions, the correlations between Bouguer gravity anomaly estimates and the terrain elevations are commonly studied. In particular, positive correlations between the terrain and Bouguer gravity anomaly estimates indicate deficient (i.e., too low) reduction densities, whereas negative correlations mark excessive (i.e., too high) reduction densities. Thus, to help visualize these feature correlations, the datasets can be normalized to dimensionless values with common statistical properties and mapped with a common set of graphical parameters [e.g., axes, colors, contours, etc.]. However, the feature correlations can be located even more explicitly using additional simple manipulations of the normalized coefficients.

### 12.5.2 LOCAL FAVORABILITY INDICES

To identify correlated features in signals **B** and **D**, the coefficients $b_i \in \mathbf{B}$ (e.g., the Bouguer gravity anomalies) and $d_i \in \mathbf{D}$ (e.g., the digital elevation model or DEM) are assumed to be co-registered with common independent variable coordinates. [**69**] introduced the concept of local favorability indices (LFI) for mapping out occurrences of positively correlated features in co-registered datasets from their standardized coefficients. [**96**] generalized the concept to normalized coefficients and also extended it to map out the negatively correlated features.

Specifically, to map out directly correlated features, the summed local favorability indices ($SLFI$) are estimated from the point-by-point addition or stacking of the normalized coefficients

$$SLFI_i = n_i(\mathbf{B}) + n_i(\mathbf{D}). \tag{12.13}$$

The coefficients satisfying $SLFI_i > 0$ emphasize the positive features in the two datasets that are correlative (i.e., peak-to-peak correlations), whereas the coefficients satisfying $SLFI_i < 0$ map out the correlative negative features (i.e., valley-to-valley correlations). In general, the $SLFI$ coefficients tend to emphasize occurrences of directly correlated features and suppress the occurrences of inversely and null correlated features in the datasets.

To map out inversely correlated features, on the other hand, the differenced local favorability indices ($DLFI$) are evaluated from the point-by-point differences in the normalized coefficients

$$DLFI_i = n_i(\mathbf{B}) - n_i(\mathbf{D}). \qquad (12.14)$$

The coefficients satisfying $DLFI_i > 0$ emphasize positive features in **B** that correlate with negative features in **D**, whereas the coefficients $DLFI_i < 0$ map out the negative features in **B** that correlate with positive features in **D**. In general, the $DLFI$ coefficients tend to emphasize occurrences of inversely correlated features and suppress the occurrences of directly and null correlated features in the datasets.

***Example 12.5.2a*** in **Figure 12.9** illustrates the simple utility of local favorability indices for mapping out feature correlations. Signals $A(x)$ and $B(x)$ normalized to $\sigma_{\mathbf{N}} = 1.394$ and $\mu_{\mathbf{N}} = 0$ as shown in respective panels **(b)** and **(c)** clearly have many correlated features. However, the overall correlation coefficient between them is zero because the respective left and right signal halves correlate directly and inversely.

The SLFI and DLFI coefficients in the respective panels **(a)** and **(d)** reveal the feature associations in considerable detail. The shaded pattern delineates the region between $\pm 1\sigma_{\mathbf{N}}$, whereas the SLFI coefficients that are presented in solid black and white fill respectively above and below the shaded region mark nearly all of the peak-to-peak and valley-to-valley feature correlations. The DLFI coefficients were obtained by subtracting signal $B(x)$ from signal $A(x)$. Thus, the cross-hatched coefficients map out positive features in signal $A(x)$ that correlate with negative features in signal $B(x)$, whereas the diagonally-ruled coefficients mark the negative features in the signal $A(x)$ that correlate with positive features in signal $B(x)$.

In general, considering the LFI coefficients from larger to smaller magnitudes allows the correlation analysis to proceed from the relatively simpler representations that involve the strongest feature correlations to increasingly complex representations which include the weaker, more subtle feature correlations. The coefficients also facilitate isolating feature correlations in the spectral correlation filtered outputs that are synthesized at all coordinates of the data grids.

## 12.6 PRESENTATION MODES

Gridded data can be presented in numerous forms that assist the analyst in evaluating and interpreting the data. Traditionally data are shown in profile or contour form, but the availability of high-speed schemes for computer processing and presentation has greatly increased the alternatives to these graphical formats. Data observed along transects are particularly amenable to presentation in profile form, although profiles can be extracted from gridded data as well and data values connected by line segments positioned by intermediate

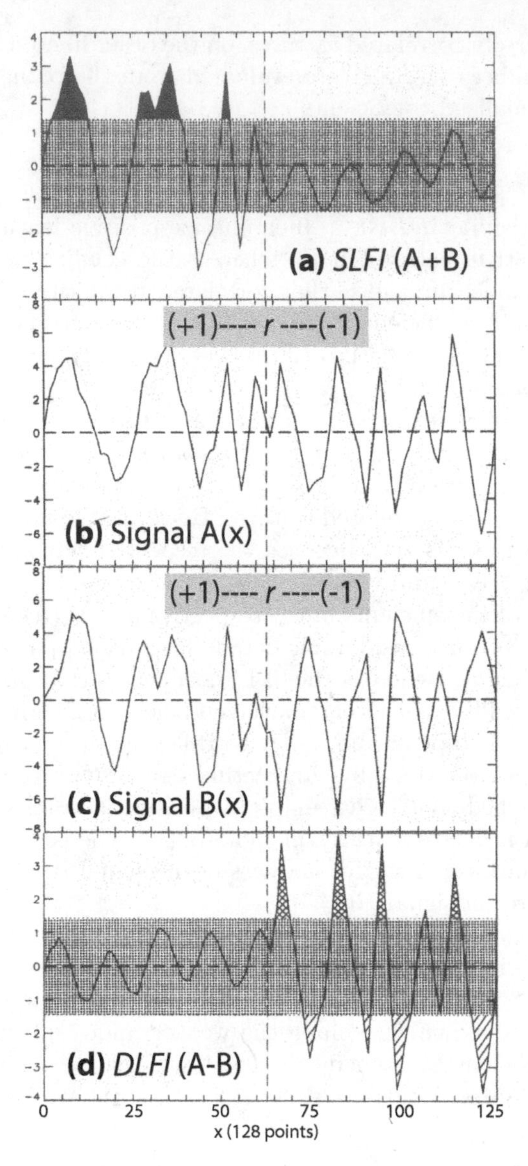


**Figure 12.9** Signals $A$ and $B$ of 128 coefficients each in respective panels **(b)** and **(c)** with correlation coefficient $r = 0.0$ have feature associations that are mapped out using the $SLFI$ and $DLFI$ coefficients in the respective panels **(a)** and **(d)**. [Adapted from [**43**] after [**96**].]

values using an appropriate interpolation scheme as described previously. Profiles are generally oriented orthogonal to the direction of the strike of elongate features of the dataset for interpretation. They can be stacked in their relative spatial position based on a level of the variable representing the position of the profile. The resulting stacked profile map gives a three-dimensional view of the dataset focusing on the higher wavenumber components that may be lost in the normal contouring process. This is particularly true of data observed along the profiles.

Contour maps of datasets and their various derivatives are the most common way of presenting two-dimensional data. Machine contouring of equispaced, orthogonally gridded data has reached a high degree of accuracy and essentially total acceptance. Numerous schemes have been developed to locate the position of a contour value on a two-dimensional representation based on interpolation along the sides of the grid squares using one of the methods described above. To improve the map presentation, the basic grid squares are divided into smaller sub-grids and the interpolation process repeated until an esthetically pleasing presentation is achieved.

The contour interval of the data map is one of the critical decisions in the map making process dependent on the purpose of the map, the map scale, and the range of variable over the mapped area. Generally, the contour interval lies between one-tenth and one-twentieth of the total range and is constant, but exceptions occur. For example, a logarithmic contour interval can be used if the range is extremely large or contour intervals may be small over the range of special interest for the map's purpose with higher or lower values given saturation levels - i.e., no contours are presented above or below the specified levels. Care is required in visual analysis of maps with non-uniform contour intervals because of distortion in the observed gradients. Color contouring, which is becoming increasingly dominant, wherein the space between adjacent contour levels is filled by a unique color, provides for rapid evaluation of amplitudes, patterns, and trends. Generally, the colors are shaded as in the electromagnetic spectrum, but with blue colors representing the lows and red colors the high values. Finally, it should be noted that contour maps should be annotated with information on the grid used in the presentation and, if possible, the location of the original point values that were used to develop the grid.

A useful alternative to the contour map which improves visualization of the anomalies is the *isometric* or *three-dimensional perspective* view. It pictures the map from a specified azimuth and angle above the horizon with variable amplitudes plotted as heights as in a three-dimensional view of topography. Alternatively, it may view depths for subsurface features in a manner comparable to heights. Complementary to the contour map is the shaded relief map which is useful for enhancing residual data variations or anomalies for visual inspection interpretation. This technique is analogous to viewing a topographic surface from an elevated position. This form of presentation is based on viewing the calculated reflection from the data surface due to a fictitious light source at a specified position [48].

In the shaded relief presentation the assumption is made that the viewing position is at a great distance from the surface and that the intensity is independent of the angular relation between the normal to the surface element and the viewing site [**24**]. As a result the brightness or intensity is proportional to the cosine of the angle between the light-source direction and the surface normal. The light source direction in terms of both horizontal angle, declination or azimuth, and vertical angle, inclination or elevation, is specified in the calculation. The surface normal, in terms of both declination and inclination, is calculated directly from the slope of the data surface at each grid point from the immediately adjacent grid values. The results are dependent not only on the data surface, but also the specified position of the fictitious light source. The appropriate position of the light source is determined by evaluating the results of iteratively employing a range of light positions, but the optimum inclination is generally of the order of $30^o \pm 15^o$ depending on the amplitude of the variations of interest and the range of amplitudes of the data surface and the declination is positioned perpendicular to the trend of the variations of interest. Presentations are often enhanced by increasing the grid density by interpolation of the original dataset and increasing the vertical exaggeration of the shaded relief intensity. Variations on the shaded relief presentation are used for enhancing specialized features. As ***Example 12.6a***, the modification described in [**19**] improves mapping of circular data features of varying radii.

High gradients and areally small data variations are enhanced, whereas regional variations are suppressed by shaded relief maps. These maps also are useful to isolate potential errors in the datasets caused by such things as isolated spurious data values and datum level problems at the join between datasets. These can be observed as geologically unrealistic local variations and linear trends. Shaded relief maps commonly are presented in gray shade images which may carry no information on the magnitudes of the variations. Alternatively, color shaded relief maps can be prepared by modulating the color contours of the data intensity map with the gray level intensity representing the reflected illumination. Numerous environmental science and engineering examples of the graphical formats and uses of color and shaded relief for presenting and analyzing data are available in the print and online literature.

## 12.7 KEY CONCEPTS

1. Graphical displays of the data and data processing results are essential elements of modern data analysis efforts. Graphics are effective for conceptualizing and implementing data analyses and presenting their results.
2. Data maps over areas of $6^o$ and larger are commonly presented using the universal transverse Mercator projection, whereas the transverse Mercator projection is used at smaller scales to roughly $0.5^o$. At even smaller scales, the simple Cartesian $[x, y, z]$-projection usually suffices for data mapping.
3. Electronically generated maps are based on data typically interpolated to grids. Data gridding schemes range from simple linear interpolation, to cubic spline interpolation that preserves the continuity

of first and second derivatives at each datum, minimum curvature with tensioned cubic splines, and kriging and least squares collocation based on the data's covariance properties.

4. In gridding, the most critical choice is the grid interval which must be at least one-half the wavelength of the smallest data feature of interest to the analysis. At larger grid intervals, the data feature is represented as a longer wavelength alias or gridding error that corrupts the analysis.
5. Effective data graphics list statistical and other attributes that summarize the key properties of the displayed data. These attributes include the range of amplitudes [AR], and amplitude mean [AM], standard deviation [ASD], and units [AU], as well as the contour interval [CI] and grid interval [GI].
6. Removing the mean and dividing by the standard deviation standardizes the data into dimensionless values with zero mean and unit standard deviation. Thus, standardized data from different datasets, including those with disparate units, may be graphically compared using the same plotting parameters [e.g., axes, contour intervals, color scales, etc.].
7. Normalized data are standardized data adjusted to arbitrary mean and standard deviation values. Nomalized data from different datasets, including those with disparate units, also may be graphically compared using common plotting parameters.
8. Standardized [or normalized] data of two datasets [including those with disparate units] may be combined by addition to yield summed local favorability indices [SLFI] that emphasize directly correlated variations. Thus, mapping out the SLFI $> 0$ coefficients tends to highlight the correlative local maxima, whereas the SLFI $< 0$ coefficients tend to map out the correlative local minima between the datasets.
9. Combining the standardized [or normalized] data of two datasets by subtraction yields differenced local favorability indices [DLFI] that emphasize inversely correlated data variations. Thus, mapping out the DLFI $> 0$ coefficients tends to highlight the relative maxima in a dataset that are correlative with the relative minima in the subtracted dataset, and vice versa for the map of DLFI $< 0$ coefficients.
10. Digital graphics are essential for conceptualizing and promoting possible solutions of the fundamental inverse problem. In general, 1-D profile, 2-D map, and 3-D isometric and shaded relief presentations are commonly used for quantifying data variations in applicable environmental parameters. Valuable insights also can result from animating the data in time and other variables.

# References


1. R.J. Adcock. A problem in least squares. *Analyst*, 5:53–54, 1878.
2. T.M. Apostol. *Calculus.* Blaisdell Publ. Co., NY, 1964.
3. J. Arkani-Hamed and D.W. Strangway. Magnetic susceptibility anomalies of the lithosphere beneath Europe and the Middle East. *Geophysics*, 51:1711–1724, 1986.
4. G. Backus and F. Gilbert. The resolving power of gross earth data. *Geophys. J.R. Astr. Soc.*, 16:169–205, 1968.
5. V.V. Belousov and M.V. Gzovsky. Experimental tectonics. *Prof. Pap. C.E.*, II:409–498, 1970.
6. P.R. Bevington. *Data Reduction and Error Analysis for the Physical Sciences.* McGraw-Hill Book Co., 336 pp, 1969.
7. P.R. Black. Seismic and thermal constraints on the physical properties of the continental crust. Ph.D. dissertation, Purdue University, 1978.
8. L.W. Braile. Comparison of four random to grid methods. *Computers & Geosciences*, 4:341–349, 1978.
9. P.W. Bridgman. *Dimensional Analysis.* Yale Univ. Press, New Haven, 1922.
10. I.C. Briggs. Machine contouring using minimum curvature. *Geophysics*, 39:39–48, 1974.
11. E.O. Brigham. *The Fast Fourier Transform.* Prentice-Hall, Inc., 252 pp, 1974.
12. E. Buckingham. On physically similar systems: Illustrations of the use of dimensional analysis. *Physical Review*, 4:345–376, 1914.
13. J.P. Burg. Maximum entropy spectral analysis. Ph. D. dissertation, Stanford University, 1975.
14. J.F. Claerbout. *Fundamentals of Geophysical Data Processing with Applications to Petroleum Prospecting.* Blackwell Scientific Publications, 274 pp, 1976.
15. A.K. Cline, C.B. Moler, G.W. Stewart, and J.H. Wilkinson. An estimate for the condition number of a matrix. *SIAM J. Numer. Anal.*, 16:368–375, 1979.
16. D.F. Coates. *Rock Mechanics Principles.* Mines Branch Monog. 874, Dept. of Energy, Ottawa, CN, 1970.
17. R.L. Coons, G.P. Woollard, and G. Hershey. Structural significance and analysis of mid-continent gravity high. *Am. Assoc. Petr. Geol. Bull.*, 51:2381–2399, 1967.
18. G.R.J. Cooper. Gridding gravity data using an equivalent layer. *Computers & Geosciences*, 26:227–233, 2000.
19. G.R.J. Cooper. Feature detection using sun shading. *Computers & Geosciences*, 29:941–948, 2003.
20. J.C. Davis. *Statistics and Data Analysis in Geology.* J. Wiley & Sons, 646 pp, 1986.
21. W.C. Dean. Frequency analysis for gravity and magnetic interpretation. *Geophysics*, 23:97–127, 1958.
22. W.E. Deming. *Some Theory of Sampling.* Wiley, New York, 1950.

23. M.B. Dobrin and C.H. Savit. *Introduction to Geophysical Prospecting.* McGraw-Hill, 867 pp, 1988.
24. S.D. Dods, D.J. Teskey, and P.J. Hood. The new series of 1 : 1,000,000 scale magnetic anomaly maps of the Geological Survey of Canada: Compilation techniques and interpretation. In *The Utility of Regional Gravity and Magnetic Anomaly Maps*, pages 69–87. Society of Exploration Geophysicists, 1985.
25. J.J. Dongarra, J.R. Bunch, C.B. Moler, and G.W. Stewart. *LINPACK Users'Guide.* Society for Industrial and Applied Mathematics, 367 pp, 1979.
26. L.M. Dorman and B.T.R. Lewis. Experimental isostasy 1. Theory of the determination of the earth's isostatic response to a concentrated load. *J. Geophys. Res.*, 75:3357–3365, 1970.
27. N.R. Draper and H. Smith. *Applied Regression Analysis.* Blaisdell Publ. Co., 407 pp, 1966.
28. A.L. Edwards. *An Introduction to Linear Regression and Correlation.* W.H. Freeman and Co., San Francisco, 1976.
29. M.R. Foster and N.J. Guinzy. The coefficient of coherence: Its estimation and use in geophysical prospecting. *Geophysics*, 32:602–616, 1967.
30. B.D. Fuller. Two-dimensional frequency analysis and design of grid operators. In *Mining Geophysics, II*, pages 658–708. Society of Exploration Geophysicists, 1967.
31. C.F. Gerald and P.O. Wheatley. *Applied Numerical Analysis.* Addison-Wesley Publ. Co., 679 pp, 1989.
32. J.W. Gibbs. Fourier series. *Nature*, 59:200, 1896.
33. H.K. Goyal, R.R.B. von Frese, and W.J. Hinze. Statistical prediction of satellite magnetic anomalies. *Geophys. J. Int.*, 102:101–111, 1990.
34. W.R. Griffin. Residual gravity in theory and practice. *Geophysics*, 14:39–56, 1949.
35. J.C. Griffiths. *Scientific Method in the Analysis of Sediments.* McGraw-Hill, Inc., New York, 1967.
36. D. Gubbins. *Time Series Analysis and Inverse Theory for Geophysicists.* Cambridge Univ. Press, 255 pp, 2004.
37. P.J. Gunn and R. Almond. A method for calculating equivalent layers corresponding to large aeromagnetic and radiometric grids. *Exploration Geophysics*, 28:72–79, 1997.
38. V.K. Gupta and N. Ramani. Some aspects of regional-residual separation of gravity anomalies in a precambrian terrane. *Geophysics*, 45:1412–1426, 1980.
39. R. Haining. *Spatial Data Analysis in the Social and Environmental Sciences.* Cambridge University Press, New York, 1993.
40. J.W. Harbaugh. A computer method for four-variable trend analysis illustrated by a study of oil-gravity variations in southeastern Kansas. *Kansas Geological Survey Bull.*, 171:58 pp., 1964.
41. C.N. Hewitt. *Methods of Environmental Data Analysis.* Springer Verlag, 1992.
42. C.R. Hicks. *Fundamental Concepts in the Design of Experiments.* Holt, Rinehart & Winston, New York, 1964.
43. W.J. Hinze, R.B. von Frese, and A.H. Saad. *Gravity and Magnetic Exploration: Principles, Practice, and Applications.* Cambridge University Press, 512 pp, 2013.

44. W.J. Hinze, R.R.B von Frese, and D.N. Ravat. Mean magnetic contrasts between oceans and continents. *Tectonophysics*, 192:117–127, 1991.
45. P.G. Hoel. *Introduction to Mathematical Statistics.* John Wiley & Sons, Inc., New York, 1962.
46. A.E. Hoerl and R.W. Kennard. Ridge regression. Applications to nonorthogonal problems. *Technometrics*, 12:69–82, 1970.
47. A.E. Hoerl and R.W. Kennard. Ridge regression. Biased estimation for nonorthogonal problems. *Technometrics*, 12:55–67, 1970.
48. B.K.P. Horn and B.L. Bachman. Using synthetic images to register real images with surface models. *Commun. Assoc. for Comput. Mach.*, 21:914–924, 1978.
49. M.K. Hubbert. Theory of scale models as applied to the study of geologic structures. *GSA Bull.*, 48:1459–1520, 1937.
50. D.D. Jackson. Interpretation of inaccurate, insufficient and inconsistent data. *Geophys. J. Roy. Astr. Soc.*, 28:97–109, 1972.
51. G.M. Jenkins and D.G. Watts. *Spectral Analysis and its Applications.* Holden-Day, 525 pp, 1968.
52. M.B. Jones. Correlative Analysis of the Gravity and Magnetic Anomalies of Ohio and Their Geological Significance. MSc. thesis, The Ohio State University, 1988.
53. M. Kanevski and M. Maigan. *Analysis and Modelling of Spatial Environmental Data.* EPFL Press, 2004.
54. W.M. Kaula. Geophysical implications of satellite determinations of the earth's gravitational field. *Space Sci. Rev.*, 7:769–794, 1967.
55. K.A. Kermack and J.B.S. Haldane. Organic correlation and allometry. *Biometrika*, 37:30–41, 1950.
56. W.C. Krumbein and F.A. Graybill. *An Introduction to Statistical Models in Geology.* McGraw-Hill, Inc., New York, 1965.
57. C. Lanczos. *Linear Differential Operators.* Van Nostrand-Reinhold, 564 pp, 1961.
58. L.L. Lapin. *Statistics for Modern Business Decisions.* Harcourt, Brace, Jovanovich, Inc., NY, 1982.
59. C.L. Lawson and R.J. Hanson. *Solving Least Squares Problems.* Prentice-Hall, Inc., 340 pp, 1974.
60. Y. Li and D.W. Oldenburg. Separation of regional and residual magnetic field data. *Geophysics*, 63:431–439, 1998.
61. R.O. Lindseth. *Recent Advances in Digital Processing of Geophysical Data - A Review.* Society of Exploration Geophysicists Continuing Education Manual, 1974.
62. K. Mallick and K.K. Sharma. A finite element method for computation of the regional gravity anomaly. *Geophysics*, 64:461–469, 1999.
63. D.W. Marquardt. Ridge regression, biased linear estimation, and nonlinear estimation. *Technometrics*, 12:591–612, 1970.
64. J.H. Mathews and K.K. Fink. *Numerical Methods Using Matlab.* Prentice-Hall Inc., 696 pp, 2004.
65. A.J.W. McDonald. The use and abuse of image analysis in geophysical potential field interpretation. *Surveys in Geophysics*, 12:531–551, 1991.

66. E.A. Mechtly. *The International System of Units, Physical Constants, and Conversion Factors.* SP-7012, NASA, 1971.
67. C.A. Mendonca and J.B.C. Silva. Interpolation of potential field data by equivalent layer and minimum curvature: a comparative analysis. *Geophysics*, 60:399–407, 1995.
68. W. Menke and J. Menke. *Environmental Data Analysis with MATLAB.* Elsevier, New York, 2011.
69. D.F. Merriam and P.H.A. Sneath. Quantitative comparison of contour maps. *J. Geophys. Res.*, 7:1105–1115, 1966.
70. H. Moritz. *Advanced Physical Geodesy.* H. Wichmann Verlag, 500 pp, 1980.
71. N.F. Murphy. *Dimensional Analysis.* Bull. Va. Polytec. Inst., v. XLII, 1949.
72. W.R. Ott. *Environmental Statistics and Data Analysis.* CRC Press, New York, 1994.
73. R.S. Pawlowski. Green's equivalent-layer concept in gravity band-pass filter design. *Geophysics*, 59:69–76, 1994.
74. R.S. Pawlowski and R.O. Hansen. Gravity anomaly separation by wiener filtering. *Geophysics*, 55:539–548, 1990.
75. R.A. Penrose. A generalized inverse for matrices. *Proc. Cambridge Phil. Soc.*, 51:406–413, 1955.
76. R.S. Peterson and M.B. Dobrin. A pictorial digital atlas. Presented at 36th Annual SEG Meeting, United Geophysical Corporation, Houston, 55 pp, 1963.
77. W.H. Press, B.P. Flannery, S.A. Teukolsky, and W.T. Vetterling. *Numerical Recipes, The Art of Computing.* Cambridge University Press, 989 pp, 2007.
78. J.E. Reed. Enhancement/isolation wavenumber filtering of potential field data. M.Sc. thesis, Purdue University, 1980.
79. C. Riemann, P. Filzmoser, R. Garrett, and R. Dutter. *Statistical Data Analysis Explained: Applied Environmental Statistics with R.* Wiley, New York, 2008.
80. A.H. Saad. Spectral analysis of magnetic and gravity profiles: Part ii - maximum entropy spectral analysis (mesa) and its applications. Research Rept. 4226RJ002, Gulf Science and Tech. Co., 1978.
81. J.K. Sales. Crustal mechanics of cordilleran foreland deformation: A regional and scale-model approach. *AAPG Bull.*, 52:2016–2044, 1968.
82. C.E. Shannon. Communication in the presence of noise. *Proc. Institute of Radio Engineers*, 37:10–21, 1949.
83. G.B. Shelly and T.J. Casham. *Introduction to Computers and Data Processing.* Anaheim Publishing Co., CA, 1980.
84. R.E. Sheriff. *Encyclopedic Dictionary of Exploration Geophysics.* Society of Exploration Geophysicists, 266 pp, 1973.
85. W.H.F. Smith and P. Wessel. Gridding with a continuous curvature surface in tension. *Geophysics*, 55:293–305, 1990.
86. J.P. Snyder. *Map Projections – A Working Manual.* U.S. Geological Survey Prof. Paper 1395, 383 pp, 1987.
87. N.C. Steenland. Structural significance and analysis of mid-continent gravity high: Discussion. *Am. Assoc. Petr. Geol. Bull.*, 52:2263–2267, 1968.
88. K. Steinbuch. *Taschenbuch der Nachrichtenverarbeitung.* Springer Verlag, Berlin, 1962.

89. C.J. Swain. A Fortran IV program for interpolating irregularly spaced data using the difference equations for minimum curvature. *Computers & Geosciences*, 1:231–240, 1976.
90. T. Szirtes. *Applied Dimensional Analysis and Modeling*. Elsevier, New York, 2006.
91. G.I. Taylor. The formation of a blast wave by a very intense explosion. ii. the atomic explosion of 1945. *Proc. Roy. Soc. London, Series A, Math & Phys. Sci.*, 201:175–186, 1950.
92. S.K. Thompson. *Sampling*. John Wiley & Sons, New York, 2002.
93. J.B. Thurston and R.J. Brown. The filtering characteristics of least-squares polynomial approximation for regional/residual separation. *Canadian Journal of Exploration Geophysics*, 28:71–80, 1992.
94. T.J. Ulrych and T.N. Bishop. Maximum entropy spectral analysis and autoregressive decomposition. *Rev. Geophys. Space Phys.*, 13:183–200, 1975.
95. US-EPA. *Guidance on Choosing a Sampling Design for Environmental Data Collection*. Office of Environmental Information, EPA/240/R-02/005, Washington, D.C., 2002.
96. R.R.B. von Frese, M.B. Jones, J.W. Kim, and J.H. Kim. Analysis of anomaly correlations. *Geophysics*, 62:342–351, 1997.
97. R.R.B. von Frese, H.R. Kim, P.T. Taylor, and M.F. Asgharzadeh. Reliability of champ anomaly continuations. In *Earth Observation with CHAMP Results from Three Years in Orbit*, pages 287–292. Springer Verlag, 2005.
98. R.R.B. von Frese, D.N. Ravat, W.J. Hinze, and C.A. McGue. Improved inversion of geopotential field anomalies for lithospheric investigations. *Geophysics*, 53:375–385, 1988.
99. R.E. Walpole and R.H. Myers. *Probability and Statistics for Engineers and Scientists*. The Macmillan Co., New York, 1972.
100. A.B. Watts. *Isostasy and Flexure of the Lithosphere*. Cambridge University Press, 458 pp, 2001.
101. N. Wiener. Generalized harmonic analysis. *Acta Mathematica*, 55:117–258, 1930.
102. N. Wiener. *Extrapolation, Interpolation, and Smoothing of Stationary Time Series*. John Wiley and Sons, Inc., 163 pp, 1949.
103. R.A. Wiggins. The general linear inverse problem: Implication of surface waves and free oscillations for earth structure. *Rev. Geophys. Space Phys.*, 10:251–285, 1972.
104. D. York. Least-squares fitting of a straight line. *Canadian J. Phys.*, 44:1079–1086, 1966.
105. D. York. The best isochron. *Earth Planet. Sci. Lett.*, 2:479–482, 1967.
106. D. York. Least-squares fitting of a straight line with correlated errors. *Earth Planet. Sci. Lett.*, 5:320–324, 1969.

# Index